NUMBER WORDS AND NUMBER SYMBOLS

A Cultural History of Numbers

Karl Menninger

Translated by Paul Broneer
from the revised German edition

DOVER PUBLICATIONS, INC., New York

*Tolle numerum omnibus rebus
et omnia pereunt.*
Take from all things their number,
and all shall perish.
ISIDORE OF SEVILLE (CA. 600)

Translated from the German with permission of the Vandenhoeck & Ruprecht
Publishing Company, Göttingen, Germany.
English translation copyright © 1969 by The Massachusetts Institute of Technology.
All rights reserved.

This Dover edition, first published in 1992, is an unabridged, unaltered republica-
tion of the English translation (published by The MIT Press, Cambridge, Mass., 1969)
of the revised German edition of *Zahlwort und Ziffer: Eine Kulturgeschichte der
Zahlen*, published by the Vandenhoeck & Ruprecht Publishing Company, Göttingen,
Germany, 1957–58.
The Dover edition is published by special arrangement with Vandenhoeck &
Ruprecht GmbH & Co. KG, Theaterstrasse 13, Postfach 3753, D-3400 Göttingen,
Germany, and with The MIT Press, Massachusetts Institute of Technology,
55 Hayward Street, Cambridge, Massachusetts 02142.

Library of Congress Cataloging-in-Publication Data

Menninger, Karl, 1898–
  [Zahlwort und Ziffer. English]
  Number words and number symbols : a cultural history of numbers / Karl
Menninger ; translated by Paul Broneer from the revised German edition.
      p.   cm.
  Translation of: Zahlwort und Ziffer.
  Originally published: Cambridge, Mass. : M.I.T. Press, 1969.
  Includes index.
  ISBN-13: 978-0-486-27096-8
  ISBN-10: 0-486-27096-3
  1. Numerals—History.   2. Numeration—History.   3. Abacus—History.
I. Title.
QA141.2.M4513   1992
513.5—dc20   91-42709
                                                                      CIP

www.doverpublications.com

Preface

The concept of number exerts a dual enchantment. When it uncovers the rich relationships between pure magnitudes to the scientist, it fills him with the satisfaction of intellectual insight; for the sake of these relationships the sovereign structure of mathematics rests upon the concept of number. For others it exerts a fascination by its deep interconnection with the daily life of the people. Is it not true that every tribe spoke and noted down numbers, that every tribe had to calculate whenever it faced life on this planet? Did man's relationship to his environment not also necessitate his relationship with numbers?

This book tries to trace this interweaving of numbers and human life. It strives to demonstrate how the clear concept of number adopts a multiple image when it grows with a people. Together the history of knowledge and of culture create a single picture: the first establishes the boundaries, the second fills the spaces with color. And all along the wealth of detail, no matter how much it takes up the reader's attention, permits the music to be heard of the eternal stream at whose banks the fate of humanity is fulfilled.

Number takes shape both in spoken and in written language, in "number word and number symbol." Just as a child learns to count one, two, three long before it learns to write or calculate with the numerals 1, 2, 3, the spoken number language precedes the written language; at least the two are so essentially separated in character that it seemed justified to separate the cultural history of number into two series of chapters, the first covering Number Sequence and Number Language and the second covering Written Numerals and Computation.

This book is written for the lover of intellectual and cultural history, but the professional historian will discern many things in it not previously expressed. Of course the plain knowledge of the number series and number symbols of practically all cultures, the Greeks, Romans, Egyptians, Babylonians, Chinese, Indians, and so on, has to be cited first, but ethnology and ethnography contribute a most colorful addition: the history of language, of culture, and of politics. There are few things of this world in which these branches of research meet each other in such an exciting and fertile manner as the concept of number. The area of its symbolic and mythical interpretation is not even included.

With this overwhelming wealth of detail it became difficult to pursue the great art of following the threads in this closely woven fabric, to separate them without destroying the fabric itself. The many illustrations and the form of presentation are intended to permit the reader to arrive at his own judgment so that he can participate intimately in the stimulating spectacle in which number has found form both in spoken and written language.

It is not often that the lover of numbers becomes acquainted with the intrinsic connection of his special area with cultural history; it is equally rare that the friend of the history of culture becomes aware of the relationship between his field and the life of numbers. I hope that the present work will serve both groups to gain the insight and the joy which comes from all knowledge of creative intellectuality in the diversity of men and peoples.

It is a very personal need for me to express my thanks for the support which I found wherever I turned. In the course of the frequently discouraging detail work of collecting, it was a happy feeling of intellectual interplay when each bit of information obtained was backed up by friendly readiness.

Above all I am grateful to the Hesse State Library in Darmstadt for laboriously securing sources for me, and also to the University Library in Heidelberg, the manuscript collection of the Bavarian State Library, the Municipal Archives in Augsburg, and the Hessian State Archives in Darmstadt. Among state collections I am particularly obligated to the Deutsche Museum in Munich, the Ethnographic Museums in Frankfurt, Stuttgart, and Munich, the Museum for Ethnology in Berlin, the numismatic collections in Munich, Karlsruhe, and Darmstadt, the Historic Museums in Basel and Vienna, the British Museum in London, the State Museums in Athens, Dublin, Helsinki, Naples, Ostia, and Rome, the Cabinet des Medailles of the Bibliothèque Nationale in Paris, the Musée de l'Oeuvre Notre Dame in Strasbourg, and the former German Institute of Archeology in Cairo.

For personal support I want to express my sincere thanks to G. Beaujouan of Paris, H. F. Deininger of Augsburg, E. Delp of Bad Nauheim, W. Eilers of Marburg, P. Hammerich of Dinkelsbühl, H. Horst of Tehran, A. König of Frankfurt, Nobuko Yokota of Tokyo, O. Neugebauer of Providence, Rhode Island, J. Rak-Sell of Singapore, the late J. Ruska of Berlin, R. Schlösser of Hannover, and K. Vogel of Munich. I am particularly beholden to Mr. Wolfram Müller of Frankfurt and Tokyo.

Heppenheim an der Bergstrasse   KARL MENNINGER
Spring 1958

List of Phonetic Symbols

ś represents a sound between s and sh

š is pronounced like sh in English "shoe"

č is pronounced like ch in English "church"

ǝ represents a suppressed or murmured e, like the last syllable in "America"

þ in Gothic and Old Nordic is pronounced like th in English "three"

ě corresponds to Russian ye

l, m, r represent syllabic sounds, without vowels

q is pronounced like k

u̯ is a semivowel, between u and w

Contents

NUMBER SEQUENCE AND
NUMBER LANGUAGE

Introduction

What image of numerical values is evoked by the curious succession of the words "one," "two," "three," and so on, which we call the number sequence? Going back deep into history, we find that the number sequence did not spring into existence fully formed, but rather that it evolved stepwise from one numerical boundary to the next. Such "rudimentary" first stages explain a series of otherwise unintelligible peculiarities inherent in full-fledged, "mature" number sequences.

These early difficulties have been overcome by an analysis of the number sequence. In the groupings that precede the sequence and in the gradation of numbers we discern the two basic laws governing both the number sequence and the written number symbols. The question of how these rules of succession are observed by the individual languages opens up a wide range of possibilities — such as succession by size, overcounting, ciphering, specification, and many others — possibilities that bear witness to the astonishing inventiveness of primitive man but also to the conceptual difficulties with number sequences encountered by him.

The key to all such investigations lies in our own words for numbers. How did *our* number sequence originate, what was its origin in the dawn of intellectual and linguistic growth, which continues to its present form in the full daylight of recorded history? For an answer we have the vast panorama of the whole family of related Indo-European languages. For our own number system does not stand alone; it is closely interwoven with those of many other European and Asiatic languages. The latter thus shed light on ours and illuminate much that would otherwise be impenetrably obscure, while in turn our number words throw light on those of others.

One question still remains unanswered here: Have the Babylonian numbers in any way influenced our own? The "great hundred," the frequent use of the number 12, and the peculiar, characteristic break that in many Germanic number sequences follows 60, all call to mind the Roman duodecimal fractions and suggest the sexagesimal order that dominates Babylonian spoken numbers and written numerals.

At the close we let the hidden number words — that is, words whose root is a number word although so distorted that it is scarcely discernible — once more reveal the richness and color of the world of number language. And once again we are conscious of how a primordial sense of order evolved, shrouded in the mystery of early dawn, and continues to influence our every-day life.

A more specific idea of the subjects covered may be gleaned from the Table of Contents.

The reader is advised to consult the complete number sequences (pp. 92 ff.), even when this is not specifically referred to, whenever he encounters a foreign number word; this will allow him to become better acquainted with the language.

Foreign number words had to be transliterated whenever the language could not be written with our own characters (as in the case of Greek, Russian, Arabic, and Chinese) or when the alphabetic symbols have phonetic values differing from those of our own (as in the case of Czech or Polish). Linguistics employs for each individual language a separate system of transcription designed for that language alone, except when it writes in the original script. But since the present book is addressed to a wider circle of readers who may become confused by an excess of phonetic symbols, I have followed a compromise method, using a transcription that reproduces approximately the sounds and symbols of *all* languages (see the list of phonetic symbols on p. vii). Where, however, some special pronunciation is called for, the reader will find it in the complete number sequence of the respective language (as, for example, in the case of Gothic, on p. 92). Symbols for long and short vowels are omitted. The reader is assumed to be familiar with English, French, Italian, and German pronunciations.

In this book it was not possible to use a pure and rigorous system of transliteration (such as the scientifically accepted $\theta\vartheta{:}'ti{:}n$ for *thirteen*), because this would completely distort the orthographic picture; even a modified system, such as the use of the Indo-European *kwetwores* for *qu̯etu̯ores* (four), for example, would throw the important connection of this word with the Latin *quattuor* out of focus.

The Number Sequence

The Abstract Number Sequence

How do we count today?

Before we go into the historical development of our words for numbers, let us first determine *how* and *what* we actually count and what "counting" really is.

Before us lies a heap of peas, which we wish to count. How do we go about this? We arrange the peas in a row, physically or mentally, touch the first one and say "one," then touch the second and say "two," touch the next and say "three," ... touch the last and say "twenty-two"; there are 22 peas in all. What have we actually done? We have *assigned* a word to each individual pea. Counting thus constitutes *assigning* words to things.

To what are these words assigned? To the things we are counting — in this case to peas. At other times we may be counting houses, trees, people, or fingers. Can we also count things of different natures: a pen, a desk, and a cat, for example? Yes, these are three "objects." Can intangible things be counted as well, such as the conclusions of a proof, or the thoughts embodied in a thesis? Yes. Even a person's traits: intelligent, slender, lively, generous, and so on, can be *enumerated.* In short, any distinguishable entity, tangible or intangible, identical or different, can be counted. These distinguishable things, considered together, constitute a *set* and are themselves the *elements* of this set.

Thus, we now say: A set can always be counted by assigning number words to its elements.

But the number words themselves form a set, the elements of which are the words "one," "two," "three," and so on. In the process of counting, the elements of the set of number words or of the number sequence, whichever we prefer to call it, are assigned uniquely to the elements of the set of things to be counted: uniquely, because only a single number word is attached to each pea.

If we think of the separate terms of the number sequence as small boxes, labeled 1, 2, 3, and so on, we can conceive of the counting process as follows: In each box, starting with the first, we place a single pea, the first in box 1 and the last in box 22. Then 22 boxes of our number sequence are full, while all the remaining boxes from 23 on remain empty.

Now we can explain the heading, "The Abstract (Empty) Number Sequence." So long as there is no counting, it is merely there, detached from all concrete objects, unused but ready. But as soon as we count, then according to our first image the number words become assigned to the objects, and according to the second one the objects are placed in the empty boxes of the number sequence. The last number word (or the last box) indicates the *cardinality*, or *number*, of the set.

This insight is as important as it is simple. We shall see, in fact, that the "abstraction" of the number sequence from the things counted created great difficulties for the human mind. We need only ask ourselves: How would we count if we did not possess this sequence of remarkable words, "one," "two," "three," and so on? Yet there was a time when it did not exist!

[7]

Thus one achievement of our number sequence is its independence of the things themselves. It can be used to count *anything*.

But can it also be used to count arbitrarily large sets, even the sands of the sea? Yes, even these sets "without number" can be counted by using our number sequence — this is its other achievement. To each successive grain of sand it assigns a number word, tirelessly and inexhaustibly. And when the last particle of sand has been counted, it still has "infinitely many" number words with which to go on counting.

The number sequence could go on counting, even though we cannot. Yet we know for certain that it would do so correctly and in the proper sequence. We hear of three million inhabitants of a city: does anyone count them one after another, 1, 2, 3? Still we are quite sure that if this were to be done, we would eventually come to inhabitant No. 2,999,974, No. 2,999,975, No. 2,999,976 . . ., and finally to No. 2,999,999 and No. 3,000,000.

Whence this certainty, which we never gained from experience? We know that our number sequence embodies the law of infinite progression; we know that every number has a successor; and we also know how that successor is formed from its predecessor.

Hence, our number sequence is not a motley collection of words arbitrarily gathered together, but an ordered creation of the mind. It incorporates the law of infinite progression, by force of which we acknowledge the countability of sets even though we ourselves cannot actually do the counting.

A finite and amazingly small quantity of number words is enough for this purpose, for the number sequence uses these words over and over again, in their proper order and context. And it is completely independent of the objects it counts: it is abstract. Therefore it can count anything.

This is our modern number sequence, the number sequence in its highest state of development. And now that we are familiar with it, the question becomes especially absorbing: was it not always so?

The Number Sequence
Used Concretely

Haven't people always counted as we do today?

We shall find the answer to this question if we descend the ladder of culture down to the very lowest steps, scarcely above the level where mind could not rise above its environment. Early man counted, too, whenever he merely gathered fruits or hunted, whether he grew his own food by more or less primitive methods of cultivation or drove his herds from pasture to pasture or whether, like many tribes living near the coast, he sought to earn his living by trade. His way of life taught him to count, the nature of his economy determining the extent of his number sequence. Why should a pygmy people, living in isolation in the primeval forest, need to count beyond 2? Anything over that is considered "many." But the cattlebreeder must count his herd, head by head, up to 100 or even more. For him "many" is something far greater, something that no longer has economic meaning to him. Thus early man's environment determined his thinking and actions, and also his counting.

So that we can understand his number sequences, which we shall now consider, let us dwell for a moment on the way primitive man perceives the world around him. It still impinges on him directly, in all its myriads of colors and forms. Things have not yet been "cooled off" for him by his intellect, which sifts them and orders them and separates them, filing their elements away in the gray, colorless pigeon-holes of concepts. On the contrary, in their immediate, hot-blooded, many-colored uniqueness they touch his innermost heart. Thus they are not objects to him, things which are alien to him and stand "outside" himself — *here* am I and *there* is the world — rather they are completely absorbed in his own life. He is a part of them, as they are of him. He is woven into the very fabric of the universe by powerful strands of religion; he does not, like "modern" man, like ourselves, stand before it in wonderment, in calculation, or indifference.

Yet some remaining fragments of that early perception of the world still loom up in our own. Many a superstition, many an oddity, survives unrecognized in the midst of the intense consciousness of our own culture. Who today knows or cares much about the number 7? Though for us it has lost its supernatural content, though it offers not the slightest advantage in measuring and reckoning time, the seven-day week still governs our whole external life. From that early interpenetration of man with the world arises the infusion of objects and numbers with mystic significance, and hence the "holiness" of the numbers 3 and 7 and the auspiciousness or bad luck of the number 13. It is the task of mythology to uncover the concepts that led early man to impart supernatural significance to certain numbers.

Our own purpose, however, is to understand how early man gave expression to things and events, with all their profusion of kaleidoscopic detail, in his own primitive language. "A man has killed a rabbit" — an American Indian would never say this in such a colorless way. His statement, broken down into its verbal component, says: "The man, he, one, animate, standing, has purposely killed by shooting an arrow at the rabbit, animate, him, one, sitting." The Indian does not put it thus because he wants to express the event in a specially picturesque manner, as our highly developed speech could also do by adding words and phrases; he *cannot* say it any other way, because that is how he has experienced the event, and he cannot free himself from its uniqueness. Generalization into pale concepts is completely foreign to him. His language proves this, since it achieves its colorful expression not as does ours, by using auxiliary words and phrases, but rather through the inflection of its words and through particles prefixed, suffixed, or incorporated in them. Just as we (in German) can indicate only the tense and the mood of a verb (*gibt, gab, gäbe* — "gives," "gave," "would give") by inflection and phonetic shift (*i—t, ä*), so can the Indian express the gender, number, intention, and detailed manner of the act of killing in one single, inflected word. Whereas we, especially in scientific language, stress only the essential aspects of an observation, shedding all incidentals and compressing the main point into a general concept, primitive man puts as many of the details observed by him as he can into his speech. We would have to reassemble a large number of general concepts in order to express the "death of the rabbit" as the Indian does. The abundant inflectional potentialities of early language and its completely different vocabulary testify to early man's keener observation and to his more intimate involvement with the world. The Lapplander has twenty different words for "ice" and twice as many again for "snow"; he can also describe thawing and freezing by a single word in almost as many various ways. How dull, by contrast, are the word forms of English, for example! The four grammatical cases *Mann, Mannes, Manne, Mann* which we express in German by inflecting the word, are all simply "man" in English and must be specified by the auxiliary prepositions "of" or "to," as in Chinese. Chinese has no inflections at all and therefore expresses the relationships among words almost exclusively by their position in the sentence. But Chinese is by its very nature an uninflected language, whereas English over the course of time abandoned its inflections, just as inflected languages generally lose their inclination and eventually even their ability to inflect.

It is also worth noting, however, that while Lithuanian, for instance, has different words for the "gray" of geese and of horses, or wool, or human hair, and so forth, it has no separate word for the generic concept of "gray," which is abstract, or "empty," and must be embodied or "filled" by actual, concrete objects. So powerfully does the idea of the unique, the real, thing persist in the mind of early men.

After this brief linguistic discussion, let us now consider the early number sequences.

NUMBERS WITHOUT WORDS

It was related by a missionary to the Abipones, a tribe of South American Indians compelled by a shortage of food to migrate (in the 18th century): "The long train of mounted women was surrounded in front, in the rear, and on both sides by countless numbers of dogs.

From their saddles the Indians would look around and inspect them. If so much as a single dog was missing from the huge pack, they would keep calling until all were collected together again. I have often since wondered how they, without knowing how to count, would tell at once, in spite of the confused throng, that one dog was missing." Yet they had only three number words and showed the strongest resistance to learning the number sequence from white men. They would indicate the size of a herd of horses by stating how much space the horses occupied when standing next to each other.

We can understand both these phenomena if we remember the far closer relationship of these people with the world around them: the keen observation that unhesitatingly notes the absence of a single animal and can say which one is missing, and the translation of a number that cannot be visualized into a clearly perceived spatial form.

The term *number sense* may be applied to the first of these manifestations. Animals show it when they immediately detect the absence of one of their young. Men also have this latent sense and can develop it. Many a teacher distinctly senses the absence of a pupil when he faces a class doing calisthenics.

NUMBERS AS ATTRIBUTES

Here I must ask the reader to focus his attention on a more subtle distinction expressed in language and offering a wealth of insight into the origin of number words.

Numbers as attributes — is a number not, indeed, an attribute, a trait? "Two cows" — "two" precedes the word "cows" just like, for example, the adjective "beautiful." But we must not let ourselves be deceived by this. "Two" is not a characteristic of the cows themselves, for one cow cannot be two; the Two could be at best an attribute of the entity "two-cows." If, however, we regard "two-cows" as a single unit, we of course no longer need to feel the "two-ness" as a particular attribute, since it is part of the essence of the concept "two-cows." Thus we see that Two is not an attribute in the same sense as "beautiful." Hence, Two is not an adjective. What is it, then? It is a special kind of word — a number word.

Nevertheless, primitive man at first always felt the number to be an adjective. We shall now demonstrate that this is so.

The Number Word in the Noun

Some primitive peoples have completely fused the number and the object into a single entity. The Fiji Islanders, for example, call 10 boats *bola*, 10 coconuts *koro* and 1000 coconuts *saloro*. Naturally this does not hold for any arbitrary number (such as 5 nuts or 23 nuts); yet in contrast to other authorities I see in these words a designation of quantity, although admittedly, tied to the object counted. In German it would be like saying *ein Malter*, a "bushel," or *eine Mandel* (15), thereby implying potatoes in the first instance and eggs in the second. We have a parallel example in the German word *Faden*, "thread." One would think that this was something in the nature of a filament, but it is not: It is a measure — "full fathoms five ..." — applied to yarn. It is as much as can be encompassed by a man's outstretched arms. Today a *Faden* has come to mean a yarn of about this length, and hence is a measure coupled to a specific object.

The examples given show that the primitive people of the Fiji Islands have no number sequence, at least not an extensive one, that has been consciously and clearly detached from objects and thus become abstract.

The Grammatical Double Number (the Dual)

The absorption of the number into the object led to remarkable word forms for the double (the dual), the triple (the trial) and, in some languages of the South Pacific islands, even the quadruple (the quaternal) number. Besides the singular, the German language has only the indefinite plural (the multiple number): *der Mann — die Männer.* If, to make up a hypothetical example, *Manna* (in German) were to mean "two men," then this would be a grammatical dual, an inflected word form indicating the specific number two. This embodiment of the number word into the noun itself is reminiscent of the primitive incorporation of all the details of the "rabbit death" (see p. 10) into a single word. Specific grammatical number words, such as "dual," thus belong to an early stage of civilization.

The ancestral Indo-European language had a form of dual which gradually disappeared from the individual languages of the Indo-European family, surviving here and there only in vestigial form. Classical Greek, for example, had a very ancient but rare dual form:

ho phílos –	*tò phílō –*	*hoi phíloi –*
"the friend"	"both friends"	"the friends"
hē cheír –	*tò cheîre –*	*hai cheîres –*
"the hand"	"both hands"	"the hands."

The ancient Semitic languages, such as the Hebrew of the Bible and also Arabic (until around A.D. 700)* had a very pronounced dual, e.g.:

Arab. *radjulun –* "man"   *radjulani –* "two men"   *ridjalun –* "men."

Surviving forms like these are very suggestive evidence of man's earliest steps beyond the number One. Two has a special place among numbers. Primitive man is intensely impressed by the geminate, or paired, condition he observes in his own body or his immediate environment: their 2 (both) eyes, hands, arms, and legs. He transfers this duality to fixed couples, such as a team of horses or a yoke of oxen (Greek *hippō, bóe*), but also to siblings, friends, and deities whom he sees or wants to see together, like the pair of goddesses (*tò theố*) Demeter and Persephone. The Sanskrit *ahani,* "the day," is grammatically a dual because it includes the night; the Turkish *valid* means "the parent," whereas the dual *valid-eijn* are "the parents"; the Indo-European *nasō* means literally "both noses (nostrils)." Sometimes a pair (like the eyes) is so strongly felt to be a single whole that one of the two (one eye) is counted, as in Chinese, by a special number word (*chih* instead of *i*), or else is indicated as a "half-eye," as in the Irish *súil,* "eye," *di súil,* "the (two) eyes," *leth-súil,* "half (= one) eye." This transference of the geminate condition of the

---

* *Editor's note:* Classical written Arabic maintains a dual form to this day.

body to any pair is likewise beautifully exemplified in Chinese, where the ideogram for "pair" in the general sense is a picture of two hands.

Two trees, two people, formed not with the dual but with the number word Two, denotes merely a fortuitous and not an intrinsic or willed duality.

I — Thou. The first step beyond the One, however, was taken at a still lower level of thought. To the awakening consciousness the world is confronted with himself; the I is opposed to and distinct from what is not I, the thou, the other. Linguistically, too, it is not unlikely that the Indo-European number word *duuo* is in some way related to the German *du* and the English *thou*. In the Sumerian number sequence, "one" and "two" have the meaning "man" and "woman," respectively.

In this primeval dichotomy of the mind, what was One before now breaks apart into One and Two. To man the Two is at first another man, a living You with whom he becomes involved in address and response, and thus in spite of the severance still feels a bond. This is echoed in the fact that the grammatical dual survived much longer in personal pronouns than in other classes of words.

"*Habt's a Geld?*" ("Do you have any money?"), asks the Bavarian peasant — here the Middle High German dual *ez* (Old High German *iz*), "both of you," still lingers on, abbreviated, in the '*s*; the genitive case is *enker*, dative and accusative *enk* (*Wir bitten enk* — "we beg you"). *Es Vogerln tragt's mein Gruss zu ihr*, "You little birds, bear her my greetings," sings the Tyrolean, too, but not the Swabian or the Swiss. Once this dual existed alongside the universal plural *ihr* ("you"); today the plural represents it. But Icelandic still distinguishes between *vid* and *þid*, "we two" and "you two," as opposed to *vjer* and *þjer*, "we all" and "you all." The Old-Frisian dual form for all three persons still survives on the island of Sylt (*wat*, "both of us"; *at*, "both of you"; *jat*, "both of them"), along with the other cases *unk*, *junk*, and *jam*. In Gothic there was *ik*, "I"; *wit*, "both of us"; *weis*, "we"; and *þu*, "thou"; *jut*, "both of you"; *jus*, "you." *Ugkara*, "of us both," and *ugkis*, "us both" or "to us both," exist alongside *unsara* and *uns* (the Gothic *ugk-* is pronounced "unk-"). The Gothic verb has a singular, a dual and a plural: *baira*, "I bear"; *bairos*, "we both bear"; *bairam*, "we all bear". There are also corresponding forms for the second person. In Old Norse *wit Hrafn* means "we two Hrafn," that is, "Hrafn and I." Thus the personal dual lived on and still survives in languages and dialects, while the inanimate dual has long since ceased to exist.

Two as Unity. In the Two we experience the very essence of number more intensely than in other numbers, that essence being to bind many together into one, to equate plurality and unity. Our mind divides the world into heaven and earth, day and night, light and darkness, right and left, man and woman, I and you — and the more strongly we sense the separation between these poles, whatever they may be, the more powerfully do we also sense their unity. To divide the unified, to unify the divided — this opposition between separation and unification has been fixed by languages, time and again, in compounds formed from the word Two, as in the German words *Zwist*, "discord," and *Zwirn*, "twine," in *diploma* and *dispute*, in the English *twin*, which also means "to separate" (see p. 172), in the Turkish *ikiz*, "twins," and *ikilik*, "a dispute" (from *iki*, "two," see p. 113). Old Armenian has an especially appealing example: from the number word *erku*, "two," are derived the opposed pair *erkin*, "heaven," and *erkir*, "earth."

The fact that the Two may also contain what is evil and despised is rooted in that primeval antithesis: twilight is not an auspicious light, the *Doppelgänger*, the "supernatural double," is a sinister companion, nor is Faust content with the two souls that haunt his breast. To impart a pejorative connotation to the meaning inherent in Two, French sometimes uses a phonetic change, as from *bis*, "twice," to *bes-* (*ba-* or *ber*). For example, *bis-sac* is the double sack which the mendicant monk carries over his back and his chest, whence *besace*, "beggar's satchel," and *besacier*, "beggar" (see p. 174).

Number word and Dual. Quite naturally, many languages express their number 2 in the grammatical dual, thus supplementing its meaning by the grammatical form (as in the Indo-European *du̯o*). Thus the Greek *dýo*, "two," and *ámpho*, "both," as well as the corresponding Latin words *duo* and *ambo* have the dual ending *-o*, so that *duo* expresses not so much the abstract number word Two as the inclusive Both, meaning "the one as well as the other." The sense of the word Both excludes counting beyond, whereas Two implies it. The Indo-European *ambhō* gave rise to the Sanskrit *ubháu*, the Greek *ámpho*, and the Latin *ambo*; dropping the first syllable yields the Gothic *ba*, *bai*, expanded to *bajoþs*, whence are derived the English word *both* and Old High German *be-de*.

Words for numbers higher than 2, such as those for 8, 20, 200 and 2000, which retained their dual form carry us deep into early man's numerical conceptions. The Indo-European *okto(u)*, Latin *octo*, is discussed elsewhere (see pp. 23 and 147). The Latin 20, *viginti* < *du̯i-viginti*, 2×10, is an old dual ending in *-i*, whereas the higher tens, like *triginta*, 30, end in *-a* (see p. 150). The Greek *eí-kosi*, 20, is also clearly distinct from the following tens such as *triákonta*, 30, etc. What does all this signify? Here, too, we see the first step beyond unity, but now the unity is no longer 1 but 10, the first cardinal number reached by early man in his counting.

How strongly the Ten was felt to be a new unit is engagingly documented by cases in which 20 is expressed simply by an indeterminate plural of "ten" ("many tens") rather than by "two tens," just as though there were only ten and then twenty but no more tens beyond that. This is what happens in Semitic languages: Hebrew *'eser*, 10, by the suffixing of the plural ending *-im* becomes *esrim*, 20; similarly in Arabic *'asrun*, 10, becomes *isrunq*, 20. Subsequent tens are formed as plurals of the single numbers, as in Hebrew *šaloš-im*, 30, "the many threes," from *šaloš*, 3 (see p. 115). The counterpart in our family of languages occurs in Danish, where the plural of *ti*, 10, which is *tyve*, means 20; The subsequent tens are formed from *tyve* and not *ti*; for example *fyrretyve*, 40, is literally 4×20 and hence 80 (see p. 65).

Not only the Ten, but also the Hundred is thus felt to be a new unit. From 100 as the new One the first step is made toward Two:

100 in Slavic is *sto*, whose indefinite plural is *sta*, but 200 is called *dve ste* (Russian *dvě sti*, Czech *dvě stě*; see p. 98). *Ste* is the dual "double hundred," so that *dve ste* is actually a redundancy. In just the same way the Lithuanian *du šimtu*, 200, is contrasted with the form *šimtai* in the subsequent hundreds (p. 93). Sanskrit uses the old dual form *dve śate* along with the common form *dviśatam*.

This principle is repeated in exactly the same manner with the next order of magnitude, a thousand: in Russian *týsjača*, 1000, *dvé týsjače*, 2000 (dual) and, oddly enough, this form of thousand is used through 4000 but then followed by *pjatj týsjač*, "five thousands" and so on (see p. 23 for the cause of this peculiar Russian word formation).

As was shown for "ten," Hebrew again forms from *elef*, 1000, the indefinite plural *alpayim*,* "thousands," for 2000, but which at the same time (literally) also serves as a round number for "thousands" and "many thousand." This ambiguity reflects the primordial opposition in thought between One and Not-one or between One and Two. The very same ancient notion is also the reason for the Latin *amb-*, "toward *both* sides," taking on the additional meaning, "at, around," that is, "toward *all* sides" — as in *ambire*, "to walk around," and for the doubled character "man" in Chinese to be read as "every man," and for "east" ("sun behind tree") as "everywhere" (Fig. 1).

 Fig. 1 Chinese "everywhere," written "east-east" — that is, "sun behind tree."

Two as the Limit of Counting. All this evidence shows that Two has a special status and is not just a number like any other in the number sequence, but instead is that extraordinary number attained by early man's first hesitant step toward counting. A hesitant step, indeed — for it is not as though man had taken all the following steps at once, running through and building up the whole number sequence. On the contrary, he stopped to catch his breath, as it were. The number 2 is a frontier in counting, the first and oldest of the many we shall encounter.

This is proved not only by the grammatical dual form, or dual number, which we have just discussed, but by other fascinating pieces of evidence as well. In Arabic, for example, the numbers 1 and 2 are adjectives that modify the object counted and testify to the early stage at which number was regarded as an attribute of the thing counted, equivalent to "beautiful" or "big," and not yet as an abstract general concept independent of the object counted. But the numbers following 3, 4, 5, and so on, are nouns (p. 81).

As further evidence we can summon the peculiar formations for 11 and 12, which in many languages are constructed differently from

* *Editor's note: Alpayim* is grammatically a dual form in Hebrew; the plural is *alaphim.*

the following numbers 13 ... 19 (see p. 84); or the formation of number words by backward counting, as in the Latin words for 18 and 19, "two and one, respectively, from twenty" (see p. 74); or the Finnish and Ainu 8 and 9, "two and one, respectively, from ten" (see pp. 75 & 69). It is always the two numbers 1 and 2 that enjoy this special status.

This becomes quite obvious in the case of the ordinals "first" and "second." In many languages these do not have the same relationship to the number words for 1 and 2 as do the subsequent ordinals "third," "fourth," etc., to "three," "four," and so on. The Indo-European preposition *pro*, "before," in the superlative degree "foremost" or in the comparative "in front of," became the Greek *prōtos*, "first"; similarly, Old-Latin *pri-* for *prae*, "before," by way of *pri-or*, "the one in front," led to *pri-mus*, "foremost, first"; whence, naturally, French *premier* and Italian *primo*. Through sound shifts from *pr* > *fr* and *pro* > , "before, for," arose such Germanic forms as the Gothic *frum-ists*, Anglo-Saxon *form-est*, English *fore-st* > *first* and German *Fürst*, "lord, prince." From the superlative degree, Gothic *air*, "early" — *airiza*, Old High German *eriro*, "sooner, earlier" — Old High German *eristo*, comes the German *der Erste*, "the first," with its original meaning of "early morning," for this word is derived from the Indo-European root *ai*, "burn, shine."

Excellent examples from non-Indo-European languages are the Hebrew *rišon*, "the first" < *reš*, "head," and the Egyptian "that which is on the head"; in modern German the word *Haupt*, like the Middle-Latin *capitaneus* > French *chef*, "captain," is restricted in meaning to the leader of a team or group.

"Second" now occasionally appears with the meaning of "the other," Old High German *andar*, Gothic *anþar*, English *other*, Latin *alter*, Lithuanian *antras*, all from the Indo-European root *anteros*, "the one of two"; and sometimes as "the following," as in the Latin *secundus* < *sequi*, "to follow," Greek *deúteros* < *deúomai*, "to lag, stay behind," and similarly in Sanskrit and Anglo-Saxon.*
Among the non-Indo-European languages I may mention Finnish, in which *yksi* is one and *kaksi* is two, but *ensimäinen* means "the first" and *toinen* "the second" (the other); in Egyptian, "the second" is called "brother." Whether these formations originate from counting on the fingers or from some other concepts is less important for our purposes than the fact that 1 and 2 clearly occupy a special place in the number sequence and that there was hence a pause after 2, early man's first step in counting. We recognize, moreover, that One itself assumed, and still holds, a unique position relative to the other numbers; we shall speak of this later (p. 19).

The Step to Three. With Three a new element appears in the concept of numbers. I — You: The I is still in a state of juxtaposition toward the You, but what lies beyond them, the It, is the Third, the Many, the Universe. This statement, in which psychological, linguistic, and numerical elements come together, may perhaps roughly paraphrase early man's thinking about numbers. "One — two — many": a

---

* *Translator's note:* In Swedish the common word for "the second" (of or more) is *andra*.

curious counting pattern, but it is exactly mirrored in the grammatical number forms of the noun, singular — dual — plural, as in the Greek *phílos, philō, phíloi*, where the third number form is thus the plural. An old Sakai in Malacca, on being asked his age, replied, "Sir, I am three years old." To him 2 was the You, the near and familiar with which he lives, to which he feels related and with which he interacts, but this is no longer true of the It, the 3; for him that is the Many, the Alien, the Unknowable. A magnificent confirmation of this is the ancient Sumerian number sequence, which begins: "Man, woman, many, ... " (p. 165).

Many investigators suspect, with good reason, that in the number words "one, two, and three" are latent the roots of the personal pronouns "this, the, and that," or even that the primordial forms of these pronouns became the first three number words.

Three is the "beyond," the "trans-." It has been thought that the Latin *tres*, 3, is related to the Latin *trans*, "across, beyond" (the root of *trare*, "to penetrate"; cf. *intrare*, "to enter by force"); and correspondingly the French *trois*, 3, to *très*, "very," the English *three* to *through* and thus the Indo-European *trejes* to *tre-*. Even though this theory cannot be proved with certainty, it does have in its favor the striking linguistic resemblance and the possible interpretation of Three as the number transcending the old numerical barrier after number 2.

The writing of the Egyptians and the Chinese, however, has perpetuated the early conceptual stage of three = many in a remarkable fashion (Figs. 2 and 3). To express the "many," the plural, of an idea, they write the character for it three times (just as the Babylonian number word *eš*, "three," became the plural ending). An ancient Egyptian inscription, for example, reads: "To the King thousands

were sacrificed and hundreds offered up (see Fig. 8, p. 42)."

The number sequence as such actually begins with three: "three, four, five, ..., etc." When a tribe of South Sea Islanders counts by twos, *urapun, okasa, okasa urapun, okasa okasa, okasa okasa urapun* (that is, 1, 2, 2′1, 2′2, 2′2′1), we distinctly feel that they have not yet taken the step from two to three. And we realize with astonishment that these people can count beyond two without being able to count to three.

The step to Three is the decisive one, which introduces the infinite progression into the number sequence. We recognize it from its action in reverse; Two is stripped of its unique position and is recognized as a number just like any other; the grammatical dual of the original Indo-European language disappears. Furthermore, from the cardinal numbers 3, 4, 5 are formed the ordinals "third, fourth, fifth," and then by analogy, going backwards, from "the other" is developed the "second." The world of numbers has entered the personal world (two = You) through the back door. Only "the first" still holds out: There is no such thing as a "firstest." Whenever such a form does occur, as in the Turkish *bir-inzi* < *bir*, "one," it

Fig. 2 Three as the plural in Egyptian: (1) flood = heaven with 3 water jugs; (2) water = 3 × wave; (3) "many" plants = 3 × plant; (4) hair = 3 hairs; (5) weep = eye with "many" (= 3) tears; (6) fear = dead goose with 3 vertical strokes, the general plural sign, next to the ideogram.

Fig. 3 Three as the plural in Chinese: (1) forest = 3 × tree; (2) fur = 3 × hair; (3) all = 3 × man; (4) speak endlessly ("much") = 3 × speak (mouth from which words emerge); (5) rape = 3 × woman; (6) gallop (ride "much") = 3 × horse.

indicates a number sequence that arose not out of a people's own early levels of conception but was constructed by analogy with an already developed neighboring number sequence (see p. 112).

The step to Three is a step across the threshold of darkness, before which the concept "number" was still deeply rooted in the life of the soul, out into the prosaic but clear and bright light of practical life. If this step means a waning of the power of detachment, of the ability to impart to each number its own distinctive features obtained from the object itself, it is compensated by the growing power to build up the number sequence as a useful structure with applications of undreamed-of scope. This was not accomplished at a single stroke, of course, but was advanced time and again, from one numerical boundary to the next, at each of which the sequence pauses to catch its breath, as it were, and waits until it is overtaken by the reality of life and pushed on to new and higher numbers.

Our analysis of the grammatical dual has yielded an abundance of insights. In addition to the grammatical dual, some primitive peoples have a triple grammatical number, and here and there we even find a quadruple. Beyond that, however, human speech has never formed a definite plural noun form. This suggests that a special position is also occupied by the number four which we shall now approach from a different tack.

NUMBERS AS ADJECTIVES

Now that we have seen how strongly the number ("four") of something that is counted ("four large houses") was felt at a very early stage to be one of its attributes like any other ("large"), we should not be surprised to find it appearing in grammatical form as well, that is, as an adjective. Two other aspects, however, may be more surprising: that only the first 4 number words take this grammatical form, and that over the millennia quite a number of languages of our own general culture have faithfully preserved this early state.

The first four number words. These words in the primitive original Indo-European language and in many languages derived from it, such as Sanskrit, Celtic, Greek, and Old Norse, not only had three genders but were also inflected, just like the true adjectives "beautiful" or "large"; in Gothic and Latin, Four has already been frozen into an nondeclinable number word.

"One" has still kept its ability to agree with the subject in gender and case in some modern languages, including German (*ein-en Baum*, *ein-er Frau*, "against a tree," "of, to a woman"); hence, I shall cite here only the less familiar forms for "two," "three," and "four":

Greek	Latin	Gothic
dýo	*duo, -ae, -o*	*twai, twōs, twa*
dyoîn	*duorum, -arum*	*twaddje*
dyoîn	*duobus, -abus*	*twaiþ*
dýo	*duos, -as, -o*	*twans, twōs, twa*

Greek	Latin	Gothic
treîs, tría	*tres, tria*	*þreis, þrija*
triôn	*trium*	*þrije*
trisí	*tribus*	*þrium*
treîs tría	*tres, tria*	*þrins, þrija*
téttares, téttara		
tettárōn	*quattuor*	*fidwōr*
téttarsi		
téttaras, téttara	(*unchanged*)	(*unchanged*)

Accordingly, the genitive "of the two men (women)" is: Greek, *dyoîn androîn* (gynaikoîn); Latin, *duorum virorum, duarum feminarum*; Gothic (Luke 9:16), "Then he took the five loaves and the two fishes," *fimf hlaibans jah twans fiskans*, and (Luke 9:33), "Master, it is good for us to be here: and let us make three tabernacles," *hleiþros þrins* (see also Fig. 27, p. 129). In Old Norse the inflected cases of four are *fiorer, fiogorra, fiorom, fioran*.

One in German. In German, even to this day, One is still declined and has three genders; it is thus an adjective in form. Its meaning as a number, however, has been thereby gradually attenuated, as we can see clearly in English, where the number word *one* (< Latin *unus*) has been eroded down to an article, e.g., *a tree*; only when the number is distinctly emphasized does it appear in its old and full form of "one," as in *one tree*. Indonesian provides an additional example: the number word *satu* has shriveled up into the article *se*: *se orang*, "a man," but *orang satu*, "one man."

High German — that is, the literate German language of today — can convey the numerical sense of *ein* ("one") only by inflection in speech. The same is true of other languages, such as French and Italian, while Dutch uses accents even in writing: *een boom* and *één boom*, "a tree" and "one tree." Only a few German dialects still make a clear-cut distinction. In Darmstadt, for instance, one might ask, *Habt ihr en Baum im Gadde?* ("Do you have a tree in the garden?") and be answered, *Ja, awwer nur aan!* ("Yes, but only one"). Languages that have no article, such as Russian and Latin (*homo* = "man," "a man," "the man," or "one man"), and those of most primitive peoples have naturally preserved the number word from external and internal erosion: "man" means "one man."

This attenuation of meaning is connected with the fact that One is never really felt as a number. One, as the antithesis of Many, had already taken a special position as far back as the Table of Ordinal Concepts established by the Pythagoreans. Plato constantly emphasized this: Like the Now in time and the Point in space, the One among the numbers cannot be further subdivided. Hence it conceals within itself no plurality which it collects together into unity, and since it is in this that the essence of number lies, One is not a number. Since, according to Euclid, "a number is an aggregate composed of units," One is itself not a number, though it is the source and the origin (*fons et origo*) of all numbers. Throughout the Middle Ages no one thought differently. One was designated sometimes as *genetrix pluralitatis*, the "mother of plurality," sometimes as *principium quantitatis,* "origin of plurality," and then again as *radix universi numeri et extra numerum,* "the root of every number and yet itself no number." The Salem Codex, a 12th-century manuscript highly significant for the history of our numerals, which we shall often encounter, states: "Every number can be doubled and halved, except for unity; this can, it is true, be doubled, but not halved — *in quo magnum latet sacramentum,* wherein lyeth concealed a great Mysterium [God]." As late as 1537 the German arithmetician Köbel wrote in his manual of computation:

Darauss verstehstu das I. kein zal ist | sonder es ist ein gebererin | anfang | vnd fundament aller anderer zalen.

"Wherefrom thou understandest that 1 is no number / but it is a generatrix / beginning / and foundation of all other numbers."

We are quite well able to understand this view, but precisely because we do understand it, we also recognize that One was regarded as a number only after the very concept of number had been clarified relative to the higher numbers. From these the number concept extended back to One, drained it of its philosophical content and made of it a number like the others. This is a process we have already described in the case of Two (see p. 17) and which we shall meet again later, and far more strikingly, in connection with Zero. Now we can understand why a French writer of the mid-16th century, in listing the digits, wrote:

les huicts figures 2, 3, . . . 9 — "the *eight* figures 2, 3, . . . 9."

Michael Stevin, the man who introduced the algorism of decimal fractions, was probably the first mathematician expressly to assert (in 1585) the numerical nature of One. He proved it somewhat as follows: If from a number 3 I subtract a non-number, then 3 remains; but since $3 - 1 = 2$, then 1 is not a non-number and must therefore be a number. Yet it is for this very reason that the old view persists, as Schiller has documented (*Piccolomini* Act II, Scene 1):

"*Fünf ist
des Menschen Seele. Wie der Mensch aus Gutem
und Bösem ist gemischt, so ist die Fünfe
die erste Zahl aus Grad' und Ungerade.*"

"Five is
The soul of Man. As Man is composed
Of good and evil mixed, so is the Five
The first number holding odd and even."

Thus One is not a number, otherwise $3 (= 1 + 2)$ would be the first odd number.

Even today we still often hear the question, "What is a prime number (such as 7)?" The answer generally given is: "A number divisible only by itself." This definition forgets that it is also divisible by 1. One feels that 1 is, after all, not the same kind of number as the others. Nor does it "act" on the number a as does every other number. For example, $a \times 1 = a$, an argument likewise formerly adduced against its numerical nature.

Language, however, has fortunately preserved this special status of One. We have already spoken in some detail of the peculiar quality of the first ordinal number (see p. 16); perhaps we should add here that, for example, when the Frenchman counts the days of the month, *le premier*, "the first," *le deux, le trois,...*, *le trente*, "the two, the three,...*, the thirty," he again gives special emphasis to One as the only ordinal number. For "one o'clock" the Italian says *il tocco*. The Germans, like the French, call the number One in card games and dice the *As* [ace], just as the Greeks once specially designated it as the *oiné*.

All this evidence supports the view that unity was first recognized as a number from the direction of plurality; hence, it is wrong to suppose that the number sequence must have been mere child's play to devise once the idea of unity was at hand. The idea of One did exist, to be sure, but embodied in the object as such and not as an independent idea of number, to say nothing of being a detached, abstract number word.

Two and Three in German. Although One is still inflected, Two has already lost its variability in modern German. Yet two and "three" used to be inflected:

Case	Old High German	Middle High German
nom., acc.	*zwene, zwa (zwo), zwei*	*zwene, zwo, zwei*
gen.	*zweio*	*zwei (g)er*
dat.	*zweim*	*zwein*
nom., acc.	*dri, drio, driu*	*dri, driu*
gen.	*drio*	*dri(g)er*
dat.	*drim*	*drin*

The *Nibelungenlied* (Verse 437) says:

Der schilt was under buckeln ... wohl drier spannen dicke.
"Beneath the boss the shield was three spans thick."

Until well into the 17th century, Two still generally had both gender and inflection.

Masculine gender *zween*:

Eber zeugte zween Söhne, "And unto Eber were born two sons" (Genesis 10:25);
Niemand kann zween Herrn dienen, "No man can serve two masters" (Matthew 6:24);

feminine *zwo* and neuter *zwei*:

Und stand auf in der Nacht und nahm seine zwei Weiber und seine zwo Mädge.
"And he rose up that night, and took his two wives, and his two women servants" (Genesis 32:22).

Inflection:

Durch zweier Zeugen Mund wird allerwegs die Wahrheit kund.
"Through the mouths of two witnesses the truth is always made known" (Goethe, Faust I, line 3013).

"*Zweier*" "of two" is still heard occasionally to this day. Again, vernacular dialects which are close to the people are faithful guardians of much that is traditional in speech; in Upper Hesse, for example, people still say:

zwien Osse, zwoo Käu, zwaa Kinner, "two oxen, two cows, two children."

The word *zwo* commonly used in German today serves merely to avoid misunderstanding (*zwei, drei*).

Four as an Old Limit of Counting. Let us pause here for a moment. What significance does the use of number words as adjectives have in the development of the number sequence? It represents the first detachment of the number from the object counted. A number word can now be used to count anything; it already stands free and independent in the realm of language and can associate itself with every object, although not yet purely counting, but still "describing an attitude."

But here is a striking question: Why do only the first four numbers, "one," "two," "three" and "four," appear as inflected adjectives? Why not "five" or "seven" or "twenty"?

We can very well answer this ourselves: Because they were the earliest number words (not counting the very first step forward, from One to Two, which had still taken place completely within the realm of the mind). A word that agrees with its subject in gender and case is more intimately bound up with it than one that does not; and the word does this because the concept for which it stands does, and hence because the number is, as yet, more intimately connected with the thing counted. But this is, as we have seen, a sure sign of prehistoric times.

One may then ask: Of course this is natural enough, but why should the break come just after Four? Why not after Seven? Two reasons may be given. The first is that the hand has four fingers, not counting the thumb. What happened to the thumb here was like what happened to One — it was not regarded as being equal, it was not a "finger" like the others. The handsbreadth, measured without the thumb across the knuckles, was used as a basic measure by almost all ancient civilizations.* The Greek and the Egyptian *ell*, for instance, had 6 handsbreadths, or $6 \times 4 = 24$ fingers; likewise the Roman foot (*pes*) was made up of 4 *palmae* and of $4 \times 4 = 16$ *digiti*. A second reason might be that a quantity larger than four, or even three, can no longer be directly apprehended. If we ask, "How many people were there?", the answer is "three or four," not "nine or ten," for that would already be "a large number." And in those early primitive times only clearly perceptible numbers were apprehended as words, as we have seen in the case of the dual, which expresses a distinct Two, the "other" that goes with the One.

Four, then, was certainly another limit of counting.

A series of unexpected pieces of evidence reveal this very ancient break at Four, even in complete, "mature" number sequences. In

* *Translator's note:* In English a horse's height at the withers is still measured in "hands" of 4 inches each.

the original Indo-European language, *octō(u)*, "eight," is grammatically a dual, as can be clearly recognized in the Greek *oktṓ* and the Latin *octo* (which have the *–o* ending of the dual; see p. 14). It must therefore mean a doubling of four (2 × 4), although the number word Four cannot be recognized in it linguistically. The hypothesis of this ancient counting by fours is confirmed by the startling similarity between the Indo-European terms for "nine" and "new":

"nine": Sanskrit *nava*   Latin *novem*   Gothic *niun*   Tocharian *nu*
"new": Sanskrit *navas*   Latin *novus*   Gothic *niujis*   Tocharian *nu*.

The explanation is that after Eight, when the breadths of both hands have been used up, there follows a "new" number, which is Nine. On the other hand, the distinction given to the number 13 and its superstitious burden cannot be explained by the supposition that is a "new" number after the third Four.

Of the still living languages, the Slavic tongues have preserved the age-old break at Four, in many cases quite sharply:

The Czech says (p. 98):

"one and one *are* two" *jedno a jedno jsou dvě*
"two and two *are* four" *dvě a dvě jsou čtyři*
but "three and two *is* five" *tři a dvě jest pět.*

The word *jest*, "is," is always used when the sum is greater than four — that is, from five on. We can readily see the connection, for we need only add the things counted:

1 cow and 1 cow *are* 2 cows; 2 cows and 2 cows *are* 4 cows; 3 cows *and* 2 cows *are* 5 cows, and so forth.

When the objects counted are named, the plural of the verb is invariably used. Thus up to the point where the number was always coupled with the objects counted, that is, up to 4, this peculiarity has persisted even today with abstract numbers. This is another remnant of prehistoric times and can be understood only with reference to them.

The break after Four is still more characteristic in Russian (p. 98). Whereas the number word One agrees with its noun, the things counted after 2, 3, and 4 in Russian are in the genitive singular; from 5 on, however, the number in Russian governs the genitive plural. Thus 2, 3, 4, 5, and 100 houses are:

odin dom   *dva (tri, četýrě) doma*   *pjatj (sto) domov*
1 house   2 (3, 4) of house   5 (100) of houses.

How did this strange rule come about? This form is by no means the meaningless genitive singular, "two of house," but an old dual that has now disappeared; the form now used resembles the genitive singular only superficially. One can easily understand how in speech the noun *house* was put in the dual, not only after Two, but also after Three and Four, but *only* through Four! In the case of a compound number, by analogy, the noun follows the same rule: "24 of the house," but "25 of the houses." What happened here? In Russian, as in Old Slavic generally, the number words up to and including Four were inflected adjectives; beginning with Five they became rigid, nondeclinable nouns followed meaningfully by the genitive plural ("five of houses").

This old limit of counting at four, however, is not restricted to the units; under their influence it extends to all the other ranks, through the *Děsjatj*, *sót*, and *týsjač* according to the rule are genitive plurals; see also the Czech number sequence (p. 98).

Tens		Hundreds		Thousands	
10	*děsatj*	100	*sto*	1000	*týsjača*
20	*-dzatj*	200	*-sti* (Dual)	2000	*-týsjači*
30	*-dzatj*	300	*-sta*	3000	*-týsjači*
40	(*sorok*, p. 185)	400	*-sta*	4000	*-týsjači*
50	*-děsjatj*	500	*-sót*	5000	*-týsjač*

Except for a single construction, inexplicable to the uninitiated, the ancient break after Four has persisted. To the question, "How old are you?" the Russian child will answer:

mne dva (*tri, četýrě*) *goda* — "to me [are] 2 (3, 4) of the *year*";

but thereafter:

mne pjatj (*děsatj*) *let* — "to me [are] 5 (10) of the *summers*."

Thus up to the age of four a person is "years" old, and after that "summers."

Now comes a surprise: The Romans had the very same custom. They counted the years of a person's age, "two, three, four, ... years old," as follows:

bimus, trimus, quadrimus, but then *quinquennis, sexennis, ...-ennis.*

Here again we have a distinct break after four. At the same time the languages have here preserved the ancient Indo-European counting of the years by winters; *bimus* < Indo-European *bi-himus*, "two-wintered," is akin to the Sanskrit *himas*, Greek *cheimốn* and Latin *hiems*, "winter." In the Germanic languages we have the corresponding Anglo-Saxon *anwintre*, "one-wintered," just as even today in the Lower Rhine region yearling cattle are called *Einwinter* "one-winter." The Greeks, moreover, had the same term, *chímaira*, a "one-winter" yearling, a male goat (with the body of a fish) from which the mythical chimaera arose. *Twalib-wintrus*, "twelve-wintered," is the term applied to the twelve-year-old Jesus in the Gothic Bible (Luke 2:42; see also p. 27).

The habit of counting by "years," as in the Latin *annus* < *-ennis*, is of more recent origin and begins only after the number Four. The old break after Four was also preserved in two further Roman customs: in the naming of children and in counting the months. The Roman father gave his children numerical names only from the fifth child on: *Quintus, Sextus, Septimus*. No *Quartus* or *Tertius* has been recorded in Roman history.

The Roman Calendar. The Roman year originally began on March 1. The first four months had the proper names *Martius* (31), *Aprilis*

(30/29), *Maius* (31) and *Junius* (30/29); thereafter, however, following the fourth month, there were *Quintilis*, (31), *Sextilis* (30/29), *September* (30/29), *October* (31), *November* (30/29) and *December* (30/29).

On the subject of the Roman calendar, which — somewhat modified, to be sure — is also our own, it is interesting to note that in the very earliest epoch the Romans used an old lunar year of ten months, from March through December, with $4 \times 31 + 6 \times 30 = 304$ days in all (the first numbers cited after the names of the months, cf. p. 37). The legendary king Numa Pompilius is said to have replaced this, following the Greek example, with a year of 12 months and 354 days, which was later changed to 355 because the odd number was considered sacred. The new calendar had four months of 31 and seven of 29 days each (the second number after certain of the months' names above), and one month of 28 days. Of the new months, Numa Pompilius placed *Januarius* (29) at the beginning and *Februarius* (28) at the end of the year. Then in 450 B.C. the Decemvirs for the first time placed January in the position of the 11th month, just before February, where it still is today. To make this revised lunar year consistent with the solar cycle, intercalary days were inserted at the year's end, in February (cf. *bissextile*, p. 174).

The Roman year began, significantly enough, in the spring, on March 1, on which day the two new Consuls were inaugurated. For a completely different reason, in the year 154 B.C. January 1 was adopted as the official beginning of the year, and has remained so until the present day. A rebellion broke out in Spain at the end of the year 154. To avoid a change of command on March 1, 153 B.C., the year 154 was allowed to last only ten months and the new year 153 was begun on January 1. With this shift the previous numerical names of the months became literally false and lost their meaning, so that October (the 8th month) should now have been December (the 10th) and others to correspond.

Caesar then undertook to rearrange the calendar, which had become distorted with the passage of time; in his honor *Quintilis*, the month in which he was born, was renamed *Julius* in 44 B.C. And *Sextilis* became *Augustus*, because that emperor, in the year 8 B.C., also had corrections made in the calendar, which had again fallen into disorder.

We may also point to some special kinds of number words that in several languages have preserved a distinct boundary after Four (Latin) or Three (Greek, English); for example, in the adverbial forms:

Latin:	*semel*	*bis*	*ter*	*quater*	but then *quinqu-ies*, etc.
Greek:	*hápax*	*dis*	*tris*		but then *tetrá-kis*, etc.
English:	*once*	*twice*	*thrice*		but then *four times*, etc.
German:	*ein-*	*zwei-*	*drei-*	*vier-*	*fünfmal*, etc.

Thus in German the break has disappeared from view.

Linguistically: Greek *hápax*, "once," in which *ha* < Indo-European *sm̥-*, "one" (see p. 146), because Indo-European *m̥* > Greek *a* and Indo-European *s-* before a vowel > Greek *h-*; cf. Indo-European *septm̥* > Greek *heptá*, "seven." Latin *-ie(n)s*, "-times" > *-iens*, afterwards forms *quot-iens, tot-iens,* "how many times," "so many times."

Let us conclude with two more examples from different cultures. In addition to their own number words for the numbers 1 ... 10, the Japanese also have words for the same numbers borrowed from the Chinese (see Far Eastern Systems, p. 449). Thus there are certain compound words, for example those compounded with "evening," which the Japanese counts "1, 2, 3, 4 evenings" only with his native Japanese number words, but from "5 evenings" on he uses the borrowed words. This break after Four, which occurs so often in the number sequences of primitive peoples, may also help us to understand the special status enjoyed by Four in the Turkish language. If a Turk wishes to intensify the force of an assertion, for example to say that he has worked *very, very* hard, he says that he has done it "with four hands" (*dört elle*). He can also "wait with four eyes," and when he rides at a gallop he rides "with four horseshoes" (*dört nal*). Perhaps this intensifying sense of the word Four is rooted in early astronomical ideas, but this together with the "numerical" reason may have led to the unique position of Four in Turkish.

Before we carry our discussion further, we must make the following important observation: All these pieces of evidence show that Four was a very ancient limit in counting. Will Five be the next such boundary? Five, of course, is practically offered by the human hand with its fingers. No, surprisingly enough, in our Indo-European culture this was not the case. Five was not a limit of counting but rather one of the essential members of the number sequence. This very knowledge of how to *arrange* the numbers following after Five is, indeed, the reason why beyond Five the number sequence no longer pauses but runs on continuously. The first four number words are not members of the sequence; they are the first steps forward, made gropingly and without any sense of a general plan, which — although it too contains numerical breaks — was finally attained with the number Five.

The Units as Adjectives. After this excursion on the subject of Four as a limit of counting, let us now go back to our main subject, numbers as attributes, and examine number words as adjectives.

Celtic has a remarkable manner of differentiating the units from the higher numbers in a very characteristic fashion. In the case of a compound number like 23, for instance, it pulls out the unit 3, places it as an adjective directly in front of the thing counted, and then puts the higher number after the object:

12 hours Celtic *di huair deec,* "2 hours 10"
23 sons Celtic *tri meib ar hugeint,* "3 sons and 20"
185,000 Celtic *coic mili ochtmugat ar chet,* "5 thousand 80 and 100."

This pattern expresses the ancient, intimate relationship to the units, the very earliest of the numbers; they must stand next to the noun and are governed by it, to the extent that they are still declined like an adjective. The new, "larger" numbers, which are held far more vaguely and remotely by the imagination, are merely hooked on afterward. The same phenomenon appears in Old Norse, in which language the age of 48 is expressed as follows:

hafdi atta vetre en fimfta tigar,
"he had 8 winters in the fifth decade."

We find the peculiar position of the unit again in Old High German, not quite so strikingly, but still distinctly enough (Luke 15:7):

niuni inti niunzug recte, "nine and ninety just persons,"

in which the unit 9 in the compound number 99 is governed grammatically by the noun and appears in the plural; by contrast, in Gothic we find

niuntehundis jah niune garaihtaize, "90 and 9."

Adjectives and Nouns in the Number Sequence. We have already encountered more than one change in the classification of number words, the last in the case of Four as a limit of counting, where the words "one" through "four" are adjectives but the following number words are not. This is the most convincing proof that the verbal number sequence was not built up all at once, in a single move. As an example, let us consider Latin (see p. 92).

Using the symbol "(adj.)" to indicate that the number word is an adjective, we cite the following Latin numbers in the dative case (so as to show their inflection as well):

(1) to the 3 men	*tribus viris*	Literally: "to the three (adj.) men"
(2) to the 4 men	*quattuor viris*	Literally: "to the four men"
to the 10 (100) men	*decem (centum) viris*	Literally: "to the ten (hundred) men"
(3) to the 200 men	*ducentis viris*	Literally: "to the two hundred (adj.) men"
(4) to the 1000 men	*mille viris*	Literally: "to the thousand men"
(5) to the 4000 men	*quattuor milibus virorum*	Literally: "to the four thousand *of* men."

Thus we see in this number sequence the succession of adjective (1); indeclinable number word (2); adjective (3); indeclinable number word (4); inflected noun (5) that governs the partitive genitive case in the object counted.

Let us examine these examples more closely:

1. *Uno, duo, tres,* and originally also *quattuor* were inflected (see p. 18); the last of these later became indeclinable through the influence (acting in reverse) of the subsequent number words 10, 9, ..., 5.

2. *Quinque,* 5, ... *decem,* 10, and the rest through *kentum,* 100, are all indeclinable number words. They stand "dumb," so to speak, next to their noun ("men" in our case), which alone assumes a relationship within the sentence through its inflection in the appropriate case (here in the dative). Since *decem* and *centum* are derived from the same Indo-European root *dekm* (see p. 126), both these number

words may have been formed simultaneously. Thereafter the language appears to have driven the number sequence beyond 100 into a new direction:

3. The hundreds, 200, ..., 900 are again adjectives. Why? Because of the remote effect of "two" and "three." *Centum* was felt as a new One. As later formations, however, the hundreds do not go through the change in word form after Three, but rather remain adjectives all the way through 900. And yet the old break after Three can still be seen in the sound shift from *–centi* to *–genti*: *du-* and *tre-centi*, but then *quadrin-genti, quin-genti, ..., non-genti*.

4. *Mille*, 1000, is an old obstacle.

5. In its multiples, 2000, 3000, ..., however, *mille* is not reduced to an adjective as are the hundreds; they remain nouns and are themselves counted, like any other nouns: *duo milia*, "two thousands." As a noun *mille* is itself declined, takes on a relationship to other words in the sentence, and itself takes the genitive case in the object counted: *quattuor milia virorum*, in contrast to the uninflected number words in paragraph 2. This procedure is followed through *centum milia*, 100,000

6. The higher thousands are all formed from this "last" number rank, "a hundred thousands." In the course of their formation *centum* is pluralized to *centena*, "every hundred," and the count now makes use of the "times" form:

decies centena milia, "ten times every 100 thousand" = 1 million.
vicies centena milia, "twenty times every 100 thousand" = 2 million.

The Roman, accustomed to dealing with very large amounts in public affairs, for example, pronounced the adverbial word "times" only once. In his History of the Caesars, the writer Suetonius relates that the Emperor Vespasian, upon succeeding to office, ordered the amount of money (in sesterces) in the state treasury to be counted and reported to him. The amount was colossal, but it could be expressed vaguely in only two words:

quadringenties milies, "400 times 1000 times,"

but this referred to *centena milia*, meaning $400 \times 100 \times 100,000 = 4 \times 10^{10}$ sesterces.

The Emperor Tiberius once deliberately overlooked the implied *centena milia*, thus making a "mistake" in his own favor. In her will, his mother Livia had bequeathed to his relative Galba

sestertium quingenties,

literally "of sesterces 500 times" (to be read: "times 100,000") = 5×10^8 sesterces. Tiberius, however, who was residual heir and moreover had not been on the best of terms with his mother, used the customary omission of the 100,000 to his own advantage and paid out to Galba only 500 sesterces — *quia notata non praescripta erat summa*, "because the intended sum had not been written out in full," according to the Roman narrator. In terms of Roman numerals, Tiberius read $\overline{\text{D}}$ instead of $\overline{|\text{D}|}$, suppressing the brackets (for more on this subject, see p. 44).

In this thenceforth undisturbed and consistent use of the adverbial "times" formation to express the largest number we distinctly sense the use of a conscious, synthetically created device, whereas the beginnings of the number sequence, through 1000, grew up laboriously and at the same time unconsciously, without direction, from still obscure conceptions of number.

Note: in medieval manuscripts the expression *mille milia*, "1000 thousands" = 1 million, often took the place of the Roman expression "ten times a hundred thousand"; thus *septies mille milia*, "7 million."

It is of interest to recall that Adam Riese knew only Thousand as the highest numerical rank and thus would have expressed 40 million as "40 thousand times a thousand" and reported the wealth of Vespasian's treasury as "40 thousand thousand times a thousand" (see p. 142).

With the Roman word for "sesterce" we encounter the same case of a number slipping into the guise of a noun in its grammatical form. *Sestertius* < *sem(is)tertius*, "half of the third," meaning two and a half, was abbreviated *II S*, for *duo et semis* (for Roman coins and their inscriptions, see pp. 28, 78, 355). "Hundred sesterces" was properly written *centum sestertii*, and thus "2000 sesterces" was *duo milia sestertium* (not *-orum*). Now, however, this third-declension genitive plural ending came to be felt as a neuter singular — *sestertium*, genitive singular *sestertii* — and then dropping the *milia*, "2000 sesterces" was expressed as *duo sestercia*, literally "two sesterces." Moreover "500,000 sesterces" was correctly said as *quinquies centena milia sestertium*, but after omitting the hundred thousand it becomes *quinquies sestertium*, in which the numismatic inscription was thought of as neuter nominative singular, "five times a sesterce." Thus "he bought it for 2 million sesterces" was expressed as *emit sestertio vicies*, literally "for twenty times a sesterce." As a result, the one word, depending on its form, could have three different meanings:

sestertius, -ii, —	*sestertia, -orum,* —	*sestertium, -ii*
one sesterce	1000 sesterces	100,000 sesterces.

To sum up: there is no doubt that the inflected adjective (such as *duo, tres*) is more closely tied to the things it counts than is the indeclinable number word in noun form (e.g., *decem, centum*), which does not depend on the objects counted. The noun (*milia*) made itself completely independent, and it, not the object counted, expressed the grammatical relationship within the sentence (the case). The object counted followed it as an appendage (in the genitive case). Thereafter, with the "times"-word forms (*decies centena milia*), such a high degree of confidence in dealing with numbers and also such a degree of consciousness had been attained that henceforth the number sequence ran on without disturbance or change in form. Thus we have, in the changes in the form of the number words, a remarkable means for tracing the way in which the concept of numbers was gradually refined and the number words became increasingly detached and abstracted from the objects counted.

Special Number Sequences and Classes of Numbers
Having concluded our discussion of numbers as attributes, let us
now speak of two more characteristic ways in which numbers have
been used.

There are some primitive peoples who at times employ a special
number sequence for various things — for example, for animate
objects, round things, days, and so on. Thus, for instance, a tribe of
North American Indians (in British Columbia) counts as follows:

	1	2	3
Animate things	*menok*	*maalok*	*yatuk*
Round things	*menkam*	*masem*	*yutsqsem*
Long things	*ments'a*	*mats'ak*	*yututs'ak*
Days	*op'enequls*	*matlp'enequls*	*yutqp'enequls*, etc.

In the language of this tribe, "two men" are called "*maalok* men,"
but "two days" are "*matlp'enequls* days." What a difference from
our own single, abstract number sequence, that is used for counting
everything!

This becomes understandable if we recall that early man experienced
things in his environment far more intimately and immediately than
we do, that to him a rabbit is a single, particular, real, living rabbit
and that he never thought of abstracting from his concrete reality
any general concept of "animal." Similarly the object also places
its imprint on the number word by which it is counted.

Incidentally, vestiges of this are still extant in our own speech. For
example, we still have different expressions for two of various objects.
We speak of a *yoke* of oxen, a *pair* of shoes, of a *brace* of partridge,
of *twins*, or of a *duet*, but we do not say "a duet of shoes," or "a
yoke of children," or "an ox twin," even though each of these expressions means two. Not only the number 2, but also something of
the actual object has gone into the numerical expression. Similarly,
pen nibs are bought by the gross and buttons by the dozen; in Germany eggs were formerly sold by the *Mandel* (15) and the *Stiege*,
"score" (20). One does not go to the store to buy a *Mandel* of buttons.

The same is true of words for groups of unspecified number, like a
herd of deer, a *pride* of lions, a *bevy* of girls, a *flock* of sheep, a *pack*
of wolves; we do not say "a bevy of sheep" or "a pride of deer."
The form of the number is closely bound to the concept. Ancient
measures are likewise strongly tied to the things they measured:
cloth was measured in *ells*, height in *feet*, depths are stated in German
folk tales in *Klaftern* and by seamen in *fathoms*. All these are measures
of length, of about the same order of magnitude, but there exists no
system of conversion (such as 1 ell = 2 feet). The fact that a "last"
in English (*last, load*) may have greater or less value according to the
thing specified (1 last of meal or herring amounted to 12, of salt to
18 tons, of gunpowder to 24 kegs, of bricks to 500, of tiles to 1000, of
hides to 144 pieces, of wool to 12 sacks) is fairly easy to understand,
but that the number word used as a measure should have different
values — that a "hundred" herring meant actually 120, codfish 124,
or salt 126 tons — is surprising. Other ancient traditional measures
(such as the quarter of grain, the carat, the bushel, or the scruple)

are similarly tied to the substance measured. The only universal measures are our "abstract" decimal measures, which were artificially devised at the time of the French Revolution.

Money is, in a sense, a kind of "thing-free" universal measure. Yet even here some primitive peoples have exchange media that are tied to the objects whose value they express, such as the shells that can be used to buy coconuts but not a chicken, which must be paid for with porpoise teeth, there being no system of conversion between the two forms of "currency."

An intermediate stage between the abstract number sequence and the concrete sequence tied to the things it counts is represented by the number classes, developed primarily by the Chinese (some 100 such classes) and the Japanese (about 50), but are also known in Persian (20) and Turkish (2). A number-class word of this kind is placed between the number word and the named object just like in our own expression 8 "head" of cattle. Thus it happens that a certain number word cannot be joined directly to the thing, whatever it may be, but only to a few classes of objects. Perhaps tables, for instance, cannot be counted directly, but must first be arranged in classes, in Chinese in the class *chang*, "spread," and in Japanese in the class *kyaku*, "foot":

4 *chang* tables,   tables 4-*kyaku*.

Since all cylindrical objects are reckoned in Japanese by the class *hon*, "root," and all poems by the *shu*, "head," 4 trees and 4 poems are counted as follows:

trees, 4-*hon*   poems, 4-*shu*.

We can understand this classification, in the case of some classes of objects, perhaps because of their physical form. Sometimes the part stands for the whole, as in Chinese "mouth" stands for "man"; in other instances the relationship is not at once discernible, as when the Japanese count automobiles by the "tower." Since they can also be represented by the seat or chair in which one sits and is carried, an automobile can be thought of as a "moving chair." But if, in contrast, the Chinese *t'iao*, "small twig," is the classification word for all elongated things (trousers, belts, strings), we can no longer understand why oxen and even murderers should be counted thus. Curious and cynical speculation is also aroused by the fact that both medicines and the boards used to build coffins are counted in Chinese by the class word *fu*, "gift."

The more classes of numbers there are, the more primitive and object-bound is the number system. It still isolates the world of physical objects and strongly emphasizes the uniqueness of each separate thing. When, by way of contrast, Turkish differentiates only between the two classes "man" (*nefer*) and "other than man" (*tane*, "piece"), as in

yüz nefer asker,   *yüz tane at,*
100 men soldiers,   100 pieces horses.

the only relevant question is whether the object has the attribute of being human or not human. We can see that if this last difference were also omitted, the number word would stand directly in front of the word for the object, and the number sequence could be used to count everything.

The number classes once more reveal clearly how closely the number was involved with the object in early man's conception, how strongly things dominated numbers. They also illustrate the mental obstacles that primitive man had to overcome not only in liberating the number sequence from these preliminary stages but also in building upon the earliest beginnings. In doing so, he obtained help from a different source — and that is what we shall look into next.

Expansion of the Number Sequence by Means of Supplementary Quantities

ARRANGEMENT BY ELEMENTS

On the island of Ceylon live the Wedda, a primitive tribe with a very low level of culture. If a Wedda wishes to count nuts, for example, he collects a heap of sticks. To each coconut he assigns not a number word but a stick: one nut — one stick; and each time he does so, he says, "that is one." So many coconuts, so many sticks; for he has no number-words. Does this mean he is unable to count? Not at all! He translates the pile of coconuts he has laid out into the auxiliary quantity of sticks. Can he then tell whether anyone has stolen one of his coconuts? Yes: he again arranges the nuts and sticks in order, one to one, and if there is one stick left over, one coconut must be missing. Can he say how many nuts he has? No, for he has no words for that. He can only point to his pile of sticks and say, "That many!"

Thus, we are here dealing with a translation of the counted things into a nameless, supplementary quantity. By arranging similar pebbles or sticks element by element, early man thus succeeded in doing away with the multifarious nature of things; he achieved a clearer view of what he wanted to know about them: namely, their number.

There is a very ancient law, inscribed in the original letters and words, according to which the senior Praetor on the Ides of September [the beginning of the Etruscan year] is to drive a nail into the right side of the Temple of Jupiter, where the sanctuary of Minerva is. It is said that this nail was to be a sign of the year's number, for letters and numerals were then very rare. Because numbers were invented by Minerva, this law was ordained in her temple. Cincius, a reliable authority, asserts that in Bolsena, too, a similar nail was driven into the temple of the Etruscan goddess Nortia as a means of reckoning the year.

With these words the Roman historian Livy describes a supplementary quantity in the form of nails used to number the years.

Among primitive peoples, the victor in battle of the hunter carries about with him the number of his vanquished enemies or slain animals in the form of scalps or boar's tusks. Young unmarried girls of the Masai, a warlike tribe of herdsmen living on the slopes of Mt. Kilimanjaro, each year add one heavy brass ring around their necks, so that their precise age can be known from this extraordinary necklace of annual rings (see Fig. 4).

We, ourselves, still continue to count with the aid of supplementary quantities. The innkeeper or bartender "tallied up" the bill, in earlier days, on a slate and today on the beer plate; in Spain at one time he would toss a small pebble (*chinas*) into the hood of the guest's cloak for each drink, whence the expression *echar chinas* "to throw pebbles," still means "to chalk something up to someone's account." The Greek sculptor Lysippos (4th century B.C.), according to Pliny,

Fig. 4 Masai girl wearing annual rings, which show her to be 23 years old.
Institute of Ethnology, Göttingen.

[33]

Fig. 5 The Poor Poet in Spitzweg's painting has "tallied up" something on the walls of his garret — perhaps the number of days' rent he owed. By kind permission of the Staatsgalerie, Munich.

produced a total of 1500 statues during his career; Pliny deduces this from the fact that he would always set aside one gold denarius from the money he was paid for each work of art, and his heir inherited 1500 denarii. What was the poor poet in Spitzweg's painting tallying on the wall of his garret (see Fig. 5)? Most likely not gold coins, but perhaps the number of days' rent he owed. Similarly, the soldier marks off the days to his next furlough on his calendar. Adalbert Stifter, the German poet, counted the days until he was to see his betrothed again by eating apples: "There are still 14 apples lying there, and yet they are only 14. When I wrote my last letter to you there were 21; and that's a comfort — tomorrow there will be only 13! Finally only one apple will be left, and when I have eaten that, I shall shout for joy" (excerpted from a letter). Each of us is familiar with tallies of this kind, and has often made them himself.

When supplementary quantities are illustrated graphically in this manner, the number need not be expressed at all, or can be apprehended merely conceptually. A chieftain on the island of Celebes was sentenced by the colonial authorities to pay a fine of 20 buffaloes. Someone expressed surprise at the severity of the punishment. Quite astonished, the chieftain asked: "Do you consider the fine that high?" and began to count out nuts from a pouch, one for each buffalo. Only when he had "grasped" the number in the truest sense of the word did he become incensed at his punishment.

By this means of aligning objects and translating them into a supplementary quantity, primitive people who possess very few number words are able to count far beyond the highest number they can express verbally. It is a mistake to think that a tribe that has only 3 number words can count only to 3. By physically representing the number by a quantity of pebbles arranged element by element, they can count much further than that.

The Damara in South Africa, for instance, have only 3 number words. An ethnologist tells how he bought a number of sheep from them at the price of 2 twists of tobacco per sheep. In carrying out the transaction he had to hand over 2 twists at a time, sheep by sheep; when he tried to hand over 4 twists of tobacco at once, for 2 sheep, the Damara tribesman became confused and lost track. He put down 2 twists side by side, turned to one sheep, then took the next two and turned to the second sheep. Rigorous ordering, trading, the exchange of a quantity far exceeding the verbal number sequence, even computation, if you please — and all without specifying the number!

But this is a path leading to written numerals sooner than to spoken number words. The words needed to specify numbers are not arrived at in this manner. They do, however, arise from the practice known as body-counting.

Body-counting. The term "body-counting" designates the number sequence arrived at by some primitive peoples, in which the parts of the human body — the head, the eyes, the arms, and so forth — are arranged in a certain order. A tribe on the island of Papua counts as follows:

1	*anusi*	right little finger	12	*medo*	nose
2		right ring finger	13	*bee*	mouth
3	*doro*	right middle finger	14	*denoro*	left ear
4		right index finger	15	*visa*	left shoulder
5	*ubei*	right thumb	16	*unubo*	left elbow
6	*tama*	right wrist	17	*tama*	left wrist
7	*unubo*	right elbow	18	*ubei*	left thumb
8	*visa*	right shoulder	19		left index finger
9	*denoro*	right ear	20	*doro*	left middle finger
10	*diti*	right eye	21		left ring finger
11	*diti*	left eye	22	*anusi*	left little finger

One might suppose that this was a sequence of number words extending through 22. There is, however, a more profound difference. What does "*doro* days" stand for — 2, 3, 4, or 19, 20, 21 days? This sameness and the mirrorlike repetition in reverse make it possible to understand a count only if the whole sequence is run through from the beginning or if the meaning is specified. This number system thus merely assigns names but does not organize and does not build on itself: it has no governing principle. Nevertheless, we can see in it the first step toward the kind of numbering that does lead to a true number sequence.

NUMBERING BY FINGERS AND TOES

Finger Signs and Number Words. Apart from teeth, fingers and toes are the only parts of the body which man possesses in any abundance. They are supplementary quantities that he always has conveniently "on hand." Thus it is not surprising that virtually all primitive peoples count on their fingers, in that they assign number values to their fingers, which they use as a supplementary quantity. Words are not absolutely necessary in this process; and with many primitive tribes which "point" the number, each number is actually a gap in the flow of speech that is filled by a finger gesture.

For the most part this assigning of numbers takes the form of extending or bending the fingers of one hand, just as we might reckon to ourselves how many months there are from May to September. Now we can understand how the Dene-Dinje, a tribe of American Indians, built up the following sequence of number words:

one	"the end is bent" (little finger)
two	"it is bent once more" (ring finger)
three	"the middle is bent" (middle finger)
four	"only one remains" (thumb)
five	"my hand is finished."

The verbal designations of their finger signs have become their number words: Four days are "only-one-remains days." Such is the case with the majority of primitive peoples; their number words, to the extent that they are expressed at all, refer to finger signs. More dramatic than "my hand is finished" are such frequently encountered expressions as "my hand dies" for 5, "my hands are dead" for 10, "my hands are dead and one foot is dead" for 15, and "a man dies" for 20. We can visualize the hand held out: the fingers stand upright and then sink down, one after another: they "die."

Some statements of numbers thus sound very strange. In translating the Bible for one tribe of Papuans, the passage (John 5:5): "And a certain man was there, which had an infirmity thirty and eight years" had to be expressed as "a man lay ill one man (20), both sides (10), 5 and 3." Even more picturesque is the expression for the number 99 in British New Guinea: "four men die (80), two hands come to an end (10), one foot ends (5), and 4." Very rarely, on the other hand, are the actual names of the fingers used as number words. An isolated example of this occurs among the South American Kamayura tribe: Here 3 is called "peak finger" (middle finger), and "three days" is thus "peak-finger days."

In comparison, number words containing the roots "clap" (10) or "jump across" (6) refer to number *gestures*: To indicate 10, the hand is clapped against the palm, and for 6 one hand is passed rapidly over the other. The number of possible variations here is beyond belief. Certain African tribes can even be identified and classified ethnically according to whether they begin to count with the right or the left hand, whether they extend or bend the fingers, or whether they turn the palm toward or away from the body. For our purposes, however, the important thing is that *man learned to count on his fingers*.

A charming instance from the recent past is provided by an anecdote related by the Englishman R. Mason about the last world war ("... *and the Wind Cannot Read*"):

Sabby was a Japanese girl in India, which was then at war with Japan. Her friend therefore introduced her as Chinese to an Englishman who had been living for a long time in India: "Miss Wei." "Really?" He stretched his face forward and examined Sabby from close up, as if he were nearsighted. "Nonsense," said he. "Count with your fingers! Count to five!" Sabby looked shocked; she wasn't quite sure whether this extraordinary man was joking or mad. Hesitantly she raised her hand: "One, two, three, four, five," she said uncertainly. Mr. Headley burst out delightedly: "There you are! Did you see that? Did you see how she did it? Began with her hand open and bent her fingers in one by one. Did you ever see a Chinese do such a thing? Never! The Chinese count like the English. Begin with the fist closed. She's Japanese!" he cried triumphantly.

The Romans, who were very poor at computations and therefore faithfully preserved the early methods of reckoning, used such turns of speech as *numerare per digitos*, "to count with the fingers," and *novi digitos tuos*, "I know your fingers," which means "I recognize your skill in reckoning," thus furnishing additional proofs of this (see also the quotation from Ovid, p. 37). Throughout the middle ages the units were called *digiti*, and in English they are still called *digits*.

Organization of the Number Sequence. The fundamental significance of fingers and toes in the development of the number sequence is revealed by the following consideration:

Whereas the custom of assigning number values to pebbles or to parts of the body yields only a continuous, undifferentiated supplementary quantity, the quantity derived from the fingers and toes is already classified and grouped by nature: 5 fingers make a hand, 10 is represented by two hands, 20 by hands and feet. It is quite natural that languages should also show these breaks: five "1 hand," ten "2 hands," twenty "one man"; this happens not only with primitive peoples but even in the number sequences of civilized societies. In addition to the familiar 10-group, we can also encounter remnants of an ancient 20-grouping, and we have already demonstrated the peculiar break after Four as a representation of the "finger-hand," held with the thumb concealed (see p. 25 and Fig. 12, p. 50).

What is the unexpected significance of this arrangement? It lies in the fact that now the path is clear to permit the number sequence to extend further, beyond the first few words: Once we have counted through "the man," the second such series can begin exactly like the first, followed by the third, the fourth, and so on. In this way one level is built up on the next; this classification ensures the ordered progress of the number sequence. In a very real sense, this is nature's gift to mankind.

I know of no more delightful expression of this idea than Ovid's verse (*Fasti* III) that alludes to the intimate union of the number Ten with nature, and in the final line even mentions the organization of the number sequence, although it may not be correct in every detail:

Annus erat, decimum cum luna receperat orbem,
Hic numerus magno tunc in honore fuit.
Seu quia tot digiti, per quos numerare solemus:
Seu quia bis quino femina mense parit:
Seu quod adusque decem numero crescente venitur,
Principium spatiis sumitur inde novis.

"A year was past when the number of full moons was ten,
This number was held in great honor by men in those days:
Because that is the number of fingers on which we reckon,
Or a woman gives birth to her child in two times five months,
Or the numerals increase to ten, and from there
We again being a new round."

While our attention was focused on other matters, we completely failed to notice that something else has happened, something that early man did not achieve: the number sequence has now finally become completely— we might almost say painlessly — detached from the objects counted. It is no longer things that are counted, but the now familiar supplementary quantity, to which all objects, of whatever kind, can be transferred.

At first the supplementary quantity insinuated itself between the amount of objects to which numbers were applied and the quantity of number words; once these have been formed, the supplementary quantity has done its job and disappears. The number sequence is born! To be sure, it has not yet acquired all the characteristics that we expect of it today, such as infinite progression. But it does at last possess the outlines from which it can develop to its logical conclusion.

Thus far we have traveled a very important path. We began by speaking about our abstract number sequence, which can be used to count anything; we then descended into the dawn of man's intellectual development, when mankind took its first timid steps in counting. From there we climbed back up, step by step, until we again arrived at the abstract number sequence with which we began.

This reconstruction is not to be taken to mean that every individual number sequence, such as our own, must have followed exactly this path; if that were the case, these intellectual steps would form the rungs of a historical ladder, the only one, in fact, which every number sequence would have to climb in order to achieve its full growth. No, these steps are merely stages of development which we can discern now in one and now in another number sequence. They also illustrate the How, and especially above all the What, of an evolution through which every number sequence has had to pass.

Our next task is to examine the manner in which our own number sequence is organized.

Principles of the Number Sequence

SUCCESSION AND GROUP

Ordering and Grouping

Let us return to our Wedda tribesman counting his coconuts (p. 33): he assigns sticks to them, one by one; then he gathers all his sticks into a pile and knows that he has so many sticks and hence so many nuts.

Members of the South American Bakairi tribe do it a little differently. They put down two coconuts together, three times; thus there are 3 pairs. Then they form these pairs on their own fingers, putting together the left little finger with the left ring finger, then pairing the middle and index fingers together, and then the left thumb with the right little finger. Thus through the supplementary quantity represented by his own fingers, he has grasped the number 6. Anything beyond that they no longer group in pairs but merely line up as does the Wedda tribesman.

This is a good example of the organization of counting which we shall call *ordering* and *grouping*. Up to 6 objects the Bakairi *groups* by pairs; from there on he merely *orders* them. Which then was the earlier stage? Ordering, of course: without first ordering, there can be no grouping. The Bakairi stands, as it were, with one foot on the lower step of ordering and diffidently places the other on the next higher step of grouping by twos: another picturesque instance of how numbers first became clear to primitive man through the use of a supplementary quantity.

But why grouping? Let us ask ourselves: why do we always divide a bundle of postal cards by twos? Why do we break up a pile of coins into little heaps of five? To count them better! "Better" means more clearly, more confidently, and therefore more quickly.

Just the same with primitive man. The Fiji Islander carved notches into his club to count the animals or the enemies he brought down with it. When the number of notches became too unwieldy for him to visualize clearly, he grouped them: After every 9 notches he carved one large notch. The large notch stands for 9 small ones, and thus constitutes a 10-group. Now he can envisage the number 54, even though he may not be able to express it verbally because he has no number word for it. After all, number words are not absolutely necessary for counting (see Fig. 6).

Counting by groups of three is exemplified on a sword from the Philippines, on which the native warrior has recorded the number of his victims by riveted silver nails (see Fig. 7).

The Germanic equivalent of this is *Hildebrand's Dirge* with its characteristic groups of eight:

FIG. 6   FIG. 7

Fig. 6 Fiji Island club, with number notches. 40 cm long.
Museum of Ethnology, Frankfurt/Main.

Fig. 7 Philippine sword, on whose blade the owner has recorded the number of his victims with silver nails inlaid in groups of three. 94 cm long. Linden Museum, Stuttgart.

"At my head stands	the broken shield,
eru thar taldir	*tiger ens atta*
on it are tallied	ten times eight
manna theira	*er ek at mordi vard*
proud men	whose slayer I became."

(This song is an Old-Norse version of the 8th-century Old High German *Hildebrand's Song*, which was probably brought to the north by German traders in the 12th century.)

There are innumerable examples of number grouping in our own daily lives. The bartender chalks up groups of 5 ticks each, the supplementary quantity standing for the number of glasses of beer consumed by a customer, by drawing the life for the fifth slantwise through four vertical strokes, forming a "gate." The truckman unloading sacks of merchandise, the bank teller counting money, the card or billiard player marking down the points he wins, all use the method of grouping. We can even group things with our eyes, without touching the things themselves. The set of dots : : : : can be counted as two horizontal rows of four each, as squares of four each, or as four pairs lined up in a row.

Commercial and Natural Groupings
Shakespeare's Shepherd in *The Winter's Tale* (Act IV, Scene 4) very astutely sets forth his complicated reckoning:

Clown. Let me see: every 'leven weather tods; every tod yields pound and odd shilling: fifteen hundred shorn, what comes the wool to? ... I cannot do't without compters.

He makes groups of 11 sheep each and assigns a weight of wool to each group — not 11 weights as the Bakairi tribesman would have done, but only one, a form of ordering that we shall speak of presently. But first let us take note of what the passage from Shakespeare also testifies: *For Primitive Man, computing was identical with counting!*

This is a very important insight. When the Damara tribesman (see p. 34) calculated his sheep transaction with the buyer, he was counting off, a simultaneous ordering of sheep and of twists of tobacco. Thus one sheep can, as here, be lined up with every 2 twists, and hence with a 2 group; or, as above, 11 sheep with one "tod" of wool and this in turn with one pound and some shillings. This unit exchange value need be established only once; the continuous ordering will then complete the transaction without recourse to any further calculations.

With this we come to the types of grouping that we wish to designate as either natural or commercial. Shakespeare's eleven sheep are a *commercial* grouping that translates directly to a unit of money, one pound. Eskimos tied their pelts together in bundles of four to prepare them for sale, since 4 pelts will make one fur coat. In many parts of Africa cowrie shells serve as currency; the natives string together 40 shells, so that 1 string = 40 cowries and 50 strings = 1 head.

Here the size of the groupings is determined by commercial expediency or by mutual agreement. But there is no such consideration in counting and grouping postal cards or coins. It would never occur to anyone to arrange some 97 dimes in groups of 11 or piles of 40. Here we revert to the *natural* arrangements by groups of 5, 10, or 20, and perhaps also 4 (see Fig. 12, p. 50). The 2 group, the pair, also belongs to this category.

Examples can be dispensed with, even though this new insight will later yield surprising conclusions; some of the instances cited previously can also serve for the groups of 5, 10, and 20 we are discussing. But two examples are pertinent just the same: one because it has preserved for us the former state of a now highly civilized, "mature" people. A native member of this people was asked by an ethnologist how he would count some given quantity of objects. "Dear Sir, we do it this way: When we have counted up to five, we set them aside in a small heap and start again from the beginning." Thus he "fives," just as Homer's Old Man of the Sea "divided into groups and counted" his herd of seals (*Odyssey* 4, 412):

αὐτὰρ ἐπὴν πάσας πεμπάσσεται ἠδὲ ἴδηται
λέξεται ἐν μέσσησιν, νομεὺς ὡς πώεσι μήλων.

"When he has seen them and counted them up by fives,
He will lie down among them, like a shepherd among his sheep."

What does the word *pempázein* mean? Since the Aeolic Greek *pémpe* is "five," it means nothing other than "to five," to divide up into groups of five. No word could more faithfully reveal the early stages of a people's culture, its counting by groups of 5.

The second example shows that even in this day and age the ancient custom of grouping can be used very prudently, for instance when one has to count an unwieldy quantity of something. Karl Bädecker, publisher of the world-famous red travel guides, wanted to know, for his volume on Italy, the exact number of steps leading to the Cathedral of Milan. He counted them by removing one pea from his vest pocket every 20 steps and placing it in his trousers pocket.

GROUPINGS OF HIGHER RANK

The Elements of a Sequence are Groupings
In the old days the German army was organized as follows: 120 men = 1 company, 4 companies = 1 battalion, 3 battalions = 1 regiment, 2 regiments = 1 brigade, 2 brigades = 1 division, 2 divisions = 1 army corps. What sort of counting is this? Nothing more than a *grouping* of soldiers arranged in order. Not only are men grouped in companies, but companies into battalions, battalions into regiments, and so forth. Thus they are "grouped groupings."

A grouping whose elements are again groupings is called a *grouping of higher order* or rank (second, third, . . ., etc.). All systems of measurement are actually groupings in ascending rank:

60 seconds = 1 minute	1000 mg = 1 g
60 minutes = 1 hour	1000 g = 1 kg
24 hours = 1 day	1000 kg = 1 metric ton
30 days = 1 month	
12 months = 1 year	

Measures are groupings of elements of known magnitude in a sequence. Are there also similar groupings of abstract numbers? Of course there are; there is no need to bring up examples from our daily life or from the lives of primitive peoples. A native counting sea shells puts down one small stick for every 10 shells; for every 10 small sticks he has one large stick, equivalent to 10 times 10 shells. Our currency system is nothing but a grouping or a division into groups.

Groupings and Numerals
Grouping is the important advance over the unorganized counting done by merely placing the objects to be counted in series; once this step has been taken, the unlimited progression of the number sequence is initiated: first ordering → first grouping → second ordering → second grouping → third ordering → third grouping, and so on. The significance of this progression is apparent from the important fact that the symbolic representation of these progressive groupings leads to the earliest number symbols used by various peoples.

As an example of pure 10-groupings, I have chosen the pictographic numerals of the Egyptians on temple walls and monuments; I have placed them here next to the discontinuous Roman method of counting by 5-groups (Fig. 8).

Fig. 8 Egyptian and Roman numerals compared.

Let us cite a historical example (see Fig. 9). The Egyptian king Sahure commanded the herds of cows, oxen, asses, goats, and sheep which he had taken as booty on a Libyan campaign to be represented on a large inscription. Our illustration shows a detail of the herd of asses and a fragment of the herd of goats. Even in those days large numbers made an impression; he therefore had the number of head inscribed above the last animal in each herd: 232.413 goats and 2234.. asses (the last two digits have been broken off). The tadpole is the symbol for the number 100,000 (see p. 122).

There is evidence that many other peoples formed their early number symbols in a similar fashion. Their *early* number symbols, for in many cases in the course of their development they went far beyond this simplest rule of ordering and grouping. Thus the clarity of

Fig. 9 Egyptian numbers in pictographs, 2500 B.C. We can read 223,4.. and 232,413. Each donkey is about 50 cm high.
German Institute for Antiquity, Cairo.

earlier Egyptian numerals was later obscured in the so-called "hieratic" script of the papyri, which developed from the old hieroglyphic writing. The Romans themselves later developed different rules for representing higher numbers, although for the lower numbers they continued to use their original systems.

An arresting example of the earlier Roman numerals, and one we shall refer to often hereafter, is the *Columna rostrata*, the commemorative column which C. Duilius had erected at Mylae in the year 260 B.C. to celebrate the naval victory over the Carthaginians. It was termed *rostrata* (< *rostrum*, "beak," by transfer from the ram in the bow of a warship) because it was ornamented with iron ship's rams of this type. On the fragment of the inscription illustrated here (see Fig. 10) our attention is caught by the bottom few lines with the number symbols:

(*Auro*)*m captom: numeri 1000 1000 1000 500*
(*Argen*)*tom captom: praeda: numei 100,000*
(*Omne*) *captom aes:* (more than 20 times) *100,000*

"— gold captured: 3500 coins
— silver captured: booty: 100,000 coins
— total loot in all: (more than) 2 million *aes*."

(The *aes*, "bronze," was a bar of bronze weighing one pound, the basic monetary unit in ancient Rome.)

One surprising thing is the array of 22 symbols representing a hundred thousand in a row; actually, as we know, there were even 32 of them. Thus a higher grouping for 10 hundred thousands, for 1 million, had not yet been devised. This is a very choice proof of the *limitation* of the Roman number sequence and number symbols. Pliny also confirms this:

Non est apud antiquos numerus ultra centum milia; itaque et hodie multiplicantur haec, ut decies centena milia aut saepius dicuntur.

Fig. 10 Inscription from the *Columna rostrata*, containing, among other things, more than 20 symbols for a hundred thousand. The column was set up in Rome to commemorate a naval victory over Carthage in 260 B.C. This fragment measures about 80 × 80 cm, with letters 2.5 cm high.
Palazzo dei Conservatori Rome.

"The ancient Romans had no number higher than a hundred thousand; hence, even today [in the 1st century A.D.] these numbers are multiplied by saying ten times a hundred thousand, or the like."

We know this already from our discussion of the Roman number words (p. 28); now we have the documentary proof, as it were, on the memorial column: at one time 100,000 was the "last" number known to the Romans. Later, as the Million became more current — we recall the state indebtedness under Vespasian (p. 28) — the Romans wrote it as 10 × 100,000 or $\boxed{\text{X}}$, exactly according to their number word for this sum. The frame around X thus stands for 100,000. Even if up to now we were not sure just how this came about, there is no longer any doubt: the frame is the terminating sign for the symbol for 100,000 which we see written out in full on the *Columna rostrata*. Used as an abbreviation, it now stands for the whole symbol; *decies*, "10 times," appears under it as X.

The Romans thereby created a numerical symbol that quite graphically reflects the number word, but with it they abandoned the rule governing their earlier writing of numbers: They no longer arranged the hundred thousands in a row, but rather counted them (Fig. 11).

Fig. 11 Large Roman numerals.

The amount that Livia left to Galba was $\boxed{\text{D}}$ sesterces; Tiberius, however, read the number without the frame around it (see p. 28). The 40 billion sesterces in Vespasian's indebtedness were readily expressed verbally and easily written down in words (p. 28) but not in numerals: A double frame would have had to be placed over the CCCC, and this the Romans did not do. All this is a compelling proof that number words are more powerful than number symbols and advance more readily to higher numbers.

Grouping and Language Speech

Now that we have recognized grouping as an important source of written numerals, the question arises: What significance does it have for spoken number words? Is the number sequence also built up by alternate ordering and grouping? Are number words also groupings, like the X among the Roman numerals?

Let us run through our own number sequence once more. "One, two, three, ... nine, ten, eleven, ... nineteen, twenty, twenty-one, ... thirty, ... ninety-nine, a hundred, ... a thousand ...." Is this ordering? Grouping? It is certainly not mere ordering "one, one, one ..," but "ten" and "-ty," "four hundred," "five hundred," "six" and "seven thousand" — are these not groupings? At any rate, the number sequence is not built up from one word alone.

GRADATION BY STEPS

The Number Steps

"One, two, three, ..., nine, ten" is a sequence whose individual elements have names. If a quantity of coins is divided into groups of 5 each, the coins are heaped together indiscriminately into the small piles without being first named individually. The coins are combined "on even ground," like bundles of grain into shocks whereas in counting we have the sensation of ascending the steps of a stairway: 1, 2, 3, ..., 9, 10, after which we come to a landing, where we can pause for breath. Then the steps continue on up: 11, 12, ..., 19, 20 — here again there is landing. Thus we climb higher and higher up the tower; as we look way down we see how the early numerals are always ordered and grouped on the floor; the stalks are tied together into sheaves, the sheaves are piled up into shocks and the shocks into wagonloads, but always on the mowed field of stubble (compare Fig. 14, p. 58 with Fig. 8, p. 42).

The reader should be guided by these illustrations. He will then recognize the essentially different arrangement of our verbal number sequence, which we have called its *step gradation*. The number sequence is graduated. This is achieved by giving names to the elements of the sequence; these elements no longer remain anonymous and undifferentiated but form the steps of a staircase: Seven is "higher" than three. In this staircase "ten" plays the part of a threshold, or a landing, as we have called it. So do "twenty," "hundred," "two hundred," and "thousand." In the Middle Ages the 10-step and every subsequent multiple of 10, such as 30, 80, or 1960, was called *articulus*.

But why is just 10 a threshold — why not 7 or 11? The answer is that we have ten fingers; after that, the hands are "finished." But the hands themselves are groupings: "hands" = 10 fingers. As we learned earlier, in New Guinea 99 is called 4 men and 2 hands and 1 foot and 4. Was this a gradation or a grouping? A grouping, of course.

But the early concepts of grouping *lead* to gradation. We can even say that the "landings" on our number staircase *are* the old groupings. They have merely lost the marks of their origin in the graduated number sequence; they faded out as groupings when the spaces between them were filled up with number steps.

Yet if the stair landings, in particular the steps 10 — 100 — 1000, are groupings, then this signifies that, in the composition of the number sequence, at first all the ranks are found by grouping — 10 tens make a hundred, 10 hundreds make a thousand — and not until later are the numbers between them, like 78 or 543, put into place. The ranks are set up ahead of time as a framework, 10 ... 100 ... 1000, and the steps are then built in from landing to landing. To put it in a nutshell: Early man possessed the higher number 1000 before he grasped the lower number 543, for 1000 was found through grouping, whereas 543 was found by counting in ascending steps. Our "theoretical" formation of the number sequence, proceeding from unity by steps of one ("plus one, plus one, ...") in building up a whole structure of numbers, does not correspond to historical reality. The true situation is far better understood from the analogy of the staircase of numbers whose chief pauses (the landings) were first erected and then followed by the intermediate steps, which were interpolated one by one.

Even today, if we contemplate the number sequence in our minds, 1000 seems clearer, more "available" to us than 543, a number of which we can really say only that we must certainly arrive at it if we keep counting long enough. We can visualize it directly only with difficulty, if at all. We are fascinated in recognizing the prominence of the landings and the secondary nature of the intermediate steps in the changes in word forms, such as we have seen, for example, in the Latin number words (p. 27): *kentum — mille* are indeclinable, solid "blocks" between which the more malleable, inflected step words for the multiples of 100, *ducenti*, etc., have been inserted. For a Germanic example of the same thing, see p. 47.

The old groupings, which we have come to recognize as the rule governing the early writing of numerals, also play a part in the formation of spoken number words. Let us examine this surprising fact through a few more examples.

GROUPINGS IN THE GRADUATED NUMBER SEQUENCE

The Ten-steps

If those whose native language is German or English run through the French names for the multiples of ten (see p. 99), they feel a sudden jolt at 80: the even progress of the number sequence is suddenly broken by the completely alien word formation *quatre-vingt(s)*. What has happened here? In our view the graduation of the number sequence has been interrupted by an old 20-grouping ("four twenties," not "four times twenty" — the -*s* plural ending allows no other interpretation). This is an excellent example of the grouping principle breaking into the graduated number sequence.

In Latin, "twenty" is *viginti, vi-ginti,* "thirty" is *tri-ginta*; in Greek, "thirty" is *triá-konta* (see p. 92) — words formed with the endings -*ginta* (-*ginti*) and -*konta*. Seemingly this is connected with "ten," but, as we shall see later, it is actually means "a tenned," a grouping of 10 units. *Vi-ginti* is then "two tenned"; the -*i* ending is the grammatical dual. *Triginta* and *triakonta* are "three tenned"; the -*a* ending is the inanimate plural.

These graduations of ten thus betray their origin in groupings. Once we have opened our eyes to them, we can discern them in many languages. The reader may look for them himself in the tables of number-words (pp. 92–99). Let me point them out in Gothic, for example (p. 92): *taihun* is 10, yet 50 is not *fimf-taihun* but *fimf-tigjus,* "five tens," and so on; similarly in Old Norse 10 is *tio,* but 50 is *fimm-tiger* rather than *fimm-tio.* We shall show later that the meaning "tenned from tens" is concealed in the word *Hundert* (see p. 128).

In modern German the multiples of ten are distinctly marked with the suffix –*zig.* But in the case of the higher ranks German still uses word forms that clearly emphasize the grouping: *Viele Hunderte* (*Tausende*) *von Menschen* ..., "Many hundreds (thousands) of people .... If we had said instead, *Viel hundert* (*tausend*) *Menschen,* "Much hundred (thousand) people," the semantic meaning would have disappeared along with the grammatical group form. We use the first form only for large but unspecified quantities; Latin, however, says *duo milia militum,* "two thousands *of* soldiers" (genitive case), and many other languages do the same. Until World War II, for example, the numbers from 200 to 2000 in Danish were capitalized: *hundrede* (100), but *to Hundrede, to Tusind* (200, 2000). Whereever the number-word (*milia*) is a noun and the quantity specified is in the genitive case after it, the meaning is always that of a grouping.

When after one rank of numbers (e.g., 100) the next higher one is indicated as "large," "big," or "strong," we clearly recognize the first as a grouping, for a number (100) does not actually grow by being termed "large." A significant example of this is provided by the Germanic word for "thousand": Gothic *pushundi* < *pus-hundi,* "the strong hundred" (see p. 132); the best-known example (although it is an artificial one) is *milli-one,* "the large 1000" (p. 134). From the word *šar* (= 60²) the Sumerians formed their highest rank of numbers, *šar-gal,* "the great *šar*" (see p. 165). Sanskrit also uses the "great" formation for numbers of higher ranks, as in *padma,* 10^{10} — *maha-padma,* 10^{11} (*maha* = "large"). Similar word formations occur among primitive peoples: the Hottentots say *disi* for 10 and *gli-disi* for 100, "big ten"; the Welsh gypsies say *baro deš,* "great ten" for 100 (see p. 191). A nice example of the way large size is indicated is provided by the Sioux Indians: *crabrah,* "10," *crabrah-hugh-tougah,* "10-herds-masculine," 100.

Here is one more Germanic example of the precedence of the (grouped) graduation and the subsequent insertion of the intermediate steps. In this connection the reader should look at the Gothic number sequence from 10 to 100 (p. 92). Gothic 10 is *taíhun* and 100 is *taíhunt-e-hund*, "tenned from tens" — clearly a grouping! The words for multiples of ten up to 60 are formed with *-tigjus*, and from 70 to 90 with *-tehund*. There can be no doubt that this break in the sequence of tens did not arise from a step-by-step building of the numbers into a "tower of numbers," but was due to the reverse influence of 100 (*taíhunte-hund*) working back to 60, which must have existed as a gradation (grouping) before the successively ascending steps of ten were inserted.

Collective Number Words

Some languages have the capacity of forming nouns from number words; this is conceptually equivalent to grouping. Thus Latin from the "landings" *decem*, 10, and *centum*, 100, derives the nouns *decuria*, "ten-ness" and *centuria*, "hundred-ness" or "century," the military detachments of 10 and 100, equivalent to squad and company, respectively; *quinque centuriae militum* does not stand for "500 soldiers" but "5 groupings of 100 soldiers each." But because the plural *milia* of *mille*, 1000, is already a noun that takes the genitive case in the quantity counted, *milia* is felt to be a grouping itself and thus does not form such a collective number word.

In the same way, the Greek language formed the collective nouns *dekás* ("decade"), *hekatontás*, *chiliás* and *myriás* ("myriad"), from the numerical ranks *déka* 10, *hekatón* 100, *chílioi* 1000 and *mýrioi* 10,000. The Bible makes frequent use of these words: "... and the number of them was ten thousand times ten thousand, and thousands of thousands " (Rev. 5:11):

ναί εἶν χο ἀριθμός αὐιόν μηριάδες μηριάδων ναί χιλιάδες χιλιάδον
literally: "... myriads of myriads and thousands of thousands."

In the Vulgate this passage has only the last expression, *milia milium*, since Latin lacks a word for a higher rank.

Based upon these coinages derived from the number ranks, Greek has created collective words corresponding to every number: *monás* ("monad, one-ness"), *dyás*, *triás*, and so forth. This generalization of the groupings applied to every number is palpably artificial and self-conscious, whereas the words for the numbers of successive ranks evolved out of number concepts lost in antiquity.

The French similarly use the suffix *-ain(e)* to form a collective noun for each separate number, and then readily employ the collective word as a numerical measure governing the genitive case in the object counted: *six-ain* or *six-aine*, "6 pieces, a six"; *douzaine*, "dozen"; *quinzaine*, "15 pieces" (German *Mandel*); *vingtaine*, "20 pieces"; *quarantaine*, "40 pieces" (from the latter is derived the term for the 40 days of detention ("quarantine") to which travelers coming from a diseased area must submit); *centaine*, "100 units"; etc. One man belonging to a group of thirty is called *un trentain*, "one of the thirty." We can quite clearly trace the grouping in these collective number words if we compare the expressions *une douzaine de livres*, "a dozen books," with *douze livres*, "12 books"; in the latter case the books are felt to be counted one by one, 1, 2, 3, ... 12, whereas the former groups them together so that they are felt in their totality rather than counted individually — so much so that the words *douzaine* or *dozen* often need not denote exactly 12 items.

20-GROUP "MAN"

We already touched briefly on the 20-group, *quatre-vingt*, in the French number sequence. This is not the only example. Before we go any further into this subject, let us discuss the abstract *20-Group "Man"* as it has grown out of the number sequence to form special words in a variety of folk languages. Only then can we understand why this peculiar 20-grouping has become an inherent part of the various number sequences.

It is difficult for an illiterate peasant to tell how many years he has lived; the Piedmontese, for instance, would answer that he carries *quat borla*, "4 large loads," on his bent back, meaning $4 \times 20 = 80$ years. This is far easier for him to visualize; he does not climb the steps of his years age, 77, 78, 79, 80, but remains standing on the ground and, like a mule, takes up the "loads" of his years grouped in bundles of 20. The Sicilian dialect has no special word for the 20-group, so that in Sicily the peasant says *tri vintini et deci*, "3 twenties and 10," for 70 years; thus he uses the number-word *vinti*. In some regions of Spain the monetary units (1 *duro* = 20 *reales*) are used in counting, so that a peasant there would reply that he is "3 *duros* and 10 *reales*" old.

This may remind us of Charlemagne's old coinage: He had 12 *denarii* of 20 *solidi* each struck from one pound of silver. This monetary system lasted throughout the Middle Ages into modern times; France did not give it up until the French Revolution, and it is still in use in England, where 1 pound = 20 shillings. and 1 shilling = 12 pence. The abbreviations £1 = 20s = $20 \times 12d$ are merely the initial letters of the Latin words *libra, solidus, and denarius*. To indicate shillings the long s, or $\int$ was also used; this has been shortened to a slanting stroke in expressions such as 5/6 (5s 6d).

On the whole, the English language has most faithfully and widely preserved the vernacular 20-group in the word *score*. This was originally a "notch" that the farmer would carve for every 20 animals as he counted his cattle or sheep (we shall have more to say about this in connection with exchequer tallies, p. 236), but today it has taken on the strict meaning of "twenty." The English Bible, just like the Piedmontese peasant, counts a person's age by groups of 20 years (Psalm 90:10):

"The days of our years are threescore years and ten; and if by reason of strength they be fourscore years, yet is their strength labour and sorrow."

Shakespeare also counts by scores. A passage in *The Merry Wives of Windsor* (Act 3, Scene 2) reads:

"as easy as a cannon will shoot print-black twelve scores."

Prince Hal says of his fat friend Falstaff (1 Henry IV, Act 2, Scene 4):

"I'll procure this fat rogue a charge of foot; and I know, his death will be a march of twelve scores."

Our oldest documentary evidence dates from around 1050, when the inventory of a monastery's possessions translated the vernacular expressions into Latin as in a dictionary:

V *scora scoep*   *quinquies viginti oves*   "5 × 20 sheep,"
VIII *score oecer*   *octies viginti agri*   "8 × 20 acres."

Modern English no longer has a word for the German *Schock* (a mass of sixty units), derived from the same root, and thus uses the far more expressive *threescore*. In some trades *score* serves as a unit of measure, e.g., in mining for a basketful of coal. In Ireland it also serves as a weight of 20 pounds.

In expressions like *scores of times* ("very often"), *scores of years ago* ("many, many years ago") or *scores of persons* ("dozens of people") the numerical meaning of 20 has lost its force, but the 20-grouping still distinctly comes to the surface. How strong the concept of 20 continues to be can also be seen quite picturesquely in the way numerals were written. A 14th-century inventory of the Scottish treasury writes the sum £6896 as

m   c   xx
vj viij iiij xvj — "6 thousand 8 hundred 4 *scores* 16."

As a rare example of a 20-grouping in an early system of numerals, I shall cite the Indian "Kharosti" numerals, which exhibit three groupings lower than 100: a 4-group, a 10-group, and a 20-group. These numerals could represent the French *quatre-vingt-dix* exactly, as could the Celtic series of multiples of ten (see Fig. 12 and p. 97).

The Low-German word *Stiege* (High-German *Steige*, "ladder, series of steps"), which is likewise a 20-group like *score*, has not acquired the same force. Previously, however, it was very widely used, as the equivalent word *stega* in Crimean Gothic testifies. *Stiege* were used in counting sheep and sheaves of grain in a field, but also eggs and linen cloth. An old book contains the words:

Lasst von dem Linnen mir für Geld nur eine halbe Stiege,

"Let me have just half a *Stiege* of the linen for money,"

and a 16th-century poet extolls the virtues of a marriageable girl in the words:

Zwei stieg Schilling sie auch hat,
davon sie rent kan kriegen.

"She also has two *stieg* of shillings,
from which an income she can get."

The way in which this word acquired its numerical meaning is lost in obscurity. The Middle High German *stiga* is a corral for small animals like sheep or goats; perhaps the amount of space designed to hold 20 sheep came to be thought of as a measure of quantity. The expression *Steige Salat* in the local German dialect of Hesse refers to a lightly built crate of laths with air spaces between (about 50 × 40 × 20 cm in size), which grocers use for shipping vegetables.

1	2	3	4	5	6	8	10
I	II	III	X	IX	IIX	XX	?

20	50	60	70	100	200

II ?I|122   × ????? ? II 274

Fig. 12 Indian Kharosti numerals, ca. 200 B.C., with three groupings of 4, 10, and 20. The symbol for 20 which resembles "3" consists of two 10-symbols superimposed. 20-groups occur very rarely in numeral systems. These numerals are read from right to left. The hundreds are no longer grouped but merely counted.

In contrast, the common Germanic word *Schneise*, "glade, a cutting of trees" (Anglo-Saxon *snas*, Old Norse *sneis*, Danish *snes*, Dutch *snees*), has the original meaning of a "rod" or "twig" used to line things up in a row. Fishermen in the Baltic use it to indicate the sticks on which they impale "stockfish" (such as cod, salmon, haddock, etc.) to dry. An old monastic register from the year 1186 includes in its annual income

24 *snesas anguillarum* — 24 "*Schneise*" (480) eels.

The expression for an indefinite quantity, "as many fishes as will go on a stick," very early became associated with a 20-group, "20 fishes," and then detached itself from this to become an expression for the abstract number "twenty." A *schnasse Zwiefel* (*Schneise Zwiebel*) "a string of onions" slang for a twist of straw with 20 onions attached. In addition to fish, eggs are still counted in Holland by *snees*, and the Swedish farmer often says *fyra sneser*, "four *snes*," when he means "eighty." All the examples here enumerated are connected with *Schneise*. Middle High German *sneise* is a string on which something (such as berries or pearls) is strung, and Bavarians today still speak of certain things in a long row as a *Schnaisen* without necessarily implying that there are 20 of them.

A remarkable evidence of the force of the 20-group in Scandinavia is the absorption of the Germanic *tiu*, "ten," as "twenty" (!) by Finnish, which after all does not belong to the Indo-European family of languages (see Fig. 21, p. 90); today in Finland it is used exclusively for eggs: *tiu munia* = "20 eggs." A similar case is the Russian word *sorok*, 40, which constitutes an exception within the Russian number sequence, for *sorok* goes back to the Old Norse *serk(r)*, "hide" or "pelt," since furs were tied together in bundles of $40 = 2 \times 20$ (see p. 185).

In the *Odyssey* (I, 431) Laertes buys a slave girl for 20 head of cattle.

Although all the Romance languages derived from the Latin a well-ordered number sequence based on gradations of 10 (see p. 99) — Latin does not place any special stress on the number 20 — traces of the 20-group have nevertheless survived here and there in popular speech, not as special words like *score*, *Stiege*, or *Schneise*, but in peculiar uses of the word "twenty." Thus, for example, money in Portuguese is counted in *vintens*, "twenties." The elided expression *dois cinco* is short for *dois vintens e cinco*, "2 twenties and 5" (= 45); *seis menos cinco* does not mean "six minus five" but "six twenties less five" (= 115). As we have said this phenomenon has been preserved in the vernacular speech of Romance countries, in some places as isolated cases and in others more commonly. Later, we shall have more to say about this and about the persistent survival of the 20-group in Celtic and French (see p. 64).

It is quite striking that the 20-grouping is most strongly rooted in the north of Europe, in Iceland, Denmark, and England. When we think of gestures symbolizing numbers, we might well suppose that these would be most commonly used and firmly established in areas where people go barefoot and thus can actually count on their toes as well as on their fingers. Yet the Eskimo, who very readily counts in terms of "men" (1 "man" = 20 fingers and toes) can scarcely do without shoes. The habit of counting by actually touching objects is, to be sure, one path leading to the vigesimal grouping, but not the only one. The mere knowledge of possessing an illustrative supplementary quantity consisting of 20 elements, to which things can be assigned in successive order, is also quite sufficient. It also "invisibly" dominated early man's conception of numbers. It may also be due to the favorable magnitude of the number 20 and its convenience in counting larger quantities; it is neither too small nor too large. This, rather than any "recollection" of the 20-group "man," is undoubtedly what prompted Herr Baedecker to count the steps in front of the Cathedral in Milan in groups of twenty (see p. 41). The latter, however, looms darkly in the background.

Let us conclude with two more instances of groupings by 20. When Pliny the Elder says in his *Natural History* (13, 77) that more than 20 sheets were never used for one scroll, this statement contradicts the fact that scrolls of more than 20 leaves existed. But perhaps he only wanted to imply the papermakers counted "bales" of papyrus in units of 20 leaves each. This ancient fashion of counting paper has been preserved even in our own day, for bales of paper are now reckoned at 10 reams of 20 quires each, so that the bale is made up of 10×20 quires ("folios" of 24 sheets); similarly in French 1 *rame* = 20 *mains* of paper ("hands" consisting of 25 sheets) or = 20 rolls of wallpaper. Both the words *ream* and *rame* are derived, through the Medieval Latin *risma*, from the Arabic *rizma*, "bale, bundle."

Of the many examples to be found among primitive peoples (see p. 35), I shall cite the number sequence of the Indonesian Enggano, a language spoken on the island of the same name (just south of Sumatra). This number sequence is clearly built up of groupings of 20, of which the following formations expressions are of interest:

kahaii ekaka	*ariba ekaka*	*kahaii edudodoka*
one man (20)	hand man (100)	one our body (400).

There is thus no special word for 100, and 100 is not a stage or gradation in counting — so strongly is the 20-step felt; *ariba* is derived from *lima*, "hand," which is the number word for "five" throughout the whole Austronesian linguistic region. The last number word for 400 may be explained as "so many times man as our body yields by its fingers and toes," and hence "man times man" ($20 \times 20 = 20^2$).

Let us now sum up our ideas about the rules governing the structure of number sequence. We have seen how, in counting, the principle of grouping arises from an ordering of the counted elements. The groups thus derived are themselves ordered in turn, and are again collected together in groupings of higher rank. In pure counting the natural divisions of 5, 10, and 20 — "hand, hands, man" — are predominant. The physical representation of this process leads to the earliest numerals used by primitive peoples, which we thus designate as *row-characters*.

The natural groupings also enter into spoken number words, but here they become gradations because the number sequence no longer recognizes a succession without individual names. The vigesimal grouping above all has been especially persistent among various peoples, in that the very ancient organization of "man" coincides with a component quantity convenient for use in actual counting.

Before we use this new insight to investigate various historical number sequences, we must make sure not to overlook a fruit that is now ready to fall ripe into our lap, but which we are for the most part inclined to forget without noticing it: the recognition that spoken number words and written numerals, curiously enough do not coincide. Written number symbols are definitely not mere representations of number words.

What should we expect? The sequence of number words in a culture is built up long before any mature system of numerals becomes established. Thus we may think that numerals are simply "copies" of number words.

But our researches thus far have shown us that the laws that govern early numerals, ordering and grouping, do not in fact correspond to the rule of the number sequence, which is stepwise gradation. Hence, the writing of numerals is *not* merely the representation of the number-word sequence.

And thus one of those paradoxes of cultural history comes about, in that the Romans, who possessed such a subtly and precisely ordered, flexible verbal number sequence, which differs from our own only in minor details, used a system of numerals so crude and cumbersome that it is hard to see how it could be a product of the same culture:

four	thousand	eight hundred	seventy-	nine
quattuor	*milia*	*octingenti*	*septuaginta*	*novem*
(I) (I) (I) (I)		DCCC	LXX	VIIII

as contrasted with 4879.

NUMBER LANGUAGE AND NUMBER SYMBOLS

It is a dramatic documentation of cultural history that various cultures throughout the world have always felt this discrepancy between speech and writing, and have thus strived to invent or borrow a good system of written numerals. Our modern numerals, so clearly and readily representing the large number 4879, are not the offspring of our own intellect. And although our forefathers had just as good a verbal number sequence as the Romans, they still did not at first have their own universally valid numerals.

On this whole planet, only two nations have succeeded in devising numerals that represent their verbal numbers, after first overcoming earlier levels of writing: the Chinese and the Indians; to be really precise, only the Chinese. This is because Chinese writing is a conceptual or pictorial system that provides a separate pictographic character for every object or idea (see the summarizing discussion of this subject pp. 450 ff.). If for example, we symbolically wrote numerals for the units, and T, H, and Th for tens, hundreds, and thousands, our number would be written:

4 Th   8 H   7 T   9,

and it would exactly represent the manner in which we speak this number ("ten" of course is shortened to "-ty"). This is precisely what Chinese does; it writes the number words with an ideogram that at the same time is its numeral (Fig. 13).

Fig. 13 Chinese representation of the number 4879.

We have written the number horizontally, instead of vertically in the proper Chinese fashion; the gradations or orders of magnitude are capitalized and are represented by the dots beneath the Chinese characters.

Whereas for us the written word "four" and the numeral 4 are two distinct and separate ways of representing a single concept, Chinese has only the one: Number word and numeral are identical. This accordance of spoken number words and written numerals is unique, the only known instance in world history.

The Indian numerals do not indicate the rank of a number by a special symbol but merely by its position: 4879. Thus the number ranks are not named, they are a nameless "positional" system of writing, so to speak; the Chinese numerals, on the other hand, make up a "named" system, as if we substituted symbols of numerical rank for position: 4 Th 8 H 7 T 9. Unlike the Chinese, the Indian numerals are not, at least not directly, a representation of the verbal number words; we shall see later that they were actually derived from the counting board. The essential difference between the two systems of numerals lies in their possession, respectively lack, of a symbol for the rank, or order, of a digit. To express the quantity 4079, the Indian number system must expressly designate the missing hundreds (by the symbol for nothing, zero) — something that the spoken number words do not do. The Chinese, on the other hand, can omit it altogether by writing 4 Th 7 T 9.

In regard to their internal structure, however, the two systems of numerals are alike. They follow the law of gradation, no longer that of ordering and grouping; thus they have risen to the highest state of perfection that a system of numerals can achieve.

In conclusion, let us touch upon another idea that is often overlooked: Our numerals do not have the same origin as the phonetic alphabet we use in writing and printing. Our letters are not akin to our numerals. One would naturally be inclined to suppose, in all innocence, that the human mind, when it took the trouble to record its ideas and concepts, would have devised similar systems of writing words and numbers, "seven" and "7." But this did not happen, neither in our western culture nor anywhere else in the world. The early system of writing numerals is everywhere the older of the two sisters. We shall learn this in the course of our voyage through cultural history, but even now the reader may recognize the extraordinary paths along which the human mind has traveled from one culture to another, as are epitomized in the following fact: Our language is Germanic, our writing is Roman, our numerals are Indian!

Historical Gradations

Levels of Rank

By now we have learned to understand the essence of gradations in the number sequence; that was the burden of the preceding section of this book. But before we look into some of the numerical gradations that have appeared in history, let us say a little more about the pictorial and written representation of such gradations, so as to visualize these concepts as clearly as possible.

Our number sequence "one, two, ..., ten" is a gradation of 10's; in other words the number 10 governs its general structure. Let us, as an example, write out $2463 = 2 \times 1000 + 4 \times 100 + 6 \times 10 + 3 = 2 \times 10^3 + 4 \times 10^2 + 6 \times 10^1 + 3 \times 10^0$. In this expression 10 has appeared as 10^3, 10^2, 10^1, and $10^0 = 1$; 2 thousand 4 hundred 6-ty 3. The exponents of 10 we call its *rank*; the "pure" levels of 10, 10^3, 10^2, 10^1, and 10^0, are thus its *levels of rank*. $1000 = 10^3$ is therefore the third level of rank. Then what are the ranks of the gradations of 20? In ascending order: 20^0, 20^1, $20^2 = 400$, $20^3 = 8000$, and so on. The pauses, the orders of magnitude in ascending this particular numerical ladder, are 20, 400, 8000, etc. As we climb upward, our steps become larger and larger at each successive rung.

The number sequence based on gradations of 5 has the following ranks: 5, $5^2 = 25$, $5^3 = 125$, etc.

In the various gradations named, the corresponding levels of rank are:

gradations of 5	5	$5^2 = 25$	$5^3 = 125$
gradations of 10	10	$10^2 = 100$	$10^3 = 1000$
gradations of 20	20	$20^2 = 400$	$20^3 = 8000$.

If we write the same number "eighty-nine" in terms of these various gradations, it would appear

in the 5-base system as	$3 \times 5^2 + 2 \times 5 + 4 = 324$
in the 10-base system as	$8 \times 10 + 9 = 89$
in the 20-base system as	$4 \times 20 + 9 = 49$.

Expressed in terms of pure gradations,

324 would be the 4th level on the 2nd level of rank 1 on the 3rd level of rank 2

89 would be the 9th level on the 8th level of rank 1

49 would be the 9th level on the 4th level of rank 1.

The reader may be able to visualize these strange forms of expression with the aid of the figure (see p. 58). If we were to write them in numerals, the vigesimal system, based on gradations of 20, would create some difficulties, for it would require nineteen different symbols and words in order to count the steps of each rank; our decimal system, based on gradations of 10, uses nine, the nine numerals 1, 2, 3, ..., 9; and the system based on gradations of 5 would have only four numerals. In all cases, a missing rank is indicated by the zero symbol 0.

[56]

In algebraic, or generalized, notation, a number expressed in a system with x levels would look like

(1) $\quad z = a_n x^n + a_{n-1} x^{n-1} + a_{n-2} x^{n-2} + \cdots + a_1 x + a_0,$

or

(2) $\quad z = a_n R_n + a_{n-1} R_{n-1} + \cdots + a_1 R_1 + a_0.$

Thus x is the base number of the gradation, the number on which it is built (in our case, 10); its powers or exponents, $x^1, x^2, x^3, \ldots$ ($10^1, 10^2, 10^3, \ldots$) are the ranks $R_1, R_2, R_3, \ldots$ (10, 100, 1000, $\ldots$). For example, in our decimal number sequence the third rank level $R_3 = 10^3 = 1000$. (For the sake of uniformity, the zero ranks $x^0 = R_0$ in expressions (1) and (2) have been combined in the last term, $a_0 R_0$.)

The coefficients a are the units (the digits) that give the number of each rank. There are always $(x - 1)$ units a_k, where a_k is one of the numerals $1, 2, 3, \ldots, (x - 1)$; again, in our own case, where $x = 10$, there are nine digits. When the ranks are expressed by the positional values of the digits, the zero symbol 0 must appear.

In reckoning by multiples of the units, the rank levels, $a \times R$, the individual rank groups $(a \times R) + (a' \times R')$ are counted together. In precisely the same manner our language uses one number word less for the units that count the ranks than the base number of the number system; hence in the system based on ranks of

10 there are 9 such number-words: one, two, three, $\ldots$, nine
5 there are 4 such number-words: one, two, three, four
20 there are 19 such number-words: one, two, three, $\ldots$, nine, and so forth.

Beyond 9 we would have to find additional number words, which could not, of course, be formed as compounds with "ten" (like "thir-teen").

In addition, the language has words for the ranks of ten, hundred, thousand and so on, and correspondingly for the other systems of rank.

A few additional examples should serve to make this general method of writing numbers more familiar. In the system based on gradations of 5, where $x = 5$ and $a_k = 1, 2, 3, 4, 0$, the number $z = 178$ would be written as follows:

$z = 1 \times 5^3 + 2 \times 5^2 + 0 \times 5^1 + 3$, or
$z = 1 \times 125 + 2 \times 25 + 0 \times 5 + 3$; written positionally, $z = 1203$.

In the 20-base system the number $z = 2657$ is

$z = 6 \times 20^2 + (12) \times 20 + (17)$
$ = 6 \times 400 + (12) \times 20 + (17) = 6 \, (12) \, (17)$.

Here the digits above 9, which we do not possess in our decimal system, are enclosed in parentheses. The Mayan language, however, had individual words for all the numbers in the vigesimal system (see p. 59), so that the number 2657 in Mayan would be read as

6-*bak* (12)-*kal* (17).

If there were no (12)-*kal*, the number 2417 in the vigesimal positional number system would look like this: 6 0 (17).

The Babylonians had a sexagesimal or 60-base system, yet they wrote their units 1 ... 59 by gradations of 10, as we do (see p. 167). Thus the number 1955 would appear in Babylonian as

$$z = 32 \times 60 + 35, \text{ i.e., (32) } sossos \text{ (35).}$$

The number 1920 is (32) 0; but the Babylonians had no zero and would merely have written (32), thus leaving the order or rank of (32) completely unspecified (on this subject, see p. 167).

Representations of Levels
Now let us graphically represent a number written to the base 5 and to the base 10 (see Fig. 14).

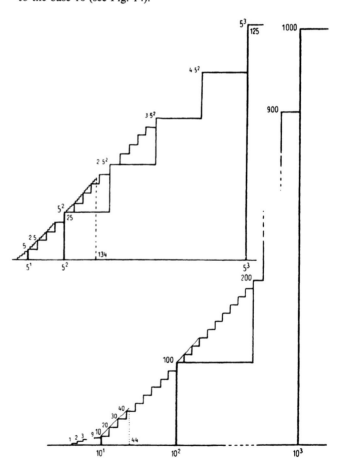

Fig. 14 Number systems based on levels of 5 and 10. The ranks $5^1, 5^2, 5^3$ and $10^1, 10^2, 10^3$ rest on the "floor." At each level the steps of the preceding rank also continue upward. The number 44 is represented in both systems: $1 \times 5^2 + 3 \times 5 + 4 = 134; 4 \times 10 + 4 = 44.$

In the 5-base system (upper diagram) we see the five unit steps ascending from the bottom up to the first rank level (5^1). From there the 5-levels of the first rank move upward by larger steps to the second rank level, 5^2; in the meantime, however, we should not forget the units still climbing by small steps from one level to the next. The diagram shows them only for the first few ranks. From the second "landing" up, the five larger steps of rank 2 ascend to the level of the third rank, 5^3, but meanwhile on each of these the small unit steps of rank 1 also continue to ascend, and so forth. Thus the number system based on gradations of 5 continues to build upward. The 10-base system ascends by much more powerful strides, as the reader can see for himself from the diagram.

The rank levels are shown from the ground up. In addition, we also see the number 44 represented graphically:

$$44 = 4 \times 10 + 4 = 1 \times 5^2 + 3 \times 5 + 4 \,(= 134 \text{ in the 5-base}$$
system).

Thus we can now understand visually what is meant by "the 4th step on the 3rd step of rank 1 on the 2nd step of rank 2." The diagrams serve to impress upon us the manner in which such a system of gradation is built up conceptually.

If we climb up once more to the threshold of the third rank on the decimal staircase and from there survey the grouping of numbers (see Fig. 8, p. 42), we can look down directly on the "expanding" essential difference between the verbal spoken number sequence and the earlier systems of numerals which keep their groupings "on level ground."

NUMBER SYSTEMS BASED ON GRADATIONS OF 20

In the following pages we shall discuss a few number sequences that have been to a greater or lesser extent based on ranks of 20. We shall encounter two somewhat disrupted but very sharply etched 20-base number systems, which constitute a remarkable convergence of the cultural histories of two pre-Columbian American peoples, the Mayans and the Aztecs of Mexico. From these we shall go on to certain peculiar number sequences of Europe (Celtic, French, and Danish) in which a 20-base structure is still distinctly traceable but could not prevail in decimal terms, although it persisted in a few noteworthy formations as vestiges of the most ancient times. Then we shall examine the number sequence of a simple and primitive people, the Ainu of northern Japan, to see how early man built up a unique and highly original tower of numbers, combining the very convenient 20-base system with a very simple 10-base system.

The Mayans

This name is applied to the Indian tribes of Central America (Yucatan and Guatemala) with their related languages, who at one time possessed a very high level of culture, including their own indigenous systems of writing and of numerals. The ruins of their enormous temples show that the Mayan priests were the real bearers and guardians of this culture. By the time the Spaniards invaded in

1524, the Mayan empire had already gone into decline, after flourishing for two periods, from the 5th to the 7th and from the 10th to the 12th centuries A.D.

The Mayan number sequence was based on the number 20. Its successive ranks

$$R_1 = 20, \quad R_2 = 20^2 = 400, \quad R_3 = 20^3 = 8000, \quad R_4 = 20^4 = 160,000,$$

were called

hun,	*bak,*	*pik,*	*calab,*

respectively, and thus had their own particular names. Quite a surprising gradation! But instead of 19 different unit designations it had names only for 1 ... 10 and used these to form the remaining necessary units 11 ... 19. Thus the 20-gradation was interrupted by a decimal gradation. Their full number sequence was as follows:

R_0 units 1 ... 19			$R_1 = 20$ to 399		
1	*hun*		20	*hun-kal*	$1'20 = 1'R$
2	*ca*			30 *lahu-cakal*	10 (of) 2'20 see below
3	*ox*		40	*ca-ikal*	2'20
4	*can*			50 *lahu-y-oxkal*	10 (of) 3'20
5	*ho*		60	*oxkal*	3'20
6	*uac*			70 *lahu-cankal*	10 (of) 4'20
7	*uuc*		80	*can-kal*	4'20
8	*uaxac*			90 *lahu-y-hokal*	10 (of) 5'20
9	*bolon*		100	*ho-kal*	5'20
10	*lahun*		120	*uac-kal*	6'20
11	*buluc*		....		
12	*lah-ca*	10'2	200	*lahun-kzl*	10'20
13	*ox-lahun*	3'10	....		
14	*can-lahun*	4'10	300	*ho-lhu-kal*	5'10'20
15	*ho-lahun*	5'10			
....					
19	*bolon-lahun*	9'10			

$R_2 = 400$ and up		
400	*hun-bak*	1'400
500	*ho-tu-bak*	5 (20) + 400 (see below)
600	*lahu-tu-bak*	10 (20) + 400
700	*holhu-tu-bak*	15 (20) + 400
800	*ca-bak*	2'400
1000	*lahu-y-oxbak*	10 (20) (from) 3'400
....		
1200	*ox-bak*	3'400
1600	*can-bak*	4'400
....		

thereafter $R_3 = 20^3 = 8000$ *pic* and $R_4 = 20^4 = 160,000$ *calab*

In this table the number words are followed by their literal translations into numbers; for example, 13 *ox-lahun* 3'10 (read "thir-teen"), 60 *ox-kal* 3'20. The reader will easily find the combinations 3 + 10 and 3 × 10. In the cases of 300 and 700 the word *lhu* is a contraction of *lahun*, 10; with 500, which should properly be *ho-kal-tu-bac* and all higher numbers, the *-kal*, 20, is omitted. For the meaning of the *-tu-* in 500, see below.

This is an unusual number sequence, one which without any doubt did not arise from the needs and experience of the people but was an artificial and conscious creation of the priests, perhaps designed for calendar computations. For what other purpose would such high numbers have been required? Certainly not in ordinary life, nor in trade and commercial transactions. From $20^2 = 400$ on up, the ranks of 20 become so vast and extensive that even we can visualize and represent them only with difficulty.

It is highly instructive to observe the stages in the growth of this number sequence. Instead of nineteen different names for the units, as we have said, we see only ten; the number words for 11 ... 19 are formed like our own. There can be no doubt that this is an ancient feature, deeply rooted in the people.

Now we may be inclined to think that these 19 units are then used to progress upward in steps of 20: 21, 22, ..., 20 + 10, 20 + 11, ... 20 + 19. But this is not what happens. A different and very remarkable mode of formation appears:

20	*hun-kal*	1'20	
21	*hun-tu-kal*		1 (on the) 20 (-level)
22	*ca-tu-kal*		2 (on the) 20 (-level)
......			
29	*bolon-tu-kal*		9 (on the) 20 (-level)
30	*lahu-cakal*	10'(2'20)	10 in the 40-interval!
31	*buluc-tu-kal*		11 (on the) 20 (-level)
......			
35	*holhu-cakal*	15'(2'20)	15 in the 40-interval!
36	*uaclahu-tu-kal*		16 (on the) 20 (-level)
......			
39	*bolon-lahu-tu-kal*		19 (on the) 20 (-level)
40	*cakal*	2'20	

Thus the method of forming number words in the interval from the first to the second twenty is not uniform. The units are added in their cardinal form with *-tu-* to the next lower twenty-step; thus 21 is the first number on the 20-step. But at 30 the next *higher* twenty-level, *cakal*, or 40, is referred to, and the same is true of 35. This is the rare "counting from above," which now no longer exists anywhere except in the Teutonic north of Europe. The Mayan concept of these numbers is thus the following: 1 ... 20 is the first and 21 ... 40 the second level; 30 lies within the second level. The level (or interval) is thus referred to in terms of its final number 40 — whence "thirty is 10 in the 40-interval" (see p. 76).

It is characteristic, moreover, that from 40 on up the Mayan number sequence uses exclusively this mode of counting from above to form its number words, and thus proceeds from 40 up without change or disturbance. As an example: 60 is *oxkal*; thus

41 is not *hun-tu-cakal*   but   *hun-tu-y-oxkal*
1 (on) 2'20   1 (in the) 60 (-interval).

The syllable -*y*- in the above is merely a connective. Accordingly,

185 is *ho-tu-lahun-kal*   and   386 is *uac-tu-hunbak*
5 (in the) 10′20-   6 (in the) 1′400(-interval).

We shall not go astray if we interpret this peculiar number sequence as follows: The units 1 ... 9 each have their own names and then combine with 10 to form the 10-level interval from 11 to 19. The ancient decimal grouping was here in use among the common people and made itself felt: The number sequence proceeds undisturbed from 20 to 40; 30 retained no name of its own. This we can readily understand; we recognize the same thing in the 60 ... 80 interval of the French number sequence, where 78, for example, is *soixante-dix-huit*, 60′10′8. But then comes a sudden break. The sequence begins to count from above, and from this point this principle governs the structure of the whole number sequence without exception; moreover, the principle of counting backward has also a reverse effect on the forms of 30 and 35. These are no longer natural outgrowths of the language but artificial and conscious verbal formations. The same thing is evidenced by the great extent of the number sequence, up to $20^4 = 160,000$. Although calendar computations may well require the use of very large numbers, they do not seem to have been the only reason for the enormously high numbers. The latter also bespeak reverence for the gods and a yearning to climb ever higher and higher, closer and closer to them, step by step. It was both a holy thing and a mystery of the initiated to ascend such a tower of numbers, to count up to such heights that numbers were no longer felt as quantities. The striving to mount ever upward was realized in the law of numbers. We shall encounter similar "holy" towers of numbers in India (see p. 137).

The Aztecs

The ancient Mexicans possessed a number sequence clearly based on ranks of 20 along with — and this was its amazing feature — a sequence of units ordered on the number 5; thus 5, 10, and 15 had their own separate names, like the levels marking the successive ranks of 20. The reader can see this for himself from the following table:

1	*ce*		11	*matlactli-on-ce*	10 + 1
2	*ome*		12	*matlactli-on-ome*	10 + 2
3	*yey*		13	*matlactli-on-yey*	+ 3
4	*naui*		14	*matlactli-on-naui*	+ 4
5	*macuilli*		15	*caxtulli*	
6	*chica-ce*	5′1	16	*caxtulli-on-ce*	15 + 1
7	*chic-ome*	5′2	17	*caxtulli-on-ome*	+ 2
8	*chicu-ey*	5′3	18	*caxtulli-on-ey*	+ 3
9	*chic-naui*	5′4	19	*caxtulli-on-naui*	+ 4
10	*matlactli*		20	*cem-poualli*	1′20
30	*cen-poualli-om-matlactli*	1′20 + 10			
40	*ome-poualli*	2′20			

......

100 *macuil-poualli*   5′20, thus not a separate word
further $20^2 = 1′400$ *cen-tzuntli* $20^3 = 1′8000$ *cen-xiquipilli*.
Example: 39 = *cen-poualli-on-caxtulli-on-naui* 1′20 + 15 + 4.

The various ranks of 20 have the following names and meanings:

1, *ce*;   20, *poualli*, "counted group"; 20^2 *tzuntli*, "hair,"

the latter a word that once, according to an ancient counting limit, had designated an unspecified multitude or "many," but later became attached to the specific meaning of 400 as a number word; and finally,

20^3, *xiquipilli*, a "pouch" containing 8000 cocoa beans.

The word *ma-itl*, "hand," gave rise to the number words for "five" and "ten," with the use of *-on-* or *-om-* "on."

There is no doubt but that this number sequence was built up more naturally and with less mental strain than the Mayan one. It contains a 5-count that appears very ancient and does not count backward from higher levels. A significant document in this connection would seem to be an old Spanish report, according to which the Indians' fighting forces were organized on the principle of this vigesimal grouping: In our terms, their army consisted of 20 battalions, each having 20 companies, each of which had 20 men — hence 20^3 men in all. From this document it seems likely that this old and convenient ordering and grouping by 20's — for this military organization just mentioned is nothing but that — gave rise to the organization by ranks of 20, in which the 20-groups became the levels of rank. Thus the Aztec number sequence provides a unique example of this very ancient procedure, already mentioned (see p. 47), by which the groupings were first set up as independent supports, between which the actual number steps themselves were inserted.

We can see from our diagram (Fig. 15) that here the 5-count is not a ranking based on 5 — it goes no further than the second rank level 5^2 — but rather that a short ladder containing three steps of 5 numbers each has been "superimposed" on the large gradations of 20.

Thus we find that no matter how many number sequences we examine, there are absolutely no pure vigesimal rankings, either in verbal number sequences or in written numerals.

Yet are there any gradations based on 5? If the size of the 20-intervals is too large for convenience in counting, the intervals of five are too small. If the human mind wishes to comprehend a larger number, it stumbles over the short intermediate steps and, once it has achieved a sequence with gradations of 10, simply climbs from one 10-level to the next (see Fig. 15). Just as there is no purely vigesimal gradation, there have also been no examples of number sequences based purely on gradations of 5.

Ever since the earliest attempts at counting, which succeeded in forming only the first four or five number words, systems based on stacks of fives have risen here and there; but it never happened that a people or culture arrived only at the second rank, 5^2 and then stopped. In most cases they continued by way of $2 \times 5 = 10$ to the larger steps of 10 (see the number sequence of the ancient Sumerians

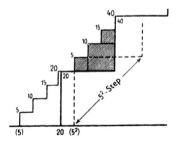

Fig. 15 Aztec numerical ranks of 20, with steps of 5 up to 20; thereafter the sequence climbs simply by large steps of 20. The fact that this is not a gradation of 5's is shown by the cross-hatched steps, which are to be seen as the second level of rank. The vigesimal ranking is clearly dominant.

p. 165). Therefore there is no justification for speaking of a "quin-ary" system or, in our terms, a gradation of 5; at most we can speak of traces or vestiges thereof. Likewise there are no number sequences based on gradations of 2. Some have attempted to see such a system in the fact, for example, that the Bushmen of South Africa have only the two number words *a*, "one," and *oa*, "two," and that they say "four" as *oa oa* and "five" as *oa oa a*. But this is an instance of mis-taking one numerical principle for another: for this is not a grada-tion but merely an ordering, a much earlier phase, especially if it occurs in speech.

Now let us look at a few peculiarities and traces of 5-gradations in mature number sequences. Irish, a Celtic language, regularly says *deich* for "ten" (< Latin *decem*; see p. 97, first column); yet it forms the interval between 10 and 20 not on the principle of 1 + 10, 2 + 10, etc., but says *oin deec, da deec* and so forth, where *deec* (< Celtic *da-coic*, 2 × 5) means not 10 but "two fives."

Moreover Cymric, the Celtic language spoken in Wales, contains an example, occurring in no other European language, of piling up units on 15 *bymthec* 5'10 (see p. 97, column two);

16 *un ar bymthec* (1 + 15)
17 *dau ar bymthec* (2 + 15)
19 *pedwar ar bymthec* (4 + 15)
20 *ugain*,

a method of counting that would be easier to understand if the units 6, 7, and 9 were formed as 5 + 1, 5 + 2 and 5 + 4, as they are in the Aztec number sequence discussed earlier.

VIGESIMAL GRADATIONS IN EUROPEAN NUMBER SEQUENCES

If we now examine number sequences in our own culture for signs of gradation or steps of twenty, we shall find some extremely re-vealing examples. The Celtic languages have maintained and dissemi-nated the vigesimal gradations with a persistence equaled by no other group of tongues; in Danish and French the vigesimal system has infiltrated in a few places into the basically decimal number sequence and has overcome the latter for considerable intervals. But once we have found that the original Indo-European language possessed not the slightest traces of any form of counting by inter-vals of 20, we naturally begin to wonder: Where did it come from, and how did it enter these particular languages?

The Celtic Vigesimal Numbers

The individual tongues that belong to the Celtic group are treated together, so far as the Celtic number sequence is concerned; the reader is asked to look these over, especially in the interval from 10 to 100 (p. 97). The adjoining map will make the over-all picture clear (Fig. 16). (No number sequence has come down to us from the now extinct language of ancient Gaul.)

Fig. 16 Where Celtic was and is spoken: 1, Irish; 2, Gaelic in Scotland; 3, Welsh; 4, Cornish (extinct); 5, Breton; 6, Gallic (extinct).

In examining these tongues, we are suprised at the regularity with which the vigesimal principle recurs. In present-day Irish (as well as in Scottish Gaelic, which differs but little from it), the sequence 20, 40, etc., runs on by intervals of 20 up through 180, and the intermediate tens, 30, 50, 70, etc., do the same up to 190:

20 *fiche*		30 *deich ar fiche*	10 on 20
40 *da fiche*	2'20	50 *da fiche's a deich*	2'20 on 10
60 *tri fiche*	3'20	70 *tri fiche's a deich*	3'20 on 10
80 *ceithre fiche*	4'20	90	4'20 on 10
. . . . . .		. . . . . .	
160 *ocht fiche*	8'20	. . . . . .	
180 *naoi fiche*	9'20	190 *naoi fiche's deich*	9'20 on 10

Examples (for the units, see the tabulation on p. 26):

deich mbliadma ocus ceithri fichit,
10 years   and   4'20 = 90 years;
deich mnaa secht fichit,
10 women   7'20   = 150 women.

Surprisingly, Old Irish did possess a distinct gradation based on 10, which was completely displaced by vigesimal gradations in modern Irish! The Old-Irish tens were formed from the words for the units:

10 *deich*	60 *sesca*
20 *fiche*	70 *sechtmoga*
30 *tricha*	80 *ochtmoga*
40 *cethorcha*	90 *nocha*
50 *coica*	100 *cet.*

A remnant of this old 10-count is the number word for 100, *ceud* (*cant, kans,* cf., Latin *kentum*), which shows its influence in the Cornish and Breton 50, *hanter-cans,* "half-100"; another inconsistency is the Breton 30, *tregont* (p. 97, fourth column).

The Danish number sequence
This is a further example from northern Europe (p. 95). Here 10 is *ti,* 20 *tyve,* 30 *tredive* and 40 *fyrretyve.* The word *tyve,* "tens," was originally the plural of "ten" and comes from an old word *tiugh,* corresponding to the Gothic *tigum,* from which the German *-zig* developed: *tre-tyve > tredive,* "thirty"; cf. *drei-ssig.* This plural is regularly used to form the first four tens; then, however, *tyve* takes on the meaning of 20:

60 *tre-sinds-tyve,* "3 × 20"   80 *fir-sinds-tyve,* "4 × 20."

Thus the same number sequence contains both these meanings of *tyve*:

tyve 20 — *fyrre-tyve* is not 4'20, but 4'10; *fir sinds-tyve,* again 4 × 20!

The meaning behind this word formation is, for example in the case of 70, "half of the fourth 20-group upon three 20-groups." In common speech the number words are shortened (see p. 95).
    The Danish sequence of tens is thus a rich deposit of primitive forms: formations based on 20, break after 4, 20-group, 20-levels

Fig. 17 The half-steps in the Danish number sequence, based on gradations of 20.

Fig. 18 Invasion of the vigesimal gradation into the French decimal number sequence.

that show up through the dominant 10-levels and, finally, the extraordinary "half-levels" (Fig. 17). These make use of a rare form of connection, that of dividing the last gradation. They correspond exactly to the German or Russian expression, "It is $\frac{1}{2}$ 4 o'clock," meaning not 2 o'clock but half of the time interval between 3:00 and 4:00. Thus in Danish "half eighty" is not "forty" but the half-step between the vigesimal levels of 60 and 80 (Fig. 18). Later we shall have more to say about this form of "overcount" (p. 76).

The Danish number sequence also provides a beautiful example of the dispossession of the vigesimal grouping *fem-sinds-tyve* (5 × 20), which is still retained in the expression for 90 = $\frac{1}{2}$(5 × 20), by the decimal gradation *hundrede*.

The French Number Sequence

Here we find evidence of one of the most fascinating cases of the infiltration of the primitive vigesimal gradation (see p. 49). The French number sequence progresses smoothly upward by levels of 10; and not until 70, *soixant-dix* (60′10) is there any intrusion of the idea of counting by twenties, which then at 80, *quatre-vingt(s)*, "four twenties" (*not* 4 × 20) clearly thrusts itself into the decimal gradation. But it is at once forcefully tied down, and as soon as it forms the next half-step, like a low stool set in front of the excessively high 20-level, it immediately becomes fully established and is then literally suppressed.

But it was not always thus. Fortunately, we can reconstruct the history of the French number sequence. Nothing is known of any Gallic numbers, but if they had possessed a strong vigesimal gradation, some traces of it ought to have been preserved in French personal or place names, or in Old French poetry. The Romans, who for centuries imposed their language and cultures on the region, had a decimal system of counting and reckoning which showed not the slightest hint of any grouping or gradation by twenties.

The 20-gradation suddenly appeared in the 11th century in northern France, spread from there to the south and until well into the 17th century formed a number sequence that ran from 60, *trois-vingt* (old form *vint*), through 120, *six-vingt*, and 140 *sept-vingt*, all the way to 360, *dix-huit-vingt* (18′20). *Les Onze-vingts*, "the eleven-twenties," was the name for a constabulary force consisting of 220 men. And the old hospital built by Louis IX in the 13th century in Paris is still called the *Quinze-vingts* from its 300 blind inmates, whence the colloquial expressions *un quinze-vingt*, a blind person, and *entrer au quinze-vingts*, "to sleep." In a few isolated local dialects (as in Savoy), the old steps of 20 are still used in common speech up to 160, *huit-vingt* (8′20). It is also remarkable that whereas Old Provençal regularly counted upward by tens, the modern dialect has adopted the vigesimal count, from *tres- vint* (60 = 3′20) up to 380, *dès-e-nou-vint*, (10 + 9)′20.

It is significant that the number sequence breaks off just at 19′20 just before the second rank level, 20′20 = 20^2, a vigesimal gradation. This shows that this vigesimal count was thought of only in terms of the popular 20-groups, without leading up to any consciously thought-out gradation of twenties, like the Mayan number sequence (see p. 59). This is confirmed by the fact that the words for 100 and 200 in French are formed with *cent*, for the most part, the expression *cinq-vingt* occurring only in vernacular speech. Above all, it is shown by the fact that from the 17th century on the widespread 20-groupings completely disappeared from all written number words. Of the original, powerful, vigesimal invasion, there now remains only the interval from 60 to 99, with 80, *quatre-vingt*, the only number word incorporating levels of twenty. The expression for $120 = 6'20$, as well as the peculiar method of writing numbers by 20s, continued in use into the 18th century. In Molière's *Le Bourgeois Gentilhomme* (Act 3, Scene 4) the creditor pulls out his account book and reads aloud to his debtor:

"... *donné a vous une fois deux cents louis*" — "*Cela est vrai*" — "*Une autre fois six-vingts*" — "*Oui*" — "*Et une autre fois cent quarante.*" "... lent you on one occasion 200 louis d'or (2 hundred)" — "That's right" — "On another occasion 120 (6 twenties)" — "Yes" — "And another time 140 (hundred 40)."

Almost certainly the second amount was written down in the account book as *six-vingt*, VI^{xx}; and IIII^{xx}, *quatre-vingt*, continued to be so written until some time in the 18th century. Did *six-vingt* survive so long because it was a French echo of the Germanic "great hundred" (120)?

Now, however, only *quatre-vingt* remains, like a large protruding boulder in the midst of a smoothly progressing number sequence; in its immediate vicinity it has displaced the forms *septante* 70, *huitante* or *octante* 80, and *nonante* 90, which still live on here and there in a few spoken dialects. The *Septuagint*, the Greek text of the Old Testament, is called *La Bible des Septante* in French (see p. 186). These all show the force of the ancient, popular 20-group, "man," in counting. Today the ground has been leveled out, and the decimal ladder proceeds unperturbed past this alien intrusion.

Now arises the question that has been on the tip of our tongues for so long: What was the origin of this rare manifestation of the vigesimal count, which so oddly broke into an alien number sequence, dominated it for a while and then disappeared almost completely? Why was the Roman-French decimal gradation suddenly put aside between the 11th and the 17th century?

The Normans

Although much still remains obscure, it seems to have been the Normans who brought the vigesimal habit of counting into the French number sequence. We have already said that the original Indo-European language contains no traces of a vigesimal grouping or gradation. But the Celtic tongues are Indo-European, and we have seen that Celtic incorporated and held on to the vigesimal count with a tenacity shown by no other language of our general culture. Were the pre-Indo-European inhabitants of the subcontinent the original bearers and transmitters of the 20-count to the Celts? We do not know.

But we do know that the vigesimal system of counting was in constant and common use in the North. We recall the groupings such

as the *Stiege*, the *Schneise*, and the *score* (see p. 49), all of them
words for groups of twenty elements which the Germanic languages
invented in addition to the regular number sequence. These special
20-forms, curiously enough, did not follow the usual Indo-European
formation with the syllable *vi-* (e.g.: Old High German *zwein-zug* as
against Latin *vi-ginti*, p. 96).

From the North, primarily from the great realm of the ancient
Danish kings, the Normans spread as raiders, conquerors, and
traders to Iceland, to the lands of the Angles and Saxons, to the
French coast and thence deep into France itself, along its rivers. In
A.D. 911, at the mouth of the Seine, they founded a kingdom of their
own, Normandy, where they adopted the French language and
French customs. From Normandy they invaded and conquered
England (1066); and they sailed along the Atlantic coast and into the
Mediterranean, where they established the Norman Kingdom of
Sicily that was to play such an important role in western history.

The Normans were not merely conquerors but also traders, whose
area of commercial activity stretched from Greenland all the way into
the Mediterranean. It is understandable that Norman customs were
introduced into areas with whose inhabitants they came into contact.
Thus the northern habit of counting by twenties came into France.
How could this custom, which had not been used for many centuries,
if at all, suddenly come into common currency, except through some
stimulus from outside? This is suggested not only by the period in
which it was introduced, the 11th century, but also by the fact that
it was thereafter more commonly used in the north of France than
in the south. But how was it able to gain a foothold at all? Because
the 20-group is so deeply rooted in the numerical concepts of primitive man that once it was given its initial impetus, it took root.
Moreover it did not have to be derived linguistically from the Norman word *skor*, "score"; the indigenous number word *vingt* could
immediately serve as the designation for a 20-group.

The Portuguese System

The Portuguese vigesimal count, which we mentioned earlier
(p. 51), is more difficult to recognize as being Norman in origin;
on the other hand, there is no doubt of the Norman influence in
Sicily, where eggs, fruits, and people's ages are still counted by
twenties in ordinary speech:

40 *du vintini* (2 twenties) — 50 *du vintini e deci*, etc., up to 100 *cincu
vintini* (5 twenties).

Moreover, the decimal number sequence was always available. For
the illiterate peasant, who has occasion to count more often than to
calculate and whose computations for the most part take the form
of counting, the 20-count, as we know, is quite convenient.

The Albanians

Against this "native" background the Albanians, whose language
is also considered to belong to the Indo-European family, have taken

the only indigenous twenty-step. An Albanian counts as follows:

10 *dhiete*
20 *nji-zet*
30 *tri-dhiete*, 3'10
40 *dy-zet*, 2'20, but from then on again by 10-steps:
50 *pese-dhiete*, 5'10
60 *gjashte-dhiete*, and so on

Albanian *-zet* < Sanskrit 20 (*vim*)-*šatih* (see p. 93).

Another endemic feature is the vigesimal counting up to 80 = 4'20 of the non-Indo-European Basque language, which we shall say more about presently (see p. 110).

THE AINU NUMBER SEQUENCE

We include the Ainu number sequence in our discussion not so much because it is a rare instance of an almost pure vigesimal count but because it has been used to build up a unique system of number words.

Today the Ainu are a dispossessed and dying people of the Far East (Sakhalin Island), who seem to belong to the white rather than the yellow race. In creating their number sequence they piled up the ancient 20-group "men" one atop another, higher and higher, not with any idea of arriving at a series of gradations of 20 with levels of 20^2, 20^3 and so on, like the Mayans and the Aztecs, but with the thought of having the numbers proceed by levels of ten each. Thus the Ainu number sequence has no special designation for the ranks 100, 1000, or 400.

1 *shi-ne*	6 *i-wan*	4'10 ('4 from 10')	
2 *tu*	7 *ar-wan*	3'10 ('3 from 10')	
3 *re*	8 *tu-pesan*	"2 steps down"	
4 *ine*	9 *shine-pesan*	"1 step down"	
5 *aschik-ne*	10 *wan*		

20 *hot-ne*	30 *wan e tu*	*hotne*	10 von 2'20	(overcount)
40 *tu hotne* 2'20	50 *wan e re*	*hotne*	3'20	(overcount)
60 *re hotne* 3'20	70 *wan e ine*	*hotne*	4'20	(overcount)
80 *ine hotne* 4'20	90 *wan e ashikne*	*hotne*	5'20	(overcount)
100 *ashikne hotne*	110 *wan e iwan*	*hotne*	6'20	(overcount)

200 *shine wan hotne* 1'10'20
300 *ashikne hot-ikashama* ("*on*')- *shine wan hotne* 5'20 *on* (1'10'20)
400 *tu-shine wan hotne* 2'(1'10'20)
500 *ashikne hot-ikashama-tu-shine wan hotne* 5'20 *on* 2'(1'10'20)

. . . . . .

1000 *ashikne-shine wan hotne* 5'(1'10'20').

A truly Cyclopean tower of numbers! The 20-group "man" (*hot*, "the whole") is the building stone that forms all the levels of rank. And the tens do not even disrupt the progression at 100 or 1000! A surprising feature is the overcount from the next step above at the odd-numbered steps, such as saying 70 as "10 from 80" in the form of "10 in the fourth twenty" (see p. 76).

Another feature peculiar to this sequence is the series of units: *shi*, "beginning"; *-ne*, "to be," helps to forms the words for 1, 5 and

20; 4 *ine*, "much," 5 *ashikne*, "hand," 10 *wan*, "two-sided" standing for "both hands." And the words for 6, 7, 8, and 9 are governed by 10. The One and the 20-group "man" are the only building blocks, and this language puts them together according to a very simple rule. To the Ainu anything more than 1000 is "many"; from here on the number words become too large and cumbersome for him.

When we compare this number sequence with our own, we can readily see the flexibility and simplicity of the latter, and the way it proceeds quickly and easily up slender number steps to the very highest numbers. Compared to ours, the Ainu number sequence stands as a rigid unbending tower, heavy and massive, but impressive because of its unvarying use of the ancient number-group as its building stones.

This insight prepares us for our next problem: How have peoples of various cultures formed the individual words of their number sequences?

The number 18, for example, occupies the same position in all number sequences, but we shall see that it is not, contrary to what the uninformed may think, generally expressed in the same manner in all languages.

First let us sum up our last discussion:

Early formations of number words lead occasionally to breaks after the number 5.

In some languages gradations based on levels of 20 still persist from the earliest grouping of numbers.

But within the framework of European culture, and as long as counting was tied to the practical life of ordinary people (unlike the Mayan), systems based on steps of 5 or 20 never took hold. The emphasis is upon gradation; that is, a clear structure erected with certain levels of rank on which the steps of the lower ranks mount upward according to some definite rule.

The gradation based on the number 10 was the only one to achieve a pure and full development. Ten lies within the natural area between intervals that are too narrow (the 5-gradation) and too wide (20-gradation); the former hinders the process of counting, the latter makes it conceptually and linguistically too hard to grasp (the Ainu and Mayan sequences). Yet one finds dramatic documentations of very early intellectual history in the way the ancient, primitive number concept of the 20-group "man," has succeeded, to a greater or a lesser extent, in breaking into the otherwise dominant decimal number sequence.

Laws Governing the Formation of Number Words

SUCCESSION OF MAGNITUDES

Thus far we have been speaking about the way number sequences were built up; now let us take a look at the formation of the number words themselves. Here, too, we shall encounter forms that seem very strange to us, but that we can understand in the light of early man's conception of numbers.

Eighteen

In the "conceptual" number sequence and hence in the succession of abstract numbers . . . 17–18–19 . . ., the number 18 stands in the same position in the minds of all peoples. It would be a mistake, however, to suppose that the number word for 18 is therefore formed in the same manner in all languages. If we glance at various verbal number sequences, we find the following:

German	*acht-zehn*	8′10
French (Classical)	*dix-huit*	10′8
Greek (Modern)	*oktō-kaí-deka*	8 and 10
Greek	*deka-okto*	8′10
Latin	*decem et octo*	10 and 8
Latin	*duo-de-viginti*	2 from 20
Lithuanian	*aštuno-lika*	8 over (10)
Irish	*ocht-deec*	8′(2′5)
Breton	*tri-ouch*	3′6
Welsh	*deu-naw*	2′9
Mexican	*caxtulli-om-ey*	15 and 3
Finnish	*kah-deksan-toista*	2 (from) 10 (in the) second (tens)
Ainu	*tu-pesan-ishama-wan*	2 climb down added to 10

Truly a bewildering assortment! We may be surprised at the variety of ways in which languages can form this one number word. Yet all these formations (except the Breton and Welsh) proceed in a natural succession from 5, 10, or 20: from one of these numbers they move either forward or back (18 = 20 − 2). The units and the rank levels are thus the building blocks; the four simple arithmetic operations of addition, subtraction, multiplication, and division are the mortar that holds them together.

The ways in which 72 is formed from levels of 20 is also fascinating:

Danish	"half (4 × 20)′2"	see p. 65
Modern Irish	"(3′20 + 10)′2"	see p. 65
French	"60′12"	
Ainu	"12 from 4′20"	see p. 69
Mayan	"12th on the 4′20(-level)"	see p. 61

We shall be able to understand these and other formations, however strange they may seem to us, if we remember that early man, who devised the first number words, was more intimately bound to his environment than modern man. Even today the peasant is more closely and tightly dependent on wind and weather, man and beast, tree and soil, than is the city dweller, to whom the world is wider, to be sure, but also more remote. And numbers are a part of the world. Early man wants to *see* numbers; they must remain visible to him, and he must be able to touch them if he is to grasp them with his mind.

For this reason he breaks down larger numbers into smaller ones, if he can. The two Celtic wordforms for the number 18, "3 × 6" and "2 × 9," are to be understood thus, as is the answer given by an aged Sicilian peasant woman when asked how old she was:

tre vvote vinti cincu anni, "3 times 20'5 years" (= 75).

In Otfried's Old High German Book of the Gospels there are the following decompositions of fairly large numbers:

zwiro sehs (jaro), $12 = 2 \times 6$
einlif stunton sibini, $77 = 11 \times 7$
thria stunton fintzug ouh thri, $153 = 3 \times 50 + 3$
thrizzug stunton zehinu, $300 = 30 \times 10$.

Units and Groupings

In addition to these, early man had yet another way of breaking down numbers. How was he to conceive a large number like 47? It is above all the units 1, 2, 3, ..., 9 — in this case 7, which are understandable; whatever is over and above the units (the 40) becomes clear as clusters or groupings. *Quat borla*, "four loads," said the Piedmontese farmer when asked his age (see p. 49); and the Old Spaniards used to say, *tres vent medidas de jarina*, "3'20 measures of grain." The visualization of numbers leads to their grouping, as we have seen especially with the 20 group "man" (p. 49).

There is a striking documentation of this: Old Norse, like Celtic, puts the counted object right after the units, and behind these places the "block" (=grouping) of the higher numbers. Thus in Old Norse, 364 days are called *fiora dagar ens fiorþa hundraþs*, "4 days in the 4th (great) hundred (120)"; in Celtic, 11 horses are called *un march ar dec*, "1 horse and 10."

The reader may also refer to the examples on pages 27 and 65.

Why did these curious turns of speech arise? Because they are easier to visualize. 47 sheep = 7 sheep and 40; 40 is a round number, a grouping whose magnitude is already familiar in the form 2'20; but here the 7 is that which must be emphasized. The number thus falls naturally into two easily visualized parts: units and groups. In contrast, the undivided number 47 is much more alien. It is merely one of so many numbers, ... 46 — 47 — 48 ..., which are met with far less often than the unit 7 and the group 40.

This interpretation also explains the reversed positions of the units and the ten-group in the German number words. If we have seen in this nothing more than a kind of inexplicable verbal caprice, now we understand the reason for it; it is just the ancient precedence of the units. As long as the units are small, this seems to make sense. With big numbers like 4385, for example, with which early man had no occasion to cope, the sudden jolt at the end, "... five and eighty," seems quite obvious to us. It is doubtless confusing, but now that we know its history we can tolerate it. This is not to say, however, that we ought not to eliminate it in this day and age.

In Swedish, for example, today one says 364, *tre-hundra sextiofyra*, in the proper order of magnitudes, having given up the Old-Norse metathesis of the units and tens. This did not happen for reasons of logic or convenience, but only because the native number words were later modeled after those of the Roman church, which were introduced into Scandinavia at the same time as Christianity (see p. 79).

The law of the succession of magnitudes explains the descending order of the levels of rank in 4385: symbolically 4 Th   3 H   8 T   5 U. Numbers are built up from the rank levels (R) of thousands Th, hundreds H, tens T and units U, together with the "pre-units" a that count them (in this case 4, 3, 8, and 5), as we may call them for the moment in order to differentiate between them and the rank of units U. Thus we shall call aR, here 4Th, a rank-group. The individual rank-groups are generally connected either without words or by the word "and":

symbolically   4—Th   3—H   8—T   5,
English   four-thousand three-hundred eighty-five,

and in Modern German with the order of the last two reversed, "... *fünf-und-achtzig*, ... five-and-eighty." Old High German, by contrast, often retained the logical and consistent order:

drizug inti ahto jar,   38 years,
zehnzug inti finfzug inti thriu,   100 and 50 and 3.

Many languages, such as Russian, Arabic, etc., reverse the order of the units and the tens, like modern German; others observe the order of magnitudes without exception, like Chinese.

The reverse succession of magnitudes or ranks, from lowest to highest, can also be documented in some languages, as in the Classical Greek expression for 56,404 talents:

téssera tetrakósia hexakis-chília pentakis-mýria tálenta,
    4   400   6 times 1000   5 times 10,000.

Number Symbols and the Succession of Magnitudes
Let us now turn our attention to the discrepancy between written numerals and the succession of magnitudes, which occurs quite commonly. This happens especially in German: we pronounce 4385 as "... 5 und 8-zig," so that a child first learning to write is likely to put down 4358. The spoken number words are older than the numerals, which have come into their domain from outside, as a stranger. In Old Norse, when the Roman numerals were introduced along with Christianity, the spoken phrase prevailed even in writing:

III *vetre oc XX*, "3 winters and 20" ($= 23$ winters),

a form of notation that follows the spoken word, exactly like the French IIII^{XX} for *quatre-vingt* (4'20).

I shall cite a little-known example from Arabic: Here, too, the units precede the tens in speech, while the normal succession is otherwise observed:

641 = 600′1′40 *sittu mi'atin wa-ahadun wa-arba'una*
                  6   100   and 1   and 40.

(Besides the usual metathesis, Arabic also uses the ascending "reverse succession," 1′40′600.)

Persian, in addition to the common Indian-East Arabic numerals, also has the so-called *Siyaq* script, which runs from right to left and which is frequently used by tradesmen and merchants in bazaars, see Figs. 109 and 110. Here one writes down the tens and units in reverse order, in our notation as 600′1′40, and thus exactly reflects the way the number is spoken (Fig. 19).

How did this unique situation in the history of writing come about? A system of writing numbers that consists of pure numerical symbols does not reverse the order of the rank-groups. The answer is surprising: The symbols in the Siyaq script are not pure numerals at all but abbreviated number words, and as such they reflect the way the numbers are spoken. Had this not been suspected for other reasons, it might have been deduced from the reverse order of the units and tens alone.

There is another curious complication in the history of these numerals: The Turks first adopted the Arabic Siyaq script in their financial administration, and along with it the reversed order of the units and tens, which the Turkish language itself does not possess. Thus the Turks would read the number 641 as *alti-yüz-kirk-bir*, 6′100′40′1, but they would write it as 600′1′40. This, of course, is the exact opposite of what happens in German: The numerals observe the proper order of the ranks and the spoken words do not, while with the Turks the language follows the correct order but the Siyaq numerals do not (only the Siyaq, not the East Arabic numerals, which is commonly used in writing).

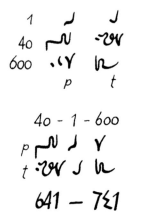

Fig. 19 The number 641 as written in Persian (p) and Turkish (t) Siyaq script. At the top are the symbols for 1, 40, and 600; below the number is written from right to left with the units and tens reversed; and at the bottom in Indian–West Arabic and Indian–East Arabic numerals.

BACK-COUNTING

As a rule, number words are formed by addition and multiplication. But the forms ("difference numbers") which arise by subtraction, or more precisely backward counting, are more interesting. Our examples show how widespread these are:

Latin 19 *un-de-viginti*, "1 from 20"
Latin 58 *duo-de-sexaginta*, "2 from 60";
Ancient Greek 58 *dyoîn déontes hexékonta*, "2 lacking from 60."

The Biblical passage, "Of the Jews five times received I forty stripes save one" (II Cor:11, 24) is, in the original Greek text,

(*pentákis*) *tesserákonta parà mían,*
5 times   40   less   1,

and in Bishop Wulfila's Gothic Bible

(*fimf sinþam*) *fidwor tiguns ainamma wanans,*
5 times   4   tens   lacking   1.

The passage (John 5:5): "And a certain man was there, which had an infirmity thirty and eight years" reads, in Old High German:

wankta zuein, thero jaro fiorzug ni was,
lacking   2,   which years   40   not was.

Moreover Anglo-Saxon 19 is *anes wona twentig*, "1 less than 20"; similarly, Old-Norse "29 nights" is

nott midr enn þritögr,
1 night less than   30,

and in a very odd fashion of counting, "54 men" is

man midr en halfr setti tögr,
1 man less than half six   tens.

Among the more remote languages, I shall cite only the examples of

Sanskrit 19 (*ek-*) *una-vimsati*, "(1) lacking (from) 20,"

and so on through all the tens, so that

Sanskrit 39 is *una-čatvarimsat*, "lacking 40."

Just as here one omits the number (1) that is subtracted, so in some Roman words for fractions one omits the number from which the subtraction is made:

$\frac{11}{12}$ *de-unx* "(as = $\frac{12}{12}$) from this 1 ounce,"
$\frac{3}{4}$ *dodrans* < *de-quadrans* "(*as*) from it $\frac{1}{4}$."

Similar formations are cited on p. 83.

Most important, however, are the examples of counting back from ten. These are to be found among many primitive peoples, sometimes even as far down as 6, as in the Ainu *i-wan* (4'10) (see p. 69), but also occur in the Finnish words for 8 *kah-deksan* (2'10) and for 9 *yh-deksan* (1'10). Perhaps the original Indo-European language may also have formed 9, 8, and 7 by counting backward from 10 (see p. 170). If we recall the Dene Dinje expression for 4, "only one is left," we have an example of back-counting with reference to the number 5 (see p. 35).

Another peculiar word formation is the Russian *děvja-no-sto*, "9 before 100" for 90, which derives from the early number concept, to distinguish it from "9 to (or before) 10". This is a beautiful example of how the ten series is basically seen as a sequence of units, but in this case "*before* 100."

How are we to interpret all these word formations? The next higher rank exerts its influence backward, dominating the numbers just preceding it, just as the full hour does the few minutes before it: "10 minutes to 6:00" rather than "5 hours 50." The rank levels, the old number-groupings, are numbers that early man could visualize. We have seen this already: 38 is just one of many numbers that are hard to grasp, but 40 is clear and palpable and so is 2; hence "2 from 40" can be understood, whereas 38 cannot. This conception of numbers as bundles is distinctly discernible in the following examples (see p. 51):

Portuguese 115 *seis (vintens) meno cinc,* "6(20) − 5,"
Sicilian   85 *cente me quinci,* "100 − 15."

On the other hand, the back-counting that occurs in the writing of numbers by some peoples using special symbols (like the Babylonians) or changes in the normal succession of magnitudes (like the Romans), is no more than a kind of shorthand:

Babylonian 19 ⟨ 𝍤 10′9 as compared to ⟨⟨𝍤 20 − 1; "less" is 𝍤 (p. 164)

Roman IIII (4) as compared to IV (1′5), or XVIIII (10′5′4) as compared to XIX (10′1′10), or 400 CCCC as compared to CD (100′500), or 89½ SXC (½′10)′100, where S stands for *semis,* "half." These abbreviations are late Roman; in medieval Latin the full form of the numeral was written out as often as the abbreviation.

A "classic" passage in which the Roman IV occurs in Goethe's *Roman Elegies* (15):

"Endlich zog sie behende das Zeichen der römischen Fünfe / und ein Strichlein davor ... (und verwischte alles wieder) — / aber die römische Vier blieb mir ins Auge geprägt."

"In the end she deftly wrote down the sign of the Roman five / And a small line before it ... (but erased ev'rything quickly) — / yet the Roman sign 'four' remains engraved in my mind."

OVERCOUNTING

Now we come to a very remarkable manner of counting, which once prevailed in two areas of the world, the Germanic north of Europe and ancient Mexico; today only traces of it remain here and there (as in the Ainu number sequence; see p. 69). This method of counting expresses 24, for example, not in the usual manner as either "4 and 20" or "4 and 2 tens," but as "4 from the 3rd ten"; for previously cited examples in Old Norse, see pp. 27 and 72, and Mayan, see p. 61.

The Meaning of Overcounting

The common method of counting places the units (in this case 4) upon the next lower rank level (in this case on the 2nd ten): this may be called counting from the lower level, or undercounting. But the other method places the units in the interval of 20 ... 30, between the two rank levels, and thus within the third interval of tens: this is counting from the upper rank level, or overcounting, because the rank-level number of the interval (the third) is determined by the next higher level, which in our case is 30.

Thus in the Old-Norse script we read:

102 men   II *menn hins ellifta tigar*
            "2 men in the 11th ten,"

in the year 969 *a niunda are hins sjaunda tigar ens tiunda hundrads*
            "In the 9th year of the 7th ten of the 10th hundred
            (*fra holdagan vars herra*)
            (after the birth of the Lord)"

If we think of the numbers shown along the line of Fig. 20 as proceeding in order from 0 running left to right, we can readily see how the method of undercounting really does build up the number from below in descending order: 9 H   6 T   9 U, whereas overcounting begins with the units 9 and then suspends these in ascending order from the next level above: 9 U in the 7th T of the 10th H. In so doing it reverses the succession of ranks. Undercounting sees the number 969 by itself, resting upon the number line; overcounting, on the other hand, regards 969 as a point within an expanse of numbers arranged one inside the other, the tens within the hundred interval. It indicates these intervals verbally by the order in which the number word is arranged.

Fig. 20 The number 969 as expressed by the methods of undercounting (UC) and overcounting (OC).

Why this peculiar way of counting? For no other reason than to make numbers (such as 969) as easily visualized as possible. It begins with the unit 9, which lies in the 7th ten-interval, and the tens in turn in the 10th hundred-interval. Instead of point-like rungs ... 967 — 968 — 969 — 970 ... on a ladder of numbers as in undercounting, the "overcounter" sees stretches of numbers (the intervals) within which the number lies. These are, in our sense, nothing but groupings: 9 units in the 7th ten-group of the 10th hundred-group. The method of undercounting 969 begins with the hundreds, a large number that the mind can scarcely grasp directly, whereas the overcounting takes hold of the number by its "nearest" end, the units, and from there climbs up to the higher 10-and 100-groups.

On the whole, therefore, overcounting has the same purpose — to get a clear view of the number — as the already mentioned fashion of placing the units in front of the object counted (see p. 72). Our example, "two men in the eleventh ten" for "102 men" unites the two, the placing of the units in front and the overcount.

The Half-count

This can be considered a form of overcounting. In German it lives on in the expression *anderthalb*, "one and a half" (literally, "half of the other") — that is, the other (second) half — and more rarely in the expressions for 2½, *drittehalb*, and 3½, *viertehalb*. These word forms for fractions using overcounting occur not only in Norse but in the Germanic languages generally; thus 2½ appears in Old Norse as *halfr þridi*, in Anglo-Saxon as *þriddehalf* and in Dutch as *derdehalf*.

In addition, verbal fractions formed in this manner also appear in other languages that otherwise form their number-words by undercounting:

Latin 2½ = *sestertius* < *semis-tertius*, "half of the third,"

from which was derived the name for the Roman coin known as the "sesterce," which originally had the value of 2½ *as* (see p. 29). Similarly *epi-deúteros* "on the second," was the Classical Greek expression for two and a half. But the Slavic language as well as Finnish and Hungarian also make use of the word formation (perhaps because they are the eastern neighbors of German) for example:

Russian 1½ *pol-tora* < *pol-vtora*, "half of the second,"
Finnish *puoli-toista*, *puoli-kolmatta*, "half of the second," resp. "third,"
Hungarian *masad-fel* (< *amasodik*, "the second" and *fel*, "half"),
Hungarian *harmad-fel* and *negyed-fel*, "half of the third," resp. "half of the fourth."

This manner of forming fractions and its general application to other fractions suggests to me that they again arise from the desire to make a difficult numerical expression (as the fraction is) easy to visualize: "one, and added to that half of the second (group)."

The "half" in this overcounting is, however, not the same as the half that is left over as a remainder in reckoning, as the latter is expressed in the Celtic *hanter-kant*; this is really "half a hundred" and thus 50; likewise, Sanskrit 50 *śat-ardha*, "100-half" = 50, or Sanskrit 25 *ardha-pañčasat*, "half 50."

On the other hand, Old-Norse *halfuhr tiunde tughr*, "half of the 10th ten," is not 50, but 95! One must pay close attention to the arrangement of the number. Similarly,

Danish *half-tred-sinds-tyve*, "half 3 × 20" or "half 60"

is not 30 but 50, in the sense of the overcount "half of the third twenty" (see p. 65). The same thing is familiar in the German way of stating the time of day: "*halb 4 Uhr*" is not "two o'clock" but 3:30. In contrast, all the Romance languages (and English) "undercount" the hours, as in the French *trois heures et demie*, English "half past three."

A very large number, such as 55,950,000, for example, which would be impossibly cumbersome if expressed by overcounting, is divided up into

½ of the 6th T × (ThTh) and ½ of the 10th HTh:
   halftum setta tigi sinna þusund þusunda ok   halft tiunda
   half of the 6th 10 times   1000 ' 1000   and half of the 10th
hundrat þusunda,
      100 ' 1000.

The Nordic Names for Fractions
These names, of which the expressions for half are a particular case,
are formed by the same procedure:

Icelandic (ca. 1300) 5⅔ hundred *tueyr hlutir   ins VI. hundrad,*
                       2   parts of the 6th hundred;
Icelandic 26⅔ ells *tveir hlutir siaunda elnar oc XX alnar,*
                 2   parts of the 7th ell and 20 ells.

(The different spellings of the same word, like *tueyr* and *tveir* in the
above examples comes about because they were taken from different
manuscripts.)

In Old Norse it was the custom, in the case of fractions whose
denominator is 1 more than the numerator, to omit the rest of the
fraction: Thus "2 parts" is understood to mean ⅔, "2 parts (of
three)." Quite surprisingly a number of other languages do the same
thing; this of course applies only to fractions commonly encountered
in daily life, not to artificial fractions that occur in computations,
such as 28/29. Thus the only ancient Egyptian non-unit fraction
⅔ was called "both parts"; the same fraction in Akkadian-Babylonian
was termed "both hands." The curious Roman name *bes* for ⅔ *aes*
is also explained in this manner:

Latin *bes* < *duessis* < *duo* (*partes*) *assis,* "2 (parts) of the *aes.*"

Modern Greek says for ⅔, ¾, and ⅘

tà dýo (*tría, téssera*) *mérē,* "the two (three, four) parts."

Old Norse then has an odd manner of expressing a fraction whose
denominator is larger than the numerator by more than 1 part:
"⅝ (of the fish caught in the River Grima belong to the Church)" is

fim hluter en þri huerta   undan,
  5   parts and 3   go underneath.

In this connection let us quote an interesting explanation of the
value of the coefficient $\pi \approx 3\frac{1}{7}$ from an Icelandic manuscript dated
about 1300:

ummaeling hrings hvers þrimr lutum lengri en breidd hans
the circumference of the circle is some 3 times longer than its breadth
ok sjaundungr of enni fiordo breidd,
and a seventh of the fourth breadth (= diameter).

The History of the Norse Custom of Overcounting
The development provides fascinating evidence of the way in which
number-words changed and migrated. Overcounting is not common
to all the Germanic languages, however. For example, Bishop
Wulfila's Gothic Bible translates the "99 just persons" (Luke 15:7)
by undercounting, as *niunte-hundis jah niune garaihtaize,* "90 and 9."

This Nordic manner of dealing with numbers gradually disappeared
under the influence of Christianity. Along with the Latinate cere-
mony and worship of the Church, the Roman method of undercounting infiltrated into the native manner of expressing numbers.

Nordic and Roman ways of writing numbers can be reliably distinguished in old manuscripts. In place of the 364 days in the Norse overcounting,

fiora dagar   ens   fiorþa   hundraþs,
   4   days   in the   4th   (great) hundred,

we find CCC *fagar oc* IIII, where only the non-Roman great hundred C (= 120) betrays the scribe's Nordic background. He has already abandoned the Norse reverse word order, "4 days and 360." The Norse habit of overcounting gradually died out between the 12th and the 14th centuries.

Thus it is remarkable that the Finns, the Esthonians, and the Lapps, neighbors of the Germanic Swedes, follow the custom of overcounting to this day. They borrowed this from their neighbors, as they did many other Germanic habits (see p. 112). In Finnish it especially dominates the second interval of tens, from 11 to 19; *yksi-toista,* "1 (in the) 2nd (tenned)" is 11, and so forth (see p. 113). Moreover Finnish has, besides the *under*counted 21, *kaksi-kymmentä-yksi,* 2′10′1, also the *over*counted 21, *yksi-kolmatta,* "1 (in the) 3rd (ten)." The overcounting formation is used especially in stating the days of the month between 20 and 30; for instance, the 27th is not the 2′10′7, but "the 7th in the 3rd."

Because of the strong influence of the Swedes on the Finns until around A.D. 700, the origin of Finnish overcounting is suspected to be due to their Swedish neighbors, among whom, however, it can no longer be found. It was the Swedes who settled Norway, where they also brought their overcounted number words, which migrated thence to Iceland (see p. 109). Here it resisted with difficulty the alien Christian fashion of undercounting and did not disappear from Iceland until around 1500.

Overcounting is now a rare phenomenon in the world. It was clearly developed by the Mayans as well (p. 61), but the Mayan overcount, of course, had no connection whatever with that of the Norse. Moreover, despite their similarity, there are very strong differences between the two (one of them being an artificial vigesimal gradation not tied to the objects counted), so that the Mayan overcounting cannot justifiably be interpreted as being founded on an attempt to visualize numbers more readily. On the other hand, this was surely the basis for the overcounting among the Ainus (p. 69).

ADDITIONAL WAYS OF FORMING NUMBER WORDS

Now that we have seen the great variety of ways in which number words are formed by overcounting and by using the arithmetic operations of adding, subtracting, multiplying, and dividing, let us look at three more instances that may show further paths that languages can follow, for example in forming the tens from 20 through 90.

Semitic

The Semitic tens, for example, are the grammatical plurals of the units. In the Arabic and Hebrew number sequences (see the complete number sequences, p. 115) we find:

Arabic 4, *arba'un* — 40, *arba'un-a,* Hebrew 4, *'arba'* — 40, *arba'im,*
Arabic 6, *sittun* — 60, *sittun-a,* Hebrew 6, *šeš* — 60, *šišš-im.*

The Arabic *-a** and Hebrew *-im* are the respective plural endings of these languages, as in Hebrew *sepher,* "book," *sepharim,* "books." In Semitic, 40 is thus literally "the forties." The word for 20 is the only exception; it is formed not from 2 but from 10 (Arabic *'asrun,* Hebrew *'eśer*); 20 *išrun-a* and *'eśr-im,* respectively, are literally "the tens." We have already indicated (see p. 14) that this is a reflection of the ancient primitive counting limit after ten. Originally the words for 20 had not a plural but a grammatical dual ending, which however was later driven out by the analogy with the other tens; this kind of linguistic "remote control" is quite common in number words arranged in order.

Although the Semitic number sequence is already remarkable for this manner of forming the tens, it is even more so in the curious gender and case inflections that characterize the number words. Each number word has both a masculine and a feminine form, e.g.:

Arabic 7 (m) *sab'un* — (f) *sab'atun*; Hebrew 7 (m) *šeba'* — (f) *šib'ah.*

But if the things counted are masculine, such as "7 men," the feminine form of the number word is used, and vice versa; thus we have 7 (f) men and 7 (m) women! Why this happens, no one knows. Have some conceptions of mythology influenced these number words? Only the Semitic words "one" and "two" are treated as adjectives agreeing in gender with the nouns they modify (as are all numbers, such as 102, that end in 1 or 2).

The changes in form of the number words in Arabic are also extremely peculiar: "one" and "two" are adjectives agreeing in gender with their nouns: "two-ish (m) men";
3, 4, ..., 10 take the genitive plural and are thus nouns: "5 (f) of the men";
11, ..., 99 take the accusative singular: "30 (f) the man";
beyond 100 the number words are again nouns and thus regularly govern the genitive case, but now in the singular: "250 of the man";
1001 nights is *alf laila wa laila,* "a thousand night and night."
Certainly an odd number sequence! How could anyone wishing to create a new number sequence artificially arrive at such a one! But the very fact that it does not proceed consistently and without disruptions shows how closely it was bound to the thought processes of its people.

Sanskrit

The Sanskrit tens and those of the related Avestan language are, from 60 up, units raised to the status of nouns (see the full sequences on p. 93):

šaš, "six"; *šaš-ti-h,* "the six-ed, six-ness, six-ity."

The number word becomes a noun by acquiring the ending *-t* (see

---

* *Editor's note:* The Arabic plural is *ūna* in the nominative case. The essential feature is the lengthening of the vowel.

p. 126). Linguistically this is the expression of a grouping, a "six-ness of tens." Sanskrit even says *daśa-ti-h*, "a ten-ness (of tens)" and hence "a hundred."

Strange things occur in the life of a language: The same *t* formation that exists in Sanskrit also occurs in Slavic from the number 5 on, as for instance the Czech *pĕ-t, šes-t, devĕ-t, dese-t*, but here it is not the tens but the units 5, 6, 9, and 10 that are treated as nouns: "five-ness, six-ness, nine-ness, ten-ness." The first four Slavic number words, on the other hand, are ancient adjectives, whereas the higher numbers are felt to be groupings; we have already spoken of this earlier (see p. 23).

In Gothic the last three tens have the form of 70, *sibunt-e-hund*, "the seven-ned ten" in the sense of "10 seven-nesses" (see p. 92).

Verbal Numeration
This is the name we shall use for the peculiar custom of an African language spoken in the Sudan, in which the words for the tens are not related in any way to those for the units and separate names are given both to the units and to the tens:

1 *telu*		6 *yeres-ko*	60 *kiyi*
2 *in-di*	20 *colo-ro*	7 *telo-ro*	70 *timba-ga*
3 *yas-ko*	30 *dere-go*	8 *ton-go*	80 *kandeln*
4 *de-go*	40 *gama-ro*	9 *inde-go*	90 *tschimbar*
5 *o-go*	50 *inde-ro*	10 *me-go*	100 *andai.*

The Turkish words for 30, *otuz*, 40 *kirk*, and 50, *elli*, likewise have no connection whatever with the names of the units (see p. 113).

Among the various known systems of numerals, such numeration was also used by the Egyptians, whose so-called hieratic script had separate symbols for each of the tens, and thus wrote the numbers up to 100 with 19 different symbols, not with 10 as we do.

Ordering is characteristically lacking in systems of number words and numerals of this kind. In their structure they have reverted to the level of counting by the parts of the body: they do not simply assign numbers to the parts of the body, to be sure, but their words and numerals nevertheless lack any internal organization (see p. 34 above). This should again caution us that our clear insight into the number sequence as we know it today — after a long succession of detours — cannot be assumed to have existed at earlier stages of cultural history.

The freezing and borrowing of number words could likewise be classified under the heading of "additional ways of forming number words." Freezing: A word with the indeterminate numerical meaning of "much" or "many" (like the Turkish *on*) is *frozen* into a definite number word with a fixed meaning ("ten"). Borrowing: if the num-ber sequence of a people stops at 100, for example, but needs a word for 1000 and then *borrows* this from a neighboring language, as the Finns took over the word *tuhat* from the Germans, this could also be considered an alternative method of forming number words. We shall encounter such freezings and borrowings time and again and where necessary shall refer to them where they occur (see p. 179).

Why do the words for "ten" deviate so much from the rule? Because "ten" is the first landing that rises above the foundation walls of the earliest number words, the units. The latter are expanded to become the tens, as it were, not just in some form of computational manner (e.g., 20 = 2 × 10 as in Old High German *zwein-zug*) but also through grammatical transformations of the words themselves (plural as in the Hebrew *eśrim*, dual as in the Latin *vi-ginti*, and 20-group "man," as in the Ainu *hot* and Japanese *hata-chi hito*, "man"). This enormous variety of ways of forming number words, which we find so intriguing, is an important proof of the slow and natural growth of the number sequence.

THE FIRST STEPS BEYOND TEN

We count ten, eleven, twelve and then thir-teen, four-teen and so on: why the odd names for just the two numbers following "ten"? Let us look into the history of these words.

Both these words are of common Germanic formation:

Gothic	*ain-lif*	*twa-lif*
Old Norse	*ellifu*	*tolf*
Anglo-Saxon	*anleofan*	*twelf*
Old Saxon	*elleban*	*twe-lif*
English	*eleven*	*twelve*
German	*elf*	*zwölf*
Dutch	*elf*	*twaalf*
Old High German	*einlif*	*zwelif*
Middle High German	*eilf* (Goethe)	*zwolf* (Luther)

Examples: A passage in the Gothic Bible, I Cor. 15:5: "And that he was seen of Cephas, then of the *eleven* (King James Version: 'twelve'; Luther: '*Zwölfen*')" reads ... *þata þaim ainlibim* (dative case!) Another passage (Mark 4:10): "... they that were about him with the twelve asked of him the parable" reads ... *miþ þaim twalibim*. From Otfried's Book of Gospels we have: "... which was one of the eleven" — Old High German *er ein thero einlifo uuas*; and Notker of St. Gall describes the wild boar as a charging beast with "teeth twelve ells long" — Old High German *unde zene sine zuuelif elnige*.

The number words we have just enumerated are combinations of "one" or "two" with the syllable -*lif*. What does this syllable mean? If we look into some related languages, we find Latin *re-linquere* (cf. "re-liqu-ary") and Greek *leipō*, "to leave remaining," Gothic *leiban* and English *leave*. The Greek word is familiar as the root of the word "ecliptic" (from Greek *ékleipsis*, "omission, nonoccurrence"), the path traveled by the sun on the celestial sphere. It is so called because the moon, when it approaches, darkens the sun so that the sun is "missing." The word "el-lip-se" also belongs to the same group (according to the Greek explanation, an ellipse is a curve that can be used to draw a desired right angle, some definite portion of which is afterwards cut off or "left out"). The Indo-European root from which all these words are derived, after appropriate phonetic shifts, is *leiqu*. Hence, eleven and twelve mean nothing more than "one-left" and "two-left." Left from what? Left over from "one-ten" and "two-ten," of course, after the ten has been subtracted! The omission of the cardinal number 10 is analogous to the omission of the denominator in the names of Roman fractions (see p. 159).

This is striking evidence that the Germanic number sequence at one time ran only as far as ten. Anything above that was "more." One and two more than ten were still counted, but anything beyond them was perhaps, as so often happens among primitive people, merely considered "many." Then along with the later clear conception of numbers, the subsequent computational number-word formations arose: 3'10 thir-teen, 4'10 four-teen, 5'10 fif-teen, and so on.

I shall cite two interesting Chinese expressions that seem to belong to this category: the phrase that says literally "not reaching 10" is used with the meaning of "incomplete," and "12 parts" has the meaning of "to be in excess." These both make sense if we consider 10 as a limit of counting: Whatever does not reach 10 is "incomplete," whatever goes beyond 10 is "excess."

Lithuanian also has fully adopted these Germanic *lif*-formations for the number words from 11 to 19; in Lithuanian these are (*lika* = *-lif*):

11 *wienó-lika*	16 *šešió-lika*
12 *dvy-lika*	17 *septynió-lika*
13 *try-lika*	18 *aštunió-lika*
14 *keturió-lika*	19 *devynió-lika*
15 *penkió-lika*	20 *dvi-dešimtis.*

There is no doubt but that in ancient times these number words migrated from the Germanic people to the Lithuanians. Whether it replaced some older Lithuanian one or whether, as may also have happened, constructed the number words of this interval for the first time in Lithuanian cannot now be ascertained. At any rate, this word formation now occurs only in Lithuanian and in the Germanic languages.

An interesting analogue of the Lithuanian number sequence is to be found in a language spoken by a Negro tribe of the Sudan, whose numbers, however, are built upon the number 5:

6 *wal-ta,* "1 and"
7 *lena-ta,* "2 and"
8 *segua-ta,* "3 and"
9 *ses-sa,* "4 and."

The Interval from 11 to 19

There are some languages that form the number words of this interval entirely without disturbances, thus without "eleven" and "twelve." Among the Indo-European languages this applies especially to those of the Slavic group, which set the units like building stones "upon" (*na*) ten:

Czech: 10 *deset,* 2 *dva;* whence 12 *dva-na-deset,* "2 on 10" contracted to *dvanást,* followed by 13 *trinást,* etc.
Russian: 10 *děsatj,* 2 *dvě;* 12 *dvě-ná-dzatj,* 13 *tri-ná-dzatj,* etc.

The "teens" are counted similarly in Albanian (*nja-mbe-dhiete*, "1 upon 10") and in Rumanian (*un-spre-zece*, "1 on 10"; see below). When the number sequences in the languages of less "mature" peoples run through this interval without disruption, one can suspect — so long as there has been no borrowing as in the case of Finnish and Lithuanian — that the number words have been formed after a pattern (as in Hungarian *tizen-egy* 10'1 and Turkish *onbir* 10'1; see p. 113). On the other hand, the Semitic sequence (Hebrew *aḥaþ'-esreh* 1'10, etc.) and the Chinese (*shih i* 10'1, etc.), to name only two, are quite independent.

We have already encountered the unorthodox Celtic examples for $18 = 2 \times 9 = 3 \times 6 = 8 + (2 \times 5)$ and the Welsh manner of counting 15, $15 + 1$, $15 + 2$, 2×9, $15 + 4$ (see pp. 64 and 71).

This is also the place to point out the unexplained break in the second interval of tens in the Romance languages, which up to or through 16 set the units in front of the "ten" but thereafter place them behind it. Classical Latin does not have this variation, although the inverted terms *decem et septem*, "10 and 7," *decem et octo*, "10 and 8," etc., have been documented in place of the more usual *septendecim* 7'10 and *duodeviginti*, "2 from 20." But the Latin spoken by the soldiers and settlers in the outlying provinces of the Roman Empire used the former words, which came from them into the Romance languages. Yet although the origin of this metathesis is thus clear enough for the case of the Romance languages, it is not explained for the case of Latin itself: why did the sequence in popular spoken Latin run to 16 according to rule and then suddenly reverse the order? Moreover, in Spanish and Portuguese the inversion takes place after 15, while Romanian lacks it altogether. In the following table the number words from 11 through 20 are given side by side in Classical Latin, Italian, French, Spanish, and Romanian, so that the reader can see this peculiar anomaly for himself:

The Interval from 11 to 19 in the Romance Languages

	Latin	Italian	French	Spanish	Romanian
10	*decem*	*dieci*	*dix*	*dies*	*zece*
11	*un-decim*	*un-dici*	*on-ze*	*on-ce*	*un spre zece*
12	*duo-decim*	*do-decim*	*dou-ze*	*do-ce*	*doi spre zece*
13	*tre-decim*	*tre-dici*	*trei-ze*	*tre-ce*	*trei spre zece*
14	*quattuor-decim*	*quattor-dici*	*quator-ze*	*cator-ce*	*patru spre zece*
15	*quin-decim*	*quin-dici*	*quin-ze*	*quin-ce*	*cinci spre zece*
16	*se-decim*	*se-dici*	*sei-ze*	*diez-y-seiz*	*šase spre zece*
17	*septen-decim*	*diciasette*	*dix-sept*	*diez-y-siete*	*šapte spre zece*
18	(*octo-decim*)	*diciotto*	*dix-huit*	*diez-y-ocho*	*opt spre zece*
19	(*undeviginti*)	*dicianove*	*dix-neuf*	*diez-y-nueve*	*noua spre zece*
20	*viginti*	*venti*	*vingt*	*veinte*	*douazeci,*
					shortened to:
				y = and	11 *unsprece*, etc.

Rom. *spre* < Lat. *super*, "above."

Underlying the break at 17 we may suspect an old 4-count, such as $2 \times 4 = 8$ (a grammatical dual; see p. 14), $3 \times 4 = 12$, and $4 \times 4 = 16$, which was then followed by the new manner of counting with the word order reversed. But it seems unlikely that the 4-count could have been pushed so high without leaving any direct traces of itself in the spoken numbers. If, on the other hand, the backward-count 10 − 1, 10 − 2, 10 − 3 for 9, 8, 7 really did exist in the Indo-European language, it could well have rubbed off on the higher interval and marked out the numbers 17, 18, 19 before the others (see p. 170). There are plenty of convincing examples of the reflection of one verbal peculiarity in a different interval of numbers; some of these have been described earlier (see p. 14).

In Classical Greek, to be sure, the difference in formation of the number words 11 and 12, on the one hand, and the remaining numbers up to 20, on the other, is not so very great, but it is still discernible:

11   *héndeka* 1′10,   12   *dódeka* 2′10,
but
13   *treîs kaì déka* 3 & 10   14   *téttares kaì déka*, "4 and 10," etc.

Thus "eleven" and "twelve" were once felt by the Greeks to be "other" or "closer" numbers than the ones immediately following them; the latter were tacked on with "and," just as two discrete concepts can always be joined together with this conjunction. Modern Greek, however, now counts 13, *dekatreîs* (10′3), 14, *dekatésseres* (10′4), and so on.

Retrospect

With this last remark, we come to the end of our investigation of the nature and origin of the number sequence. Now let us take one last look back over the road we have traveled.

At the end we directed our attention to the laws characterizing the formation of the number words in various languages. Number words had to be combined, of course, unless an impossibly large quantity of them were to be invented. What did we learn? That the combinations of number words have not followed any thought-out, clearly recognized plan, but quite often have had to be put together awkwardly from words already at hand. In addition, we found word formations that reflect the force of the very oldest number groupings. We recall how these groupings systematically invaded graduated number sequences, and we have seen how strongly they still live on in everyday popular speech.

Let us therefore return to the ideas we encountered in the course of our wanderings through linguistic history and which emerged from the ordering and grouping of the number sequence, its gradation, supplementary quantities, spoken numbers, and written numerals, and the attached number sequence, which became abstract by gradual detachment from the objects it counted — then we shall recognize what is the most significant conclusion of our study:

A people's number sequence is not a system created fresh out of the pure workings of the mind; it is rooted in the same soil as the people. Like culture itself, it grows up slowly over the millennia, and even in its mature form it reveals the history of its people through the successive deposits of the passing years.

Our Number Words

The Indo-European Family
of Languages

"Number words are among the most change-resistant words of any language; they lose their original meanings very early, and they change scarcely at all with the passage of time and the rising level of culture. If our number sequence stood isolated in the sphere of language, we should never be able to learn anything about its origin."
(See pp. 100 and 125)

Up to this point we have been speaking about the general development of the number sequence; we shall now turn to our own number words. What can they tell us about their history and the history of our own forefathers?

Our number words are linguistically related to most of the other European and to the Sanskrit-Iranian number words. Thus we can shed light on them from the widest variety of different languages and gain insights that would otherwise have been lost to us forever.

Now I must ask the reader to travel with me through the regions of linguistic history that lie beyond the sphere of numbers themselves, in order to learn something about the nature of *linguistic affinity*. Our efforts will be rewarded by allowing us to take part in one of the most magnificent dramas in the history of peoples and their cultures — one which language alone has preserved for us.

To keep the reader who is unfamiliar with linguistics from losing his way in a sea of minute details and wandering aimlessly in a dark wasteland, let us take a quick look at the route we are about to travel. We shall begin with a rapid survey of the Indo-European languages and their number words, with which the reader should gradually become familiar, so that he can perceive, test, and work out for himself the conclusions that we draw from them. Next we shall learn, from the example of the Romance languages, what is meant by linguistic affinity. Last, we shall obtain a general idea of the basic Indo-European languages and of the individual languages that grew out of it.

Thus prepared, we shall turn to the *meanings* of our number words. We shall speak of their flexibility and their fading out. We shall begin our discussion of these words with the ranks of ten, hundred, thousand, and higher numbers (as used in India, by Archimedes, and by ourselves), and continue with the units and the tens. Through them, the remarkable break in our sequence after 60 will lead us to a discussion of the great hundred (120), of the number 12, and of the Babylonian sexagesimal system. There our study of number words will end, and after a last look at *hidden number words* we shall leave not only these, but the entire subject of spoken numbers.

SURVEY OF LANGUAGES AND NUMBER WORDS

Before we arrive at the number sequences, we come first to the languages themselves, in the order in which they are treated in the study of Indo-European linguistics. They can be readily visualized from the accompanying linguistic map (Fig. 21). Here is the succession in which we shall meet them:

Kentum and *Satem* languages (p. 91) and their number words (pp. 92–93);
Germanic languages (p. 94) and their number words: old Germanic (p. 94), new Germanic (p. 95) and Modern German (p. 96);

Fig. 21 Map of the Indo-European family of languages, showing the geographic distribution of the *kentum* and *satem* branches and of a few non-Indo-European languages.

Celtic languages and number words (pp. 96–97);
Balto-Slavic languages and number words (pp. 97–98);
Romance languages and number words (p. 99).

The reader will profit greatly from a comparison of these Indo-European with non-Indo-European languages. Of the latter, we shall present:

the Basque number sequence (p. 110);
the Etruscan number sequences (p. 110);
the Finno-Ugric family of languages (p. 111) with the Finnish, Hungarian, and Turkish number words (p. 113);
the Semitic family (p. 114) with the Arabic and Hebrew number words (p. 115);
the Sumerian number sequence (p. 165);
the Chinese and Japanese number words (pp. 447 ff.).

Linguistic affinity involves not only a similarity of words, but above all identical linguistic structures. In the case of the last three of these language groups, the differences in their structures are shown briefly by means of examples, so that we can thereby recognize the peculiar characteristics of our own language.

"Affinity" of Languages

Let us begin our discussion with the number words in the Romance languages: Italian, French, Spanish, and Romanian (see table on p. 99). Even a superficial glance will reveal the similarity of these words, for example those for 5 or 40 or 100. Is this resemblance just accidental? No; even if we did not know the histories of these words, this would be impossible to believe. It is contrary to the agreement of almost all the words in these number sequences, even those of peoples who live at great distances from one another, such as the Spaniards and the Romanians.

All the languages of this group are descendants of the mother-tongue, Latin, the language of the Romans. The Roman Empire and its civilization embraced the provinces of Spain and Gaul in the west; its northern boundary ran along the Danube River, north of whose lower reaches was the Roman province of Dacia. To these frontier regions the Romans sent soldiers, traders, and settlers, who along with Roman culture also introduced their Latin folk speech. And the forms of vernacular Latin continued to live on in the far corners of the Empire, long after Rome had ceased to be the Empire's capital and its only source of culture and civilization. Left to its own devices, popular spoken Latin developed into separate languages, in both pronunciation and form: French *homme* and Italian *uomo*, both of which however, like two sisters, betray the common heritage of their Roman mother: Latin *homo*, "man" or "person." Hence, they have been called "Romance" languages.

Here we have a case of related languages or linguistic affinity, the explanation for which is provided by recorded history.

Once again, let us test the relationship of the languages listed below by comparing the corresponding number words in the tables on pages 92 and 93: in the German *drei* ("three"), for example, from the Germanic *þreis*, and in the same word in Greek, Latin, Celtic, Tocharian, Baltic, Slavic and even the Sanskrit *trayah* — in all these we see the same combination of sounds, *tr-*, which German preserves even to this day. A striking similarity!

The words for "two" and "six" also maintain their resemblance through all these various languages. Other numbers, like "eight" and "nine," are similar only between the languages listed on p. 92, while the words for "eight" in Baltic and Sanskrit seem more closely

Kentum Languages		Satem Languages
	In Europe:	
Greek*		Slavic*
Italic (Latin)*		Baltic*
Celtic*		Albanian (see p. 68)
Germanic*		
	In Asia:	
Tocharian*		Sanskrit*⎫ Aryan
Hittite		Iranian ⎭
		Armenian and other Near-Eastern languages

* Asterisks indicate languages whose number words are given in the tables following.

Kentum Languages

	Greek	Italic	Celtic	Germanic	Tocharian
		Latin	Irish	Gothic	
1	*heîs, mía, hén*	*unus, -a, -um*	*oin*	*ains*	*sas, sam*
2	*dýo*	*duo, -ae, -a*	*da*	*twai, twos, twa,*	*wu, we*
3	*treîs, tría*	*tres, tria*	*tri*	*þreis, þrija*	*tre, tri*
4	*téttares, -a*	*quattuor*	*cethir*	*fidwor*	*śtwar*
5	*pénte*	*quinque*	*coic*	*fimf*	*päñ*
6	*héx*	*sex*	*se*	*saíhs*	*sak*
7	*heptá*	*septem*	*secht*	*sibun*	*spät*
8	*oktố*	*octo*	*ocht*	*ahtaú*	*okät*
9	*en-néa*	*novem*	*noi*	*niun*	*ñu*
10	*déka*	*decem*	*deich*	*taihun*	*śak*
20	*eíkosi*	*vi-ginti*	*fiche*	*twai-tigjus*	*wiki*
30	*triá-konta*	*tri-ginta*	*tricha*	*þreis-*	*tary-ak*
40	*tettará-*	*quadra-*	*cethorcha*	*fidwor-*	*śtwar-*
50	*penté-*	*quinqua*	*coica*	*fimf-*	*pñ-*
60	*hexé-*	*sexa-*	*sesca*	*saíhs-*	*säksäk*
70	*hebdomé-*	*septua-*	*sechtmogo*	*sibunt-ehund*	*säpt-uk*
80	*ogdoế-*	*octo-*	*ochtmoga*	*ahtaút-*	*okt-*
90	*emenế-*	*nona-*	*nocha*	*niunt-*	*ñm-*
100	*hekatón*	*centum*	*cet*	*taihunt-ehund*[1]	*känt-*
200	*diakósioi, -ai,*	*ducenti, -ae,*	*da-cet*	*twa-hunda*	*we-känt*
300	*tria-* [*-a*	*tre-* [*-a*	*tri-*	*þrija-*	*tri-*
500	*penta-*	*quingenti*	*coic-*	*fimf-*	*päñ-*
800	*okta-*	*octingenti*	*ocht-*	*ahtaú-*	*okät-*
1000	*chílioi, -ai, -a*	*mille*	*mile*	*pusundi*	*wälts*
2000	*dis-chilioi*	*duo-milia*			*we-wälts*
3000	*tris-chilioi*	*tri-milia*			*tre-*
10,000	*mýrioi, -ai, -a*	*decem-milia*			*tmam*
	For 11 to 19, see p. 86 For 30–90, see p. 150	For 11 to 19, see p. 85 *c*: hard k; thus 100 is "kentum," not "sentum."	For 11 to 19, see p. 64 For 20–90 in Old Irish, see p. 65 *c*: hard k *ch*: ch as in German "*ach*"	[1]also *taíhuntaíhund* *-h*: ch as in German "*Buch*" (6,8,10,70) *ai*: short *e*, open; 10 is "teh-khun" *aú*: short *o*, open; 8 is "ah-khto" *ei*: long *ī*, pronounced like "ee" in English "bee"	*ñ*: ny as in "canyon"

Satem Languages

	Indian	Slavic	Baltic	
	Sanskrit	Old Church Slavonic	Lithuanian	Basic Indo-European
1	ekab, eka	jedinų, -a, -o	vienas	oi-nos; oi-qos sems, smiǝ, sem
2	dvi, dve	dva, dve	du, dvi	duŭo; duo
3	trayah, tisrah,	trije	trys	trejes, trie
4	čatvarah, čatasrah	četyre	keturi	qŭetŭor (es)
5	panča	pētj	penki	penqŭe
6	šaš	šestj	šeši	sŭeks; seks
7	sapta	sedmj	septyni	septm̥
8	ašta, aštau	osmj	aštuoni	oktou
9	nava	devētj*	devyni†	neŭn, eneŭen
10	daśa	desētj	dešimt	dekm̥; dekm̥t
20	vim-śatih	dva-desēti	dvi-deszimt	
30	trim-śat	tri-	trys-	
40	čatvarim-	četyri-	ketures-	
50	panča-	pētj-desētj	penkes-	⎫
60	šaš-tih	šestj-	šešes-	⎬ compounds formed with -(d) kom̥-t-
70	sapta-	sedmj-	septynes-	⎭
80	aši-	osmj-	aštuones-	
90	nava-	devētj-	devynes-	
100	śatam	sŭto	šimtas	km̥tóm
200	dvi-śatam[1]	dve-ste	du-šimtu	
300	tri-	tri sta	trys-šimtai	
500	panča-	pētj sŭtj	penki-	
800	aštau-	osmj-	aštuoni-	
1000	sahasram	tysēšta	tukstantis	sm̥-gheslo-m
	ś: light sh as in German *Stern* (10) š: full, heavy sh as in German *Schwamm* (6) č: ch as in "child" (4) v: as in "vow" (4) -im-: nasal i (20) [1]also dve sate	š and č same as in Sanskrit ē: nasal j: final palatal "y" for Slavic ь and ъ ŭ: short u for ъ *see p. 147	for 11–19, see p. 84 †see p. 147	m̥: syllabic sound (am, em, om, or um) ŭ: short u qŭ: as in English "quite" ; indicates that both variations are possible (2, 6)

Germanic Group (p. 106)

West	North	East

Anglo-Saxon (from 12th century)
  English*
 	
Frisian (from 14th century)
 	
Low German
  Dutch*
  Low Saxon ("Heliand," 9th century)
  Plattdeutsch
— 2nd Phonetic Shift (see p. 127)
High German (table on p. 96)
  Old High German (8th cent. to 1100)*
  Middle High German (1100 to 1500)*
  New High German (after 1500)*
	Old Nordic*
	Norwegian*
	Icelandic
	Danish*
	Swedish
		Gothic*
		Wulfila's Bible
		(4th century)

Old Germanic Languages

	West Germanic		North Germanic	East Germanic
	Anglo-Saxon	Old Saxon	Old Nordic	Gothic
1	*an*	*en*	*einn*	*ains*
2	*twegen, twa*	*twene, two twe*	*tveir, tvan*	*twa*
3	*þri, þreo*	*thria, thriu*	*þrir, þriu*	*þreis, þria*
4	*feower*	*fiuwar, fior*	*fjorer*	*fidwor*
5	*fif*	*fif*	*fimm*	*fimf*
6	*six*	*sehs*	*sex*	*saihs*
7	*seofon*	*sibun*	*siau*	*sibun*
8	*eahta*	*ahto*	*atta*	*ahtaú*
9	*nigon*	*nigun*	*nio*	*niun*
10	*tyn*	*tehan*	*tio*	*taihun*
11	*endleofan*	*ellevan*	*ellefo*	*ain-lif*
12	*twelf*	*twelif*	*tolf*	*twa-*
13	*þri-tene*	*thri-tehan*	*þrettan*	*(þreis-taihun)*
19	*nigon-tyne*	*nigen-*	*nitian*	*(niun-)*
20	*twen-tig*	*twen-tig*	*tuttugu*	*twai-tigjus*
30	*þri*	*thri-*	*þri-tiger*	*þreo-*
40	*feower-*	*fiwar-*	*fjorer-*	*fidwor-*
50	*fif-*	*fif-*	*fim-*	*fimf-*
60	*six-*	*sehs-*	*sex-*	*saihs-*
70	*hund-seofontig*	*ant-sibunda*	*siau-*	*sibunt-ehund*
80	*-eahtatig*	*-ahtoda*	*atta-*	*ahtaút-*
90	*-nigontig*	*-nigonda*	*nio-*	*niunt-*
100	*-teontig*[1]	*hunderod*	*tio-*	*taihunt-*[1]
110	*-endleofantig*		*ellefo-*	
120	*-twelftig*		*hundraþ*	
200	*tu hund*	*twe hund*	—	*twa hunda*
1000	*þusund*	*thusundig*	*þusund(raþ)* = 1200	*þusundi*
	[1]also *hun(dred)*			[1]also *taihun-* *taihund ai, aú* (see p. 92)

Modern Germanic Languages

	West Germanic		North Germanic		
			Northwest	Northeast	
	English	Dutch	Icelandic	Danish	Swedish
1	one	een	einn	en	en, ett
2	two	twee	tveir	to	tva
3	three	drie	þrir	tre	tre
4	four	vier	fjorir	fire	fyra
5	five	vijf	fimm	fem	fem
6	six	zes	sex	seks	sex
7	seven	zeven	sjö	syv	sju[1]
8	eight	acht	atta	otte	åtta
9	nine	negon	niu	ni	nio
10	ten	tien	tiu	ti	tio
11	eleven	elf	ellefu	elleve	elva
12	twelve	twaalf	tolf	tolve	tolv
13	thir-teen	der-tien	prettan	tret-tan	tret-ton
19	nine-	negen-	nitjan	ni-	nit-
20	twen-ty	twin-tig	tuttugu	tyve	tjugo[2]
30	thir-	der-	þra-tiu	tredive	tret-tio
40	for-	veer-	fjörn-	fyrretyve	fyr-
50	fif-	vijf-	fim-	halvtres[1]	fem-
60	six-	zes-	sex-	tres	sex-
70	seven-	zeven-	sjö-	halvfjers	sjut-
80	eigh-	tach-[1]	atta-	firs	at-
90	nine-	negen-	niu-	halvfems	nit-
100	hundred	honderd	hundrad	hundrede	hundra
1000	thousand	duizend	þusund	tusind	tusen

ij: ey as in "hey!" (5)
z: s (6)
ui: öi (1000)
[1](see p. 152)

[1]For the full forms of 40–90, see p. 65

[1]"shoo"
[2]"chügu"
å: oh (8)

German (and Gothic)

	High German			Gothic (for comparison)
	New	Middle	Old	
1	*eins*	*eins*	*ein*	*aina; ain, ainata*
2	*zwei*	*zwene, zwo, zwei*	*zwene, zwa, zwei*	*twai, twos, twa*
3	*drei*	*dri, driu*	*dri, drio, driu*	*þreis, þria*
4	*vier*	*vier*	*fior*	*fidwŏr*
5	*fünf*	*fünf*	*fimf; finf*	*fimf*
6	*sechs*	*sehs*	*sehs*	*saihs*
7	*sieben*	*siben*	*sibun*	*sibun*
8	*acht*	*ahte*	*ahto*	*ahtaú*
9	*neun*	*niun*	*niun*	*niun*
10	*zehn*	*zeben*	*zehan (zehen)*	*taihun*
11	*elf*	*eilf; einlif*	*einlif*	*ainlif*
12	*zwölf* (p. 83)	*zwelf*	*zwelif*	*twalif*
20	*zwan-zig*	*zweinzic (-zec)*	*zwein-zug*	*twai-tigjus*
30	*drei-ßig*	*dri-*	*driz-*	*þreis-*
40	*vier-zig*	....	*fior-*	*fidwor-*
50	*fünf-*	....	*finf-*	*fimf-*
60	*sech-*	....	*sehs-*	*saihs*
70	*sieb-*	....	*sibun-zo (-zug)*	*sibunt-ehund*
80	*acht-*	....	*ahto-*	*ahtau-*
90	*neun-*	....	*niun-*	*niun-*
100	*hundert*	*hundert*	*zehan-zo*	*taihun-taihund*
200	*zwei-hundert*		*zwei hunt*	*twa hunda*
1000	*tausend*	*tusent*	*dusunt; thusunt*[1]	*þusundi*
	β: ss (30)		[1]also *tusent* older forms also 200 *zwiro zehanzug* 500 *finfstunt zehanzug* 1000 *zenstunt zenzech,* "10 times 10-ty"	For pronunciation, see p. 92

Celtic Group (see p. 104)

MAINLAND CELTIC:	*Gallic* (now extinct)
ISLAND CELTIC:	*British* (from the 8th century)
	Walisian (Welsh)*
	Cornish*
	Brythonic (Breton)
	Gaelic
	Irish*
(see Fig. 16)	Gaelic (in Scotland)

Celtic Languages

	Irish	Welsh	Cornish	Breton
1	oin	un	un	eun
2	da	dau	dow	diou
3	tri	tri	tri	tri
4	cethir	þetwar	þeswar	þevar
5	coic	þimþ	þymþ	þemþ
6	se	chwe	whe	chouech
7	secht	seith	seyth	seiz
8	ocht	wyht	eath	eiz
9	noi	naw	naw	nao
10	deich	dec; deg	dek	dek
11	oin deec	un ar dec	ednack	unnek
12	da-	dour ar dec (deudec)	dewthek	daou-zek
13	tri-	tri ar dec	trethek	tri-
14	cethir-	þetwar ar dec	þuzwarthak	þevar-
15	coic-	hymthec	þymthek	þem-
16	se-	un ar þymthec	whettak	choue-
17	secht-	dou ar-	seitag	seit-
18	ocht-	deu naw	eatag	tri- (ch)ouech
19	noi-	þedwar ar þym- thec	nawnzack	naou-zek
20	fiche	ugeint	ugans	ugent
30	deich ar fiche*	dec ar ugeint	dek warn ugans	tregont
40	da fiche	de-ugeint	deu ugens	daou ugent
50	deich ar da fiche	dec ar de-ugeint	hanter-cans	hanter-kant
60	tri fiche	tri ugeint	try ugens	tri ugent
70	deich ar tri fiche	dec ar tri-ug.	dek warn try ugens	dek ha tri ugent
80	ceithri fiche	þedwar-ugeint	þeswar ugens	þevar ugent
90	deich ar ceithri [fiche	dec ar pedwar-u.	dekwarn þesw. ug.	dek ha pevar ugent
100	cet	cant	cans	kant
1000	mile	mil	myl	mil

The differences in spelling are due to the different ages and origins of the documents. For the numbers 20 to 90, see p. 65.

ch: ch as in German "ach"
c: hard k* (pp. 65 & 150)

ou: oo
ch: ch as in German "ach"

Balto-Slavic Group (see p. 105)

PRUSSIAN (died out in 16th century)
LATVIAN
LITHUANIAN (see table on p. 93)

South Slavic
   Old Church Slavonic (extinct; = Old Bulgarian, 9th century; see table on p. 93)
   Serbo-Croatian
East Slavic
   Russian*
West Slavic
   Polish
   Czech*

Slavic Languages

	West Slavic	East Slavic
	Czech	Russian
1	*jeden, -na, -no*	*odin, -ná, nó*
2	*dva, dvě*	*dva, dvě*
3	*tři*	*tri*
4	*čtyři*	*četýrě*
5	*pět*	*pjatj*
6	*šest*	*šestj*
7	*sedm*	*semj*
8	*osm*	*vósmj*
9	*devět*	*dévjatj*
10	*deset*	*děsatj*
11	*jede-náct*	*odin-na-dzatj*
12	*dva-*	*dvě-*
13	*tři-*	*tri-*
19	*devate-*	*děvjat*
20	*dva-cet*	*dvá-dzatj*
30	*třia-*	*tri-*
40	*čtyři-*	*sorok*[1]
50	*pa-desát*	*pjatj děsjatj*
60	*še-*	*šestj-*
70	*sedm-*	*semj-*
80	*osm-*	*vósemj-*
90	*deva-*	*devjanósto*[2]
100	*sto*	*sto*
200	*dvě stě*	*dvě sti*
300	*tři-sta*	*tri-sta*
400	*čtyři-*	*četýrě-*
500	*pět set*	*pjatj-sót*
600	*šest-*	*šestj-*
1000	*tisíc*	*týsjača*
2000	*dva tisíce*	*dva týsjači*
5000	*pět tisíc*	*pjatj týsjač*

ě: "ye" (2)	*č*: ch (4)
ř: rsh (3)	*v*: v as in "vow" (2)
š: sh (6)	*ě*: ye (2)
c: ts (11, 20)	*-j-*: Russian ь, "soft sign," indicating that final consonant is palatalized.
	[1](see p. 185)
	[2](see p. 75)

Romance Group (see p. 91)

Italian* and Sardinian (spoken on Sardinia)
French* and Provençal
Spanish*, Portuguese, and Catalan
Rhaeto-romanic ("Churwelsh" in the Upper Rhine region)
Romanian*

Romance Languages

	Italian	French	Spanish	Romanian
1	*uno, una*	*un, une*	*uno, una*	*uno*
2	*due*	*deux*	*dos*	*doi, doua*
3	*tre*	*trois*	*tres*	*trei*
4	*quattro*	*quatre*	*cuatro*	*patru*
5	*cinque*	*cinq*	*cinco*	*cinci*
6	*sei*	*six*	*seis*	*šase*
7	*sette*	*sept*	*siete*	*šapte*
8	*otto*	*huit**	*ocho*	*opt*
9	*nove*	*neuf*	*nueve*	*noua*
10	*dieci*	*dix*	*diez*	*zece (< diece)*

11–19	(see p. 85)

	Italian	French	Spanish	Romanian
20	*venti*	*vingt*	*veinte*	*doua-zeci*
30	*trenta*	*trente*	*treinta*	*trei-zeci*
40	*quaranta*	*quarante*	*cuar-enta*	*patru-*
50	*cinquanta*	*cinquante*	*cincu-*	*cinci-*
60	*sessanta*	*soixante*	*ses-*	*šase-*
70	*settanta*	*— -dix*	*set-*	*šapte-*
80	*ottanta*	*quatre-vingt*	*och-*	*opt-*
90	*novanta*	*— — -dix*	*nov-*	*noua-*
100	*cento*	*cent*	*ciento*	*o suta*

	Italian	French	Spanish	Romanian
1000	*mille*	*mil*	*mil*	*o mie*
2000	*due mila*	*deux mils*	*dos mil*	*doua mii*

	Italian	French	Spanish	Romanian
	c before *i* and *e* pronounced ch (5, 10,100)	**huit* < Old Fr. *oit, wit*	*c* and *z* before i and *e* pronounced like Engl. *th* (100)	*c* as in Italian
			c before a, o, u is hard: k (5 = "thinko")	*z*: voiced s
			ch: as in "child" (8)	*š*: sh; Rom. *ş*

related; the words for "nine" in Baltic and Slavic, however, do not fit the pattern at all. Likewise in the case of "ten" and "hundred" we can find some individual languages that resemble each other more closely (Greek *déka*, Latin *decem*; Sanskrit *daśa*, Slavic *desętj*), and others in which these words do not seem to be even remotely related: Germanic *taíhun* and Sanskrit *daśa*; Latin *centum* and Slavic *sto*.

But these similarities overlap inconsistently: one number word seems to be almost the same in two given languages, while some other is quite different. The question becomes increasingly complicated: Are these languages related to each other, or are they not?

They are indeed! When their similarities were first noticed, of course, other words besides those of the number sequences were compared as well. If a certain word is the same or nearly the same in several languages, we speak of a "word-equation": Sanskrit *pitár* = Greek *patér* = Latin *pater* = Gothic *fadar* = English *father* = German *Vater*. This particular "word-equation," however, does not extend into Baltic and Slavic. Only in a very rare cases does such an equation hold true for all the languages, but these include most of the number words. A few exceptions, like the Lithuanian "nine," can be readily explained (see p. 93).

Number words are among the words of a language that most strongly resist change; they lose their original meanings very early, and they change scarcely at all with the passage of time and the rising level of culture. Consider the Gothic word for 5, *fimf*, on p. 96, a word dating from the 4th century A.D.: The Modern German word *fünf* has remained virtually unchanged for more than a millennium and a half. Thus it is the number words that directly reflect the relationships of the Indo-European languages in their purest form. Now let us turn to these relationships.

Kentum and Satem Languages

From our survey (tables on pp. 92 and 93) we saw that the relationship between these languages is not quite so obvious as the affinity of the Romance languages to each other (p. 99). A much graver difficulty, however, is that recorded history has nothing whatever to tell us about the relationships between the Indo-European languages. Which was the ancestor of this great variety of languages?

Since history remains silent, linguistics must take the whole burden up its shoulders. Students of comparative philology have succeeded in discovering the phonetic laws behind such differences as between the Latin *centum* and the Sanskrit *śatam*, "hundred." One of the most important criteria for classifying the Indo-European languages has turned out to be the fact that some languages have softened the hard consonant *k* to a sibilant *s* (or *sh*).

This division is manifested in the words for "hundred" just cited above: Latin *centum* (*k!*) — Sanskrit *śatam* (*sh!*). This ties together a whole cluster of differences we noted earlier: Latin "four," *quattuor* (*kw!*) versus Sanskrit *čatvarah* (*ch!*); Greek, Latin, and Celtic "eight," *ok-* (*k!*) versus Sanskrit and Baltic *aš-* (*sh!*); Greek, Latin, and Celtic "ten," *dek-* (*k!*) versus Sanskrit, Slavic, and Baltic *des-* or *deš-* (*s, sh!*). Using this consistent phonetic feature, linguistics has differentiated the *Kentum* from the *Satem* languages (*satem* is the Old Iranian word for 100), which we find, arranged accordingly, on the tables on pages 92 and 93.

Germanic is numbered among the *Kentum* languages, although its word for "hundred" begins with a *t* and "four" with an *f*. We shall find the explanation for this later.

The languages belonging to the two main branches are:

Kentum group: Greek, Latin, Celtic, Germanic, Tocharian, Hittite;
Satem group: the "Aryan" languages of the Indo-Iranian group, the languages of the Baltic and Slavic groups, and Albanian and Armenian (see map on p. 90).

Now that he can readily cross the *Kentum-Satem* bridge from one branch to the other, the reader should be able to recognize the over-all linguistic similarity of the Indo-European languages; and I ask him to do just that.

THE ORIGINAL INDO-EUROPEAN LANGUAGE

If all these languages are sisters, they must have a common ancestor, an original language from which they have developed. But we know of no people that spoke or wrote such a mother language nor have we any direct evidence or written documents concerning it. Thus comparative philology has had to build up the basic Indo-European language, going back into prehistory, just as if Latin had completely disappeared and had had to be reconstructed from French, Italian, and Spanish. This was done by applying all the laws of phonetic and linguistic change that have been discovered and to which some-times one, sometimes another language has contributed. Languages that have become extinct, such as Sanskrit, the language of the Indian scholars and priests, as well as languages that were either left or pushed aside and have escaped attrition by being far away from the cultural mainstream, such as Lithuanian, preserve the old Indo-European features and forms more faithfully than those that are still living and changing. Which of us, for instance, could understand Old High German without specially studying it?

The basic mother language was named after its two geographic end members: Indian (Sanskrit) in the east and Celtic in the west. But since the existence of the Celtic group had not been fully established at the time this great discovery was made, and Germanic was then thought to be the westernmost group of this family of languages, the Indo-European languages are still known in German as *Indo-germanisch*, "Indo-German." Today, moreover, it is known that Tocharian was the easternmost member of the family, so that we should properly speak of "Tochario-Celtic" when referring to the primeval mother language by the names of its outermost descendants.

A language is used and developed by people. Who spoke the original language? The "Indo-Europeans," of course. But this is purely a name for the people who spoke the language. Students of Indo-European prehistory have quite readily inferred the occupations, customs, and way of life of the Indo-Europeans from their reconstructed vocabulary, but so far it has not been possible to locate their place of origin with any degree of certainty. According to one older view, they (initially) inhabited a long belt extending from the Central Asian highlands across the steppes north of the Caspian Sea into Europe itself; another view confines their initial home to the regions of Central Asia; and according to a third hypothesis there are solid reasons for supposing them to have originated in the region between the Baltic and the Black Seas.

Let us take another look at our linguistic map (Fig. 21), which shows the present geographic distribution of the *Kentum* and *Satem* languages, both of which derive their common origin from the Indo-European language. It does *not* show — and I ask the reader to take careful note of this — the migrations of the Indo-Europeans, about which we know nothing whatever. All we do know is a little about the wanderings of some individual peoples during historical times. Our map is thus a purely linguistic one, and at the same time a kind of family tree of the Indo-European languages; when so used it will, taken together with the table of Indo-European languages on p. 91, clarify their geographical distribution.

We do know, to be sure, that the Aryans, who were the first Indo-European people to appear on the stage of history, migrated around the middle of the third millennium B.C. from the northeastern regions through Persia (Iran) toward India, where they partly displaced and partly intermingled with the aboriginal inhabitants of the subcontinent. That the Greeks and the ancient Italian tribes migrated southward to their present homes from the north has also been ascertained. To that extent and with respect to these peoples, the arrows on the linguistic map can also be regarded as general routes of migration.

THE INDIVIDUAL INDO-EUROPEAN LANGUAGES
Now that we have gained at least a general picture of the *Kentum* and *Satem* languages, we can turn to the individual languages of the Indo-European family and try to learn who once spoke them or still speaks them the periods from which their first linguistic documents originate.

Indic Languages
The language of the oldest collection of hymns in existence, the Rig Veda (second millennium B.C.) is Vedic. In a broader sense this forms a part of Sanskrit, the written language in which the great popular epics (the Mahabharata and the Ramayana) were composed. The *Sakuntala*, by the Indian dramatist Kalidasa (5th century B.C.) was also written in Sanskrit. Sanskrit then gradually became a purely learned language, like Latin in western Europe during the Middle Ages, and thereafter no longer took part in the evolution of the vernacular (see table on p. 93). In addition to the languages already mentioned, this group included many common vernaculars, from which Prakrit and Pali, the written language of Southern

Indian Buddhism, were derived. The language spoken by the gypsies can also be included among the innumerable dialects of modern Indic (see p. 190).

Iranian Languages
These are so closely related to the Indic that they have been classified together with them in an "Aryan" group. The language of the writings concerning Zarathustra's religion is Avestic. Somewhat younger is Old Persian, which during the period from Cyrua (as well as Xerxes, Darius and others) to Alexander the Great was the official language of the ancient Persian Empire and has been preserved in inscriptions written in cuneiform characters. Middle Persian, the so-called Pahlavi, was the language of the second great Persian empire (of the Sassanidae), which lasted until it was overthrown by the Arabs in A.D. 641. Afghan (see p. 191) is also one of the Persian dialects. Since the Iranian tongues are so very similar to Indic, only the Sanskrit number sequence has been cited here (see p. 93).

Near Eastern Group
We have likewise omitted the number sequences of this group, which among others has included Armenian (spoken south of the Caucasus), Phrygian (interior of Asia Minor), and Hittite. The latter is one of the *Kentum* languages, and was written in cuneiform on clay tablets. Excavations by German archaeologists (carried on since 1906) have uncovered the archives of the Hittite kings of the middle of the second millennium B.C.; thus, Hittite is today considered, along with Sanskrit, to be the oldest preserved Indo-European language. The Hittite Empire was overthrown by Phrygian invaders in the 12th century B.C.

Tocharian
We did, however, present the number sequence of the easternmost Indo-European language, discovered around the turn of the century by German scholars. The written documents of this extinct people come from the areas of Kashgar and Turfan in eastern and northern Turkestan. It has been possible to distinguish two related yet sharply differentiated Tocharian dialects A and B, which are thought to have flourished in the first millennium A.D. The encroachments of the Turkic peoples put an end to Tocharian, which belonged to the *Kentum* and not to the *Satem* branch, as did the Aryan languages. Thus was destroyed the hypothesis that all Indo-European languages of the interior of Asia belonged to the *Satem* branch (see p. 92).

Greek
Among the Indo-European languages of Europe, this is the oldest that has been documented (dating from the 8th century B.C.). It consisted not of just a single, common language but of a series of dialects, of which the most important are *Ionian-Attic* (Ionian on the west coast of Asia Minor was the language of Homer, Hesiod, and

Herodotus, while the Attic spoken in Athens and elsewhere was used by the classical poets and tragedians and by Thucydides, Plato, and Aristotle; "four" is *téttares* in Attic, *tésseres* in Ionian); *Aeolian* northern Asia Minor (written by Sappho of Lesbos; here "four" is *péssyres*); and *Doric*, spoken in the Peloponnese, Crete, Sicily, and southern Italy ("four" is *tétores*).

Only after Alexander the Great and the political overthrow of Greece did the Attic and Ionian dialects give rise to the Hellenistic common language known as the *Koiné*, which continued to be the literary, political, and legal language of the Eastern Roman Empire until its final conquest by the Ottoman Turks in 1453. *Koiné* is also the language of the New Testament (see p. 92).

Latin

Latin has been known since about 300 B.C.; as the language of the Romans, it won out very early over the other dialects of the ancient Italic branch (Oscan, Umbrian, Sabellian and others). Classical Latin is still in use today as the language of the Roman Catholic Church and was until quite recently the universal language of scholarship; it was not this, however, but the Vulgar Latin spoken in the early Middle Ages which gave birth to the modern *Romance languages* (see p. 91).

Celtic

This is both linguistically and geographically connected with Italic. Around the middle of the first millennium B.C., when Greece and Italy had "settled down" to begin their political and cultural dominance of the then civilized world, the mounted Celtic nomads in the north restlessly wandered and raided here and there. Their Late Iron Age culture (Latène) was distinguished for the richness of its ornamentation. Whether Britain and Gaul were their original homes cannot now be determined; at any rate, from there they migrated to Spain, where they mingled with the aboriginal Iberians to form the Celtiberian culture. In 390 B.C. they sacked and destroyed the city of Rome; a century later they plundered the Greek sanctuary of Delphi. One group of Celts migrated to Asia Minor, where they still live, surrounded by alien peoples; the region of Galatia (St. Paul's Epistle to the Galatians in the New Testament) takes its name from them.

In prehistoric times the Germanic peoples invaded western and southern Europe, generally along the Danube valley. Somewhat later the Celts abandoned these sites, and the Germanic peoples settled there and took over the old Celtic place names: Rhine, Main, Danube, Worms, Melibokus, Mainz, Walchensee, and Chiemsee are all names of Celtic origin. Toward the end of the first millennium B.C. these Germanic movements broke the power of the Celts. The Germanic tribes penetrated to the west, where their Gaul was conquered by Julius Caesar. In the course of the next several centuries the Roman culture and language became so deeply and thoroughly rooted that today, apart from a few place names, virtually nothing is left of Gallic, the mainland Celtic language. On the other hand,

Celtic is still extant in the British Isles (Island Celtic), in the forms of Irish in Ireland, the somewhat different Gaelic of Scotland, and Walisian (the language of Wales, also called Welsh, and not to be confused with the Romance dialect known as Churwelsh, spoken in the ancient province of Rhaetia in the Upper Rhine region); Celtic also remains in use in the form of Breton in Brittany and until recently as Cornish in Cornwall. The former region was, to be sure, a part of ancient Gaul, but its inhabitants came from Britain in the 5th century A.D., as the present names of the language and the peninsula testify. Except for Cornish, which died out some 100 years ago, all three Celtic languages are still spoken today. The oldest documents (the Irish of the so-called *Ogham script*) are all from the Christian era and go back to the 8th century A.D. (see p. 97 and Fig. 16).

Slavic and Baltic
These are the eastern neighbors of the Germanic languages. The individual languages of this group and their geographic distribution may be seen from Fig. 21 and the table on p. 98.

The oldest Slavic document is the translation of the Bible by Cyril and Methodius in the 9th century; its language is known as Old Church Slavonic or Old Bulgarian (see table on p. 93). In the Orthodox Slavic countries this has become the language of the Church, as did Latin in the Roman Catholic regions.

Together with Baltic, whose principal language is Lithuanian, Slavic forms a closely interrelated major group similar to the Aryan of Persian and India; the Balto-Slavic languages (along with Albanian) are the *Satem* languages of Europe. But whereas the Aryan

Fig. 22 Time-scale of the earliest documents of the Indo-European languages. Between Lithuanian (+1700) and Sanskrit (−1500) there is a gap of more than three thousand years.

group, through the vicissitudes of history, developed further and
further away from the original Indo-European language, present-
day Slavic, in contrast, and especially Lithuanian, surpasses all
the other Indo-European languages in the profusion of ancient Indo-
European forms it has preserved. Yet the oldest Lithuanian documents
date only from the 17th century A.D.! The reader will see from the
"time ladder" (Fig. 22) that the word equations previously discussed
(see p. 100) do not all lie on the same chronological level, but from
Lithuanian back to Sanskrit run through more than three millennia.

The Germanic Languages
Figure 21 shows the Germanic group split up into the West Germanic languages spoken between the Rhine and the Elbe Rivers,
the East Germanic along the northeastern shore of the Elbe, and the
closely related North Germanic tongues. They are all wedged tightly
between Celtic to the west and south (see p. 104) and the Balto-
Slavic group to the north and east on the continent. Nevertheless
they are sharply differentiated from both these neighboring language
groups. And if we refer to their number words and compare them
(see p. 92) — "ten": Celtic *deich*, Lithuanian *dešimt*, Slavic
desētj, but Greek *taíhun* — we find that we cannot at first glance
assign the Germanic languages to either the *Kentum* or the *Satem*
branch. According to the Roman historian Tacitus, the Germanic
tribes intermingled less than any others with alien peoples who had
occupied the areas they inhabited since the earliest times. Since there
is also further evidence to this effect, it is at least not impossible that
the Germanic peoples initially acquired an Indo-European language
as a primitive people and that they were later "Indo-Europeanized,"
perhaps as a result of the dominance of the Celts, who at one time
penetrated deep into what is now Germany. This hypothesis may
at least explain the break, or gap, in linguistic relationship whose
sharpness is unique among all the Indo-European languages: The
Germanic tongue has adapted the new language to its own idio-
syncrasies, since originally it was not "suitable," just as Germans
today have trouble pronouncing Swiss dialects or Russian, and
Englishmen notoriously speak poor French and colorless Italian.
   Philology has been able to establish the rules governing this
Germanic modification of the Indo-European sounds and thus
determine which sounds have been "displaced"; for they have not
all been changed, as the Germanic words for the numbers 2, 3,
7, 8, and 9 testify (see tables on pp. 92 and 93). This "Germanic
phonetic shift" was the furnace in which the Indo-European language was "melted down" into "Germanic." If we examine the
accompanying *dekm* table (Fig. 23), we see how the original Indo-
European word *dekm* appears in the various *Kentum* and *Satem*
languages to the right and left.
   While all the other branches derive directly from the main trunk,
the Germanic group undergoes two phonetic shifts, the "Germanic"
and the "German," indicated by the circles on the diagram. As the
word *dekm* runs through the Germanic branch, it changes phoneti-
cally to such an extent that it appears in Gothic as *taíhun*.

Fig. 23 Changes of the Indo-European word *dekm*, "ten," in the *Kentum* and *Satem* languages; observe the Germanic and the German phonetic shifts (indicated by the circles).

If we pursue this path still farther, we shall come, beyond the West Germanic languages (Old Saxon, Anglo-Saxon, and English), to a second circle with yet another phonetic shift: Once again the Germanic languages show a phonetic shift (in this case from *t* to *z*, pronounced "ts"), giving rise to the Modern German *zehn*, "ten." Thus the original Indo-European *dekm* has had to undergo two phonetic shifts, stronger in Germanic and milder in German, before it finally adjusted to the German ear in the form of the word *zehn*. Later we shall discuss the specific rules that have governed these phonetic shifts (p. 127).

Thus the individual Germanic languages differ from each other only as dialects. The oldest Germanic language that has been preserved is *Gothic*. Bishop Wulfila (A.D. 310–383) translated the Bible into the language of his own people and thereby created the oldest and most significant Germanic document. Fragments of this translation, written in the characteristic Gothic script in silver on purple-colored parchment and bound in a silver cover (the *Codex argenteus*,

"Silver Book"), have come down to us and are now preserved at Uppsala in Sweden (see Figs. 27 and 28; also *Gothic Bible*, Figs. 92 and 93). Just as Boniface was the "Apostle to the Germans," so was Wulfila the Apostle to the Visigoths. Originally the Visigoths were united with the Ostrogoths in the Gothic nation, which together with the Vandals and other peoples formed the cultural community of the East Germans (see p. 94).

Land hunger and pressure from the Slavs set the Germanic tribes into motion in the 1st century A.D.: The previously quiescent sea of northern peoples now set out on the migrations that surged forcibly against the bulwarks of the Roman Empire. The strongest and boldest of these barbarian floods were the East Germans; they were the true migrants, adventurers driven not only by hunger but also by a desire to make conquests in sunny and shining faraway lands where they could settle.

In Dacia in the Lower Danube region they were stopped by the resistant wall of the Roman military. A part of them came to an agreement with the Romans, stayed on, and adopted Christianity (in the Orthodox Greek form, whose missionary was Bishop Wulfila). Other remnants of these tribes persisted for a long time on the Crimean Peninsula, where their "Crimean Gothic" language became known in the 16th century (see p. 50); they and their tongue, however, died out in the 18th century. But the rest of the East Germans, now driven on by the pressure of the Huns, broke forcibly into the world of Roman culture. For a time the Romans succeeded in deflecting the streams of Goths into Spain, where it was dammed up briefly into a Visigothic kingdom, but this ultimately collapsed, having reached the end of its restlessly squandered strength a few centuries later, under the blows of the Arab invaders.

After that, however, a second flood of barbarians broke into the very heart of the Roman Empire, halting briefly when the leader of the Ostrogoths, Theodoric the Great, consolidated his young nation on Italian soil (with his capital at Ravenna). Not until after Theodoric's death in A.D. 526 did dissensions among the Germans and their damned-up lust for adventure again overflow the ordered channels of the Gothic Kingdom; now, wave upon wave, the Gothic floods swirled onward into the enticing south. They burst forth once more at the slopes of Mt. Vesuvius, then flowed back to the broadly curving, seductive Bay of Naples, which had already absorbed so many other peoples and a scant millennium later was to be enriched by the blood of the last Hohenstaufen Emperor.

The Gothic tongue came to an end with the Gothic people. Their example shows how difficult it is to shed light on the dim past of the Indo-Europeans and to trace their way of life and the course of their wanderings; for the peoples who themselves were the bearers and transmitters of the original Indo-European language vanished without a trace.

The West Germanic peoples were more peaceful. In the West the walls of the Roman Empire held fast and quite early forced the migrating tribes to settle down and support themselves by a more highly developed agriculture. Thus at the time when, far from their northern home, their Gothic and Vandalic cousins were doomed to extinction, they gradually consolidated the nations and kingdoms which, after the Roman rule had been overthrown, were destined to become the bearers of Western culture. To be sure, there were advances and retreats among the West Germanic peoples too, but these movements were less violent and destructive. The Angles and the Saxons, who left the Elbe River area to invade England in the 6th century, ultimately combined with the later invaders, the Normans, to become the true "Englishmen," next to whom the native inhabitants (the Celts) from then on played only a subordinate role in British history. But the vacuum which the departed East Germans had created along the Elbe River could not be filled adequately by those who remained, and so a Slavic people (the Prussians) pushed their way to the Elbe itself. Deliberate and forceful policies ultimately succeeded in stemming their invasion, and in the centuries to come the territory was regained step by step.

The oldest documents of the languages of the West Germanic group (English, Frisian, Low and High German; see p. 94) date from the 8th century; they include the Low-Saxon Biblical hymns of "Heliand," of the 9th century. High German is commonly broken down into Old (to 1100), Middle (to 1500) and New High German (from 1500 on; see table on p. 96). Charlemagne spoke Old High German; the language of the *Nibelungelied* and of the Minnesingers was Middle High German.

The North Germanic subgroup lay to one side of the more rapidly flowing mainstream. With the passage of time it deviated from the remaining Germanic groups and by A.D. 700 had formed the original Nordic language, which has been preserved only in isolated runic inscriptions. Later it became divided into West Nordic (in Iceland and Norway) and East Nordic (in Denmark and Sweden); the language prior to about 1500 was known as Old Norse. The Old Germanic tongue lived on in its truest form on the remote island of Iceland (settled by Norwegians around 900 and Christianized around 1000). The Edda, a collection of Old Norse songs about gods and heroes, was first written down in the 13th century.

In the tables on pp. 94–96 the reader will find the old and modern West and North Germanic languages compared, with Gothic included in the first of these groups. He can not only compare these languages directly at a given period of time, but also learn which changes, English, for example, went through as it developed from the Anglo-Saxon.

Thus we have presented a general survey of the Indo-European languages, represented visually in the linguistic map (Fig. 21), the time-ladder (Fig. 22), the *dekm* table (Fig. 23), and the number sequences of the various individual languages, past and present (pp. 91–99).

The Non-Indo-European Languages

We have yet to discuss the non-Indo-European languages, which appear within the circles on our linguistic map (Fig. 21): the now defunct Iberian, Ligurian, and Etruscan tongues and the still spoken Basque language and Hungarian (Magyar), the latter of which, along with Finnish and Ostyak, belongs to the so-called Finno-Ugric branch, which grew out of the stem of the Uralic-Altaic family of languages from which Turkish and Mongolian have also sprouted.

The number sequences of these non-Indo-European languages spoken in Europe are given in the next few pages. To these, as examples of a broader and stronger linguistic relationship, are added the Semitic number sequences, Arabic and Hebrew, which we shall encounter repeatedly in the course of our study. (The reader will find the Chinese and Japanese number words and numerals on pp. 447 ff.)

Basque

Basque today is spoken by about a million persons and is thought to be a possible successor of the now vanished Iberian language:

1 *bat*			
2 *bi*	20 *oguey*		20
3 *hiru*	30 *oguey-t-amar*		20 + 10
4 *lau*	40 *ber-oguey*		2'20
5 *bost*	50 *ber-oguey-t-amar*		2'20 + 10
6 *sei*	60 *hirur-oguey*		3'20
7 *cazpi*	70 *hirur-oguey-t-amar*		3'20 + 10
8 *zortzi*	80 *laur-oguey*		4'20
9 *bederatzi*	90 *laur-oguey-t-amar*		4'20 + 10
10 *amar*	100 *eun*		
	1000 *milla.*		

The Basque number sequence, whose number words all the way to 100 have not the slightest resemblance to the Indo-European (except for the word for "six," which by coincidence sounds familiar), is clearly built up by levels of 20. The old Basque limit of counting was evidently 100, for the word *milla* for 1000 has been taken from the Romans.

Etruscan

The Etruscan and Ligurian peoples, who have now disappeared, were the Romans' preceptors in many things pertaining to government and religion, and maintained their highly respected special status well into the period when Rome had outwardly overtaken them. Originally the Etruscans occupied large areas of Italy; later their home was confined to Tuscany, which takes its name from them (*Tusci = Etrusci*). They were probably one of the peoples who came over from Asia Minor, appearing at the borders of Egypt with other maritime nations around the middle of the second millennium B.C.; driven away from Egypt, they crossed the sea and set foot in Italy. Their language is preserved only in the form of a few not yet reliably deciphered remnants. The Etruscan number words up to 90 were:

1 *thu*	10 *sar*		
2 *zal*	20 *zathrum*		
3 *ci*	30 *ci-al-ch*		
4 *huth*	40 *huth-al-ch*		
5 *mach*	50 *muv-al-ch*		
6 *sa*	60 *se-al-ch*		
7 *cezp*	70 *cezp-al-ch*		
8 *semph*	80 *semph-al-ch*		
9 *nurph*	90 *nurph-al-ch*		

As in the case of Basque, the words are completely foreign to us and show no similarity at all to those in any of our familiar languages. But we can see that the tens, apart from 20, were formed from the units with the suffixes *-al-ch.*

Philologists see in the *-al-* a genitive ending that serves as the equivalent of an adjective; the *-ch* then turns this into a noun: thus 30, *huth-al-ch,* is "three-like, the" and hence "the three-like." The numbers just before the gradations, such as 17 or 29, are formed by back-counting: "3 from 20," *ci-em-zathrum,* or "1 from 30," *thun-em-cialch.*

Finno-Ugric

The Finno-Ugric group of languages, which together with Samoyedic belongs to the so-called Uralic branch, is represented in Europe by its two largest offshoots, Finnish and Hungarian. Along with some smaller groups of languages it stretches from the East Slavic region as far east as the Urals. Philology has divided the Uralic group into the following branches:

1. Lapp;
2. Baltic Finnish (Finnish, Estonian, Livonian, Carelian);
3. Volga Finnish (Mordvinian and Cheremyshian, both belonging to small groups spoken along the bends of the Volga River south of Moscow);
4. Permian (Zyryanian and Votyakian) spoken in the region of Perm, now the city of Molotov, in the Central Urals;
5. Ugric (Vogulian and Ostyakian on the Ob River; Magyar or Hungarian west of the Carpathians).

All of these languages trace their descent back to an original language whose native home was somewhere in what is now eastern Russia. From the brief survey in the table on p. 113 we see the wide gap between Finnish and Hungarian, the latter of which is far more closely related to languages spoken east of the Urals. Their affinity consists in having the same or similar grammatical structure and vocabulary, and in the case of the basic words also in regular phonetic changes; there is less of the obvious similarity of words that the laymen can recognize, which we are accustomed to find among various Indo-European languages.

To mention but a few of the features peculiar to both these languages: They have no grammatical gender, inflect all nouns in the same manner, form the plural with the same symbol (Hungarian *-k-*, Finnish *-i-*, in the nominative and accusative cases *-t-*), and have the same case endings for the singular and the plural. Hungarian has 24 cases and Finnish 15, expressions that we would treat as prepositional phrases as "in, into, to, on, from the house" all being considered as "cases." A few examples will illustrate the difference between these and our Indo-European verbal forms:

Hungarian: *a haz,* "the house"   *a haza-k* "the houses"
           a haz-ban, "in the house" *a haza-k-ban,* "in the
                                         houses"
Finnish: *talo,* "the court"   *talo-t,* the courts"
           talo-ssa, "in the court" *talo-i-ssa,* "in the courts."

Both Finnish and Hungarian are agglutinating or amalgamating languages, not inflected like modern German, because they leave the root (e.g., *talo*) undisturbed and represent their grammatical indications by suffixes "tacked on" (agglutinated) to the ends of the roots (for example, *talo-i-ssa* contains the suffixes *-i-* for the plural and *-ssa* for the locative).

Now let us consider the number sequences on p. 113. Surprisingly, both the Finnish and the Hungarian number sequences are built up uninterruptedly on the basis of the number 10. All the tens are formed by adding suffixes to the units, 30 = 3'10, etc.; even 20, which in so many other languages takes a special form, is here 2'10. The numbers 11, 12, ..., 19 have no special words for "eleven" and "twelve" but run consistently straight through, in Hungarian taking the form 10'1, 10'2, etc. and in Finnish using the overcount borrowed from the Germanic languages (see p. 80). Does this testify to a rare case of a regular, undisturbed progression on the part of a highly mature number sequence possessed by the Finno-Ugric peoples? Let us examine the root meanings of the number words; we find that the 2nd and 3rd rank levels have been borrowed from Indo-European:

Finnish	Mordvinian	Vogulian	Zyryanian	Hungarian
sata	*sada*	*sat*		*szaz* < Indo-Iranian *sata*
tuhat		*sotr*	*surs*	*ezer* < Indo-European

and, to be sure, Finnish *tuhat* < Baltic *tukstantis,* Hungarian *ezer* < Indo-Iranian *hazar.*

Thus the Finno-Ugric number sequence was built up independently only up to ten. In fact all these languages have in common only the number words from 1 through 6; the Hungarian word for 7 comes from Indo-European (*setpm,* Vogulian *sät,* Hungarian *het,* whose initial *h* was created by analogy with 6, *hat*).

At the very first rank threshold, 10, the Finno-Ugric languages bcgin to deviate from each other, whereas the Indo-European languages have all the numbers through the 100s in common and do not go their own separate ways until 1000. The basic Finno-Ugric language has a word meaning "number," which in the Lapp *lokk* is the number word for 10; Finnish has it in *luka*, "to count, to reckon." Finnish 10 is *kymmenen* (Mordvinian *kemein*), whereas Hungarian borrowed the Indo-Iranian *tiz* (Syryenian *das*). Lapp also uses the word *tseke*, which means "notch, groove," for 10 — an exact analogue of the English word *score* for a group of 20.

The Permian tongues form the tens with *-mis*, Hungarian *-mints*, which we recognize as a degenerate remnant in the Hungarian 20, *hu-sz*, and 30, *harmin-c*, as well as in 8, *nyol-c* and 9, *kilen-c*, the latter

	Finno-Ugric Languages		Turkish
	Finnish	Hungarian	
1	*yksi*	*egy*	*bir*
2	*kaksi*	*kettö*	*iki*
3	*kolme*	*három*	*üč*
4	*neljä*	*négy*	*dört*
5	*viisi*	*öt*	*beš*
6	*kuusi*	*hat*	*alti*
7	*seitsemän*	*hét*	*yedi*
8	*kahdeksan*	*nyolc*	*sekiz*
9	*yhdeksan*	*kilenc*	*dokuz*
10	*kymmenen*	*tiz*	*on*
11	*yksi-toista**	*tizen-egy*	*on bir*
12	*kaksi-*	*-kettö*	*on iki*
13	*kolme-*	*-harom*	*on üč*
20	*kaksi-kymmentä*	*búsz*	*yirmi*
30	*kolme-*	*harminc*	*otuz*
40	*neljä-*	*négy-ven*	*kirk*
50	*viisi-*	*öt-ven*	*elli*
60	*kuusi-*	*hat-van*	*altmis*
70	*seitsemän-*	*hét-ven*	*yetmis*
80	*kahdeksan-*	*nyolc-van*	*seksen*
90	*yhdeksan-*	*kilenc-ven*	*doksan*
100	*sata*	*szaz*	*yüz*
200	*kaksi-sataa*	*két-szaz*	*iki yüz*
300	*kolme-*	*harom-*	*üč-*
1000	*tuhat*	*ezer*	*bin*
2000	*kaksi-tuhatta*	*ket-ezer*	*iki bin*
3000	*kolme-*	*harom-*	*üč-*
10,000	*kymmenen-*	*tiz-*	*on-*
	For 11–19, see p. 80 *(see p. 78)	*sz*: ss (20) *c*: ts (8) *g*: dj (1, 4) *z*: soft s (10)	

two having the meanings "2 from 10" and "1 from 10," respectively. The same syllable also appears in the Turkish 60 and 70.

Finnish forms these same back-counted 8 and 9 with *-deksan*. The Hungarian tens from 40 to 90 have the "10" syllable *-van* or, after phonetic shift attraction, *-ven* in 70 and 90 (Syryenian *-min*, Vogul *-men*).

The surprising thing is that with this multiplicity of words for 10, the Hungarian word form *hu-sz* < *ke(ttö)-(min)ts*, 2'10, runs through most of the Finno-Ugric languages: Ostyak *ko-s*, Vogul *ku-s*, Mordvinian *ko-ms*. This surely is due to the influence of the 20-group "man" concept, here expressed verbally not as a grouping but computationally in the form 2'10 (like the French *vingt* for the Norman *skor*; see p. 68).

What have we learned from our study of these number words? It has demonstrated clearly that an undisturbed progression of the number sequence is not due to any higher degree of maturity, but to a less developed number sequence hidden in the background. This then modeled itself on the pattern of the already developed neighboring languages (Slavic and Germanic) and thereby was able to avoid all the incomprehensible anomalies and disturbances in the progression, which it would have undergone had it developed independently.

Turkish

Similar circumstances have impressed themselves on the *Turkish number sequence*. Turkish is part of the Altaic subfamily of languages, a sister family of the Uralic, while the Finno-Ugric group belongs to the Uralic division. Thus Turkish is a "cousin many times removed." They share, however, the same clear grammatical structure. The Turkish word *on*, "many," for 10 is an old limit of counting. In the Turkish 60 and 70 we recognize the Finno-Ugric 10 ending *-mis*; *-san* (or *-sen*) is also a form of "ten." In contrast, the words for 20, 30, 40, and 50 are irregular, and no one has yet succeeded in explaining them satisfactorily; they are not formed from the units but are "numerational" and thus preserve a very early state of the number sequence that extended only to 50 (see p. 82).

Semitic

The Semitic languages, like the Indo-European, make up a large family of related tongues, of which we shall cite only the Arabic and the Hebrew number sequences.

Philologists have broken down the Semitic family into the following classification:

1. Eastern Semitic (Babylonian from the third millennium B.C.; Assyrian from 1100 B.C. on; see p. 165);
2. Western Semitic
   a. Northern Group (Aramaic and Canaanitic, including Hebrew and Phoenician)
   b. Southern Group (Arabic and Ethiopic including Abyssinian).

Semitic Languages

	Arabic		Hebrew	
	Masculine*	Feminine*	Masculine*	Feminine*
1	*ahadun*	*ihda*	*ehad*	*ahaþ*
2	*iþnani*	*iþnatani*	*šǝnayim*	*šittayim*
3	*þalaþun*	*þalaþatun*	*šaloš*	*šaloša*
4	*arba'un*	*'arba'atun*	*arba'*	*arba'a*
5	*hamsun*	*hamsatun*	*hameš*	*hamišša*
6	*sittun*	*sittatun*	*šeš*	*šišša*
7	*sab'un*	*sab'atun*	*šeba'*	*šiv'a*
8	*þamanin*	*þamaniyatun*	*šǝmonē*	*šǝmona*
9	*tis'un*	*tis'atun*	*teša'*	*tiš'a*
10	*'asrun*	*'asaratun*	*'eser*	*'aśara*
20	*'išruna*		*'eśrim*	
30	*þalaþuna*		*šalošim*	
40	*'arba'una*		*arba'im*	
50	*hamsuna*		*hamiššim*	
60	*sittuna*		*šiššim*	
70	*sab'una*		*šiv'im*	
80	*þamanuna*		*šǝmonim*	
90	*tis'una*		*tiš'im*	
100	*mi'atun*		*mē'a*	
200	*mi'atani*		*ma'þáyim*	
300	*þalaþu mi'atin*		*šaloš me'oþ*	
1000	*'alfun*		*elef*	
2000	*'alfani*		*alþayim*	
3000	*þalaþatu-alafin*		*šalošeþ'alafim*	
10,000	*'ašarat-'alfan*		1) *'aśereþ'alafim*	
			2) *alafim 'aśara*	
			3) *rǝbaba* or *ribbo*	

The case endings -*un* are no longer always pronounced. *h*: ch as in German *Buch*	The murmured vowel is almost silent: "30" is thus almost "shloshim."

' is a guttural; ' is a glottal stop: German "Post-amt," not "Pos-tamt."

* *Editor's note:* In traditional grammar, as well as in usage (both spoken and written), columns 1 and 3 are considered "masculine," and columns 2 and 4 "feminine."

Hebrew, developed since the migration of the Jews into Canaan (14th century B.C.), is the language of the Old Testament. As a result of the destruction of Jerusalem (by the Assyrians) and the Babylonian Captivity (in the 6th century B.C.), Hebrew from that time on became increasingly displaced by Aramaic, and was virtually extinguished after the great westward migration of the Jews during the Hellenistic epoch, beginning in the 2nd century B.C. Jesus and His Apostles spoke not Hebrew but Aramaic. As a ritual language, however, Hebrew has been maintained down to the present. Next to it, around A.D. 200, arose the New Hebrew of the Talmud, the compilation of Judaic laws. Finally, as a result of Zionism and the return of the Jews to modern Israel, Hebrew has again become an official and commercial language in that part of the world.

Arabic

Arabic and its many dialects were spread by the great conquests of the 7th and 8th centuries A.D., of which we shall have more to say on pp. 406 ff.; through the Koran alone, which Moslem believers may read only in the original Arabic text, it was introduced into all the lands where Islam predominates. Thus, for example, fully half the vocabulary of Persian, which as Avestan was once a purely Indo-European language, now consists of words of Arabic derivation.

The essential feature of the Semitic languages is that they attach the root meaning of the word to a framework generally of three consonants (like the English *drnk*). Words of related meanings are then created by inserting vowels (as in *drink, drank, drunk*). This vowel gradation, or *ablaut*, which as our examples show can also occur in German or English (apart from words formed by adding prefixes and suffixes), is much more important in Semitic than in the Indo-European languages.

An example from Hebrew is the word for 10,000, *rəbaba*; its root is *rbb*, from which *ribbō* is immediately derived, as the number sequence shows. But the stem *rb* is also part of the following family of words: *rab*, "much, strong: Lord" — *rob*, "multitude, size, abundance" — *raba'b*, "to be many, numberless" — *rabba'* (> Rabbi), "leader." Thus we see, as these words have documented, that the Hebrew word *rəbaba* has been condensed from the indeterminate meaning of "many" to the strict concept of 10,000. Compare the example on p. 133. Another peculiarity of the Semitic number words, their habit of changing gender and word class throughout the sequence, has already been mentioned (p. 81).

Herewith the Arabic and Hebrew number sequences, whose affinity, in contrast to the Finno-Ugric, can be readily demonstrated by the great similarity of the number words.

The Meaning of Our Number Words

IMAGES BEHIND THE CONCEPTS OF SIZE AND NUMBER

If we address ourselves directly to our number words, "one, two, three ... nine, ten, hundred, thousand," they remain strangely silent about their inner meanings; in this respect they behave like but few other words of our language. It is remarkable that in our daily lives we encounter them constantly and use them as the most reliable bearers of concepts in our possession. Yet we allow ourselves to be content with their usefulness while knowing them only "by name." They pass by us mute, like alien slaves valued only for their services, and we do not dignify them by inquiring into their "person" or their homeland. And yet they do have "personalities," chosen by early man out of his colorful, chaotic environment to be the bearers of his concepts of numbers. We are aware, of course, that the number sequence did not spring from a single brain, as a perfected system, but that it developed slowly, like a tree, keeping pace with man's gradual development. Thus we cannot view the first few number words as being artificially contrived formations invented, like Esperanto, by one person alone and then adopted by others.

But however much we may rattle and shake the word "hundred," tap it or squint at it through a magnifying glass, it remains silent. And no matter how diligently we may search among all words for its relatives, none will admit belonging to the same tribe. Anyone will reject the word "hound" as a cousin of "hun-dred." Thus we have no alternative but to seek evidence for our hypothesis that number words originally had graphic or figurative meanings. Once again we must turn to common speech and look for words that designate magnitudes, which is, of course, what number words are.

Pictorial Quantities

A German farmer does not cultivate so many square meters of acreage, but so many "mornings" or "days" — that is, as much land as he can plow in one or more days. Elsewhere he speaks of a day's work, of a *Juchert* (the amount of land a "yoke" of oxen can plow each day) and of a *Mannshauet* (as much as one man can "hew" in one day). In brief, out of the things that surround man day after day, from what he sees and knows and feels, he creates the units of measure that convey ideas of size or quantity to him. From his own body he took the *ell*, which has been used as a measure of length by virtually all civilized peoples: Latin *ulna* > French *aune*, Greek *ōlénē*, Old High German *elina*, modern German *Elle*, Italian *braccio*, "arm"; and also the foot, the handbreadth (Latin *palmus*, Greek *dóron*), the fingerbreadth (English *digit*, "¾ of an inch"); French *pouce*, "inch" < Latin *pollex*, "thumb" and "inch" or $\frac{1}{12}$ of a Roman foot); and in addition, all the measures of extent: the span, the pace, and the fathom (the distance a man can "fathom" with his two arms extended); the Swiss have a *Chlupfel*, a "*Ge-lupf*," an armful of hay, just as in Swabian a *Hämpfli* is a "handful." The sheaf also belongs to this class (< Indo-European *ghreb*, "to grasp, grip"), since it is a quantity of stalks "grasped" together into a bundle. The German *Klafter* is the English *fathom*, from Anglo-Saxon *fathmos*, "both arms stretched out"; the fathom later became specifically a measure of depth, but in northern Germany it once also measured a volume of wood, like a "cord." *Faden*, "thread"* is thus as much yarn or thread as can be measured with both arms, in sewing. From the same root comes the word *Fuder*, a large wine measure called the "cartload," which was orginally also a fathom's length, as was the related Latin *passus*, "pace," a thousand of which (*mille*) came to be called a *mile* (see p. 186).

The Greek coin known as the *drachmé* likewise was derived from a fathom or grasp measure; the word originally meant "handful" and was worth 6 pieces of iron (Greek *óbolos*), which had a monetary value in early Greece (see the story of the slave girl Rhodopis in Herodotus' *History*, II, 35). The English apothecaries' weight *dram* is a descendant of the drachma. On the other hand, it is purely fortuitous that the Russian *kopek* also derived its name, like the Greek obol, from the word for a spear or lance (Russian *kopje*), since the image of a rider carrying a lance was stamped on the early kopeks.

Since the first money used by early man did not consist of coins with commonly accepted arbitrary values, but of objects of barter — cattle, sheep, iron and so on — the names of many coins of antiquity and even of the Middle Ages were originally those of objects. Who would have thought that the French equivalent of the "dollar," the *écu*, has an animal hide as its ancestor? Yet if we trace the development of this word back to antiquity, the connection becomes quite clear: *écu*, Spanish *escudo*, Italian *scudo*, derived through the Latin *scutum*, the leather-covered long shield carried by Roman soldiers (compare Latin *ob-scu-rus*, "hidden") from the Greek *skýtos*, "leather, hide"; this animal skin was originally traded and bartered, and thus was used as money. Later on *écu* was also inscribed on the stamped *escutcheon* of the master of the mint. These designations of quantity and monetary value grew directly out of the ordinary life of early man.

Whenever a measure did not immediately confront him pictorially, he created one that did: He took a *rod* as a measure of length; the English still use the *yard*, a word cognate with the German *Gerte*, "rod." Another example is an obsolete German measure of capacity, the *Simmer* (Old High German *sümber*), literally a basket. Similarly, *Schäfflein* (a small trough), from Latin *scapha*, "boat, skiff" > Old High German *scaf*, English *skiff*, became a bushel; and an *Eimer*, "bucket," (Greek *ámē* > Latin *ama* > Middle High German *ame*) became an *ohm*, a liquid measure of about 35 gallons. The modern German word *nachahmen*, "to imitate, counterfeit," which Luther still spelled *nachohmen*, literally means "to measure again, to verify," and was later changed to "become equal to a measure." The *Malter*, "quarter" of grain, roughly equivalent to a "bushel" is a *Mahl-ung*, a "grinding" or "milling," and the *Metze*, "peck" (Old High Ger-

---

* *Translator's note: Faden* in German means both "fathom" and "thread," thus revealing the relationship between its meaning.

man *mezzan*) of grain was originally the "measured quantity" or "measure." These dry measures gradually evolved into weights. The best example of this is the "ton" (cf. archaic English *tun*, "barrel"), which like the others was originally a measure of grain; "a ton of land" was as much land as could be planted with a ton of seed grain. This later became a measure of salt, coal, and the cargo capacity of a vessel. Today it is our largest unit of weight.

But even for weights, there were things which people took over directly from nature to use as measures: the carat is the small red kernel in St.-John's-bread, which looks like a small horn: Greek *kerátion*, from *kéras*, "horn" (as in rhino-ceros, "nose-horn"). This was worth four barley- "corns" (Latin *granum*), from which our apothecaries' weight, the *grain*, is derived. The Babylonian weight measure *ma-nu*, which entered the Greek scale of weights as *mnâ*, "mina," is probably a date pit. The old German weight, the *Lot*, is related to English *lead* (löten = solder) which could be poured out in measured amounts as weights.

Even for something as hard to measure as time, primitive man drew on his observations of his surroundings: mid-day, sun-down, sun-up. The ancient Sanskrit "forenoon" was *samgava*, "the time when the cows are driven together," just as the Greek quitting time in the evening, the *bu-lytós*, was the time for "unharnessing the oxen" (< *bous*, "ox"; *lyein*, "to release"; *bulýtonde*, "in the afternoon"). An early Germanic example is the Anglo-Saxon *undorn* (going back to the Latin *inter*, "between"), the "between-meals time." The Russian peasant divides up his whole day after this fashion: first, second, and third *upovod* "between," just as he also breaks up the night into the first and second "cockcrow."

Wherever we look we see that the early measures of size and quantity were not deliberately synthesized and artificially given new names especially invented for them. Instead, early man simply took some events or objects which most conveniently designated magnitudes and introduced them as measures. Today we still make the same use of objects of fairly definite and commonly known magnitude: We speak of a "ream" and a "quire" of paper, a "skein" of wool, a "pack" of cigarettes, a "bale" of cotton; as informal measures of time, we have "before you can say 'Jack Robinson'," and a human generation.

Descriptive Number Words

But images have been substituted even for abstract numbers. We need only recall our 20-group, the *score* "notch," the *snas* "rod," and the *borla*, "load" (see p. 49), not to mention the Lapp *tseke*, "notch" (see p. 113); other examples are provided by words for the rank levels (as on pp. 63 and 141).

Primitive cultures also offer some excellent illustrations. Things that commonly come in twos are often called "wings," "eyes," or similar paired objects. In the Carib language the word for 2 has the root "break, split," indicating two created out of one; an expression still used in modern German is *ent-zwei*, "into two". In the language of one Negro tribe "four" is called "two forks"; the Abipone Indians say "rhea toes" for "four" because the kind of rhea (ostrich) with which they are familiar has four claws.

The hands and fingers have become measures and number words all over the world. We recall how the finger gestures became the Dene-Dinje number words, how the "hand" is almost universally used to indicate five, and how 10 was called "both hands," 15 "both hands and one foot," and 20 "one man" (p. 35).

Symbolic Numbers of India

In India the custom arose of expressing numbers by symbols: 1 is "moon"; 2 is "eyes" or "arms"; 4 is "throw of the dice," or "brothers" (because Rama in Indian poetry had three brothers), or "world's age"; 7 is "head," because the head has 7 apertures; 9 is "cattle," also in reference to Hindu myths, and so forth. Other cultures also provide numerous examples, such as the "Muses" representing 9 or the "Apostles," 12. For the numbers 1 and 2 there are more than three hundred different symbolic words, including even "rhinoceros."

Although the Hindus possessed a fully formed number sequence and thus had no need for such substitutes, such symbols were still readily and without difficulty understood as constituting numbers. They were used not only in poetical works, to express a "naked" number couched in symbolic terms, but predominantly in scientific texts on mathematics and astronomy. Most of these were written in verse, and in order to fit numbers into the rhythm of the lines and perhaps also to enable the reader to grasp them better pictorially, the verbally fixed number words were expressed by appropriate verbal symbols. This practice extended throughout India, Tibet, and Southeast Asia all the way to Java where in fact it is still used when a number must be expressed without the slightest ambiguity, e.g., in administrative affairs or in historical inscriptions on public monuments.

Since the 7th century the custom of expressing numbers symbolically has seen some rather remarkable uses. For example, the number 1021 is put together from symbols used in the same manner as digits, according to the place-value principle which after all had its origin in India. The only difference here is that the number is written and read in ascending order from left to right:

	one	two	zero	one
Hindi:	*śaši*	*paksa*	*kha*	*eka*
	"moon"	"wing"	"hole"	"one."

The *kha*, "hole" or "empty space," stands for zero, likewise a device of Indian invention in its written form "0." In this place-value notation the numerals 1, 2, ..., 9 were represented by the meaning of a word rather than by a fixed written symbol. A number consisting of two digits can also be given by a single word, such as "sun" for 12 (in the zodiac) or "teeth" for 32. To make up an example in our own language:

3125 = "hand" — "Apostles" — "the Fates" = 5 — (12) — 3.

Here we note for the first time that peculiarly Indian delight in numbers which is manifested in their playful habit of substituting symbolic words for numbers. It springs from a uniquely Indian gift for numbers which has often been an object of wonder and admiration.

Images as Number Symbols

We have mentioned pictorial measures and number words; now for the view that early man drew on images from his environment to express his ideas we shall call on a third witness: writing, language's younger sister. Here we shall make a distinction which has, quite surprisingly, seldom been made before: the distinction between the writing of words and that of numbers.

How many people are aware that our letter A is an oxhead and the letter G a camel's neck? The letters of our alphabet have a colorful history: they have come down to us from Rome, whither they had migrated from Phoenicia by way of Greece. Yet even Phoenicia was not their origin, but Egypt. The image painted and chiseled on the walls of tombs and temples by the scribes of the Pharaohs were the true ancestors of our alphabet. If we turn the *A* upside down and write the *G* in the form of the Greek gamma, *Γ*, we may recognize in them the horned cow's head and the neck and head of the camel. Their names make this absolutely clear, however: Hebrew *álef* > Greek *alpha*, "cattle"; Hebrew *gimel* > Arabic *djamal* and *gamal*, our word "camel."

We may also turn to China, where still today pictures of objects are considered "written" symbols, or ideograms; the character for "tree," for example, is in fact a tree with trunk, root, and crown (see Fig. 1). Thus, no further proof is needed for the assertion that the first written symbols were pictures of the objects they represented.

But numerals are much closer to us in time. They can be of various different kinds, as we shall see in the course of our discussion; those of one kind are definitely of pictorial origin. A glance at Fig. 8 will show that the ancient Egyptian symbol sign for 10 was a horseshoe-shaped stanchion for fettering animals; that for 100 a measuring cord of 100 ells, shown rolled up as depicted in illustrations of Egyptian surveyors; that for 1000 a lotus blossom; and that for 10,000 a finger. We are not now concerned with the significance of these symbols apart from their pictorial quality, but this leaves nothing to be desired.

An excellent example of the way measures, coinage, and pictorial representation are united is provided by the so-called marriage tablets from the Kai Islands, south of New Guinea (Fig. 24). On these carved tablets the groom and the father of the bride inscribed the price that had been paid for the bride years before, when the bride was purchased as a child; thereafter she remained in her parents' house until the purchase price was fully paid. On the tablets a disc stands for a bronze gong, a ring for a bracelet, a pair of "tongs" for an earring, a supine animal for a goat worth 10 Dutch guilders, and a toothed bow for a golden comb. The second tablet (on the right) shows very clearly how the bride price was paid off in

Fig. 24 Marriage tablets from the Kai Islands. The groom and the father of the bride mark the bride's price on the tablet. An example of the interconnection between measure, coinage, and pictures. Size 41 × 16 cm and 86 × 14 cm. Linden Museum, Stuttgart.

Fig. 25 Symbols for common fractions. Babylonian half and half, Egyptian half and quarter.

Fig. 26 Number beasts. *Top*: tadpole, the Egyptian symbol for 100,000; *bottom*: scorpion, the Chinese sign for 10,000 on the obverse side of a 6th-century coin.
Schlösser Collection, Hannover.

installments: the objects already paid have been cut out, leaving blank spaces.

Early signs for fractions also derived from visual symbols and developed from measures. Thus the Babylonian sign for one half was a measuring vessel whose contents are shown as cut in half (Fig. 25). The ancient Egyptian half is clearly the "half," i.e. one side of "the whole," just as our own word "half" also originally indicated "one of two side(s)" and in this manner first acquired its meaning of half; the Low German expression *up der halwe lîn* means "to lie on one's side" and hence "to be ill." The Egyptian sign for one quarter is a recumbent cross, which as a hieroglyph means both "to break" and also stands for a "doubled half," or one quarter, represented by the four parts, the four arms of the cross.

This category also includes the high "number beasts" whose pictures and names are used by some peoples as numerals or number words. Thus, for instance, the Egyptian symbol for 100,000 is a tadpole, such as wiggle by countless numbers in the mud of the Nile when the water retreats within its normal banks after the yearly flood (see Fig. 26 for tadpole symbol in the hieratic script and Fig. 9 for the same sign as a hieroglyph). Its name, *hfn*, also means "innumerable." Similarly the scorpion, because of the large numbers in which it occurs and perhaps also because of its many feet and segments ("thousand-foot," milliped), in China became the numeral, and its name *wan* the number word, for 10,000. Our illustration of its appearance on an ancient coin of the 6th century shows the creature, with its two antennae and its barbed tail, in a form that is easier to recognize than its counterpart in modern Chinese (see pp. 450 ff.).

Another fascinating instance is the reverse of the process in the Indian name for the lac louse (*Tachardia lacca*) which, like an aphid, always occurs in "countless" multitudes and whose secretions form a hard surface (shellac) on certain plants — its name derives from the Sanskrit number word *lakša*, lakh, 100,000.

To these examples we can also add the Greek word for 10,000, which from *myríos*, "countless," was condensed down to the specific number word *mýrioi* and was probably derived ultimately from the Greek *mýrmex*, "ant." The shift of the accent was made by grammarians who wanted to differentiate between the specific number word and the indefinite word "countless." We may also wonder whether the Avestan *baevar*, "beaver," which also means 10,000, acquired its numerical meaning because of the abundance of these animals.

Two number concepts grew out of the Hebrew *alef*, "ox," by coalescence. The letter *alef* (Greek *alpha*), originally an oxhead, had the value of 1 as the first letter of the alphabet (we shall have more to say about this later: see pp. 262–267, especially the table of Fig. 95). The Hebrew word for 1000 is *elef*, obviously the same root as *alef*. The connection between the two is probably this: The letter's original form was the picture of an oxhead, and so it was called *alef*, "ox." Once the indefinite "many" was visualized as a herd of oxen (a quantity), people also began to say "cattle," Hebrew *alef*, for "many," and the word after a time came to be the specific number word "thousand."

As a complement to our numerical bestiary, we have a "number plant" in the Egyptian lotus blossom, which was used as a sign for 1000 (Figs. 8 and 9).

In summary, we can say that early man, in speech as in writing, drew on his environment for symbols to indicate sizes and quantities; he did not fashion artificial words or symbols to apply only to specific numbers and measures but had no intrinsic connection with them.

Now we can hope for insight into the meaning underlying our own number words "one," "two," and "three."

SEMANTIC FADING

Eene deene Bohnenblatt,
unsere Küh sind alle satt . . .

"*Eene deene* bean and leaf
All our cows are satisfied."

Here and there German children still count things off with this jingle. But what are children's nonsense rhymes doing in a scholarly study of written numbers? We are faced with a conundrum: We know what beans, leaves, and cows are, but what does "eene deene" mean?" Some might say smilingly that this is simply childish gibberish and nothing more.

But then we find a cousin of this German rhyme in America:

Eeny meeny miny mo,
Catch a nigger by the toe.
If he hollers let him go;
Eeny meeny miny mo,

and we are forced to admit that this American children's verse is very much like the German one. And if we now look at still another example, which English children recite in India and South Africa as well as at home:

Une dune des
Catlo wuna wahna wes
Each, peach, muskydom
Tillatah, twenty-one,

then, in spite of some differences, we must again acknowledge the similarity, and we can no longer dismiss *eene deene* as mere nonsense and beneath the notice of serious students.

Perhaps we can get at the meanings of these obscure words if we look on them as number words. The English *one, two, three* do not get us very far; but let us try the French:

Une, deux(-ne), trois
Quatre-cinq, six-sept, huit-neuf, dix;

then if we rewrite the last two lines of the English rhyme as

Each speech must be dumb
Till I tell twenty-one,

we have one of the best and clearest examples of a counting rhyme.

Our English rhyme makes sense only as far as *catlo*. The following explanation may be found plausible. Because of the rhythm, *deux* has also taken on the *une* ending *-ne*; "one-ah, two-ah, three." *Catlo* has the same meter as *quatre-cinq*, but the subsequent syllables do not fit the pattern at all. Yet *wahna* has the same ring as *huit-neuf*. Then *wuna* may be understood as an altered *six-sept* changed by remote influence of the words next to it, as often happens in children's verses:

Eins zwei drei,
hicke (?) hacke hei!

Similarly, *wes* and *des* are explained as syllables exchanged for number words: *des* < *dix* and *wes* < *(tr)ois*. Now the two first lines of nonsense words reveal themselves as number words whose meanings have faded out and been lost.

What is the history of this rhyme? We know that in the year 1066 the Normans invaded England and settled there as lords over the earlier Anglo-Saxon and Celtic inhabitants, and centuries passed before they mingled completely with the people they had conquered. Thus the *un*, *deux*, *trois* came to England, where its sound remained as alien as its sense; for the English count *one*, *two*, *three*. In this way the number sequence in the children's verse became eroded to the point of unintelligibility; if this were not a counting rhyme, no linguist would read the French *deux* into the syllables *duna*.

Two more examples show the connections between general history, children's verses, and the number sequence. In the German province of Hesse, in Mainz, this verse is still in circulation:

Eene deene dus manee
Riwwele rawwele sondernee.
Ecke Brot, sonder Not,
Dusee.

The first and last lines sound French: *dus* < *douce*? ("sweet"), *manee* < *monnaie*? ("money"), and *dusee* < *douze* or *douzain*, "twelve" or "dozen" — are there twelve number words? The second line is metrical gibberish, but the third line can be understood: if a man has an *Ecke*, "corner" (piece), of bread, he will not be in want (*Not* = want). Mainz was occupied by the French many times in its history; the children heard a strange language and composed a jingle whose beginning and end were the foreign number sequence but whose middle came from their own language. Here too we see the surprising longevity of these unintelligible foreign phrases.

There is still another impressive example. In the county of Lincolnshire, which today is a purely English-speaking region in the eastern part of the Midlands, sheep in the fields and stitches in knitting are counted in a very peculiar fashion:

1	*yan tan tethera pethera pimp*	5
6	*sethera lethera hovera covera dik*	10
11	*yan a dik tan a dik — — bumpit*	15
16	*yan a bumpit — — — figitt*	20.

What is this? The numbers at the ends of the lines, 5, 10, 15, and 20, show quite clearly that this is Celtic; we need only compare these with the Welsh number sequence (p. 97). But how does Celtic suddenly pop up right in the middle of an English-speaking area? There is no doubt that Celts previously lived here who were later driven westward, into Wales. Thus the traces of a historical event, for which there is perhaps no other documentary evidence, have been tenaciously preserved in this number sequence. Here, where the number sequence was not officially preserved in its pure form but was left to the common people, we have a beautiful example of the way adjacent number words and others in the sequence work on each other: 4, *pethera* < *pedwar*, has expanded 3, *tri*, into *tethera*; the dactylic rhythm, however, has also implanted itself in the numbers 6 through 9; the vowel change from *e* to *o* in 8 then extends to 9 as well, and so on. This vernacular number ditty suggests a whole series of linguistic observations that can equally well explain the "accepted" number sequences of a language.

The old ways of counting of course live on in children's verses too, sometimes altered to the point of unintelligibility, as in

Eetern feetern penny pump,
All the ladies in a lump.

The words of the second line, which is governed primarily by the requirements of rhyme and rhythm, do not make much sense; but in the first line, *eetern* < *een—teen*, 1, 2; *feetern* < *tether-em*, 3, *penny pump* < *pethera pimp*, 4, 5.

Modern Greek provides an example of the erosion of number words that is taking place right now: In the popular speech of Greece today, the tens in the number sequence are *triánta, saránta, penénta, hexénta, hebdoménda, ogdónta*, and *ennenénta*, rather than *triákonta*, etc. as was shown in the table on p. 92.

Against the background of our larger subject, these small children's ditties are outstanding examples of how in the course of time words can have their forms and meanings eroded to unintelligibility and yet for this very reason continue to live on and move about without change, because fear of the unknown keeps them from being meddled with.

OUR INDO-EUROPEAN NUMBER WORDS

In the course of our search for the meanings of our own number words, we have found both encouragement: that images enter into concepts of number and magnitude — and disappointment: that words may gradually slough off their meanings until they become mere gibberish. The resistance shown by our number words to changes in form unfortunately tends to show an early fading out of their pictorial quality.

There is one hope for enlightenment, however, and only one: the relationships among the various Indo-European tongues. If our number sequence were a solitary orphan in the world of languages, with claim to neither ancestors nor relatives, we should be able to learn nothing at all about its origin and its past. But when we assemble all languages into a family gathering and compare them, we see a common cast to the features of each number word that we had scarcely noticed. This may now help us to discover their previously unrecognized similarities.

The First Two Ranks: the Tens and the Hundreds

Dekm-kmtóm. We must turn once more to the table on p. 107, to recall the similarity of the number word "ten" in the *Kentum* and *Satem* languages, derived from the original word *dekm*. Since two phonetic shifts intervene between *dekm* and the Modern German *zehn*, we shall eliminate the Germanic element in the table that follows (see pp. 92–93). In this comparison the reader may be surprised

	ten	hundred
	Kentum (c = k)	
Greek	*de-ka*	*(he) ka-to-n*
Latin	*de-cem*	*cen-tu-m*
Celtic	*de-c*	*can-t*
	Satem	
Sanskrit	*da-śam*	*śa-ta-m*
Old Slavic	*de-sē-ti*	*s-to-*
Lithuanian	*de-šim-t*	*šim-ta-s*
	Original Indo-European forms:	
	dé-km	*km-tó-m.*

to see the curious recurrence of the terminal syllable of "ten" as the initial syllable of "hundred."

The original Indo-European language has the root -*km*-: this consists of the laryngeal consonant *k* which is softened in the *Satem* languages to *s* or *š* (see p. 100), and the nasal *m*. The dot beneath the latter indicates that a short vowel *a*, *e*, *i*, *o*, or *u* is sounded with it: for example, in the German root *nmmt* a suppressed *i* turns the consonant *m* into a syllable and the root into the word *nimmt*, "takes." The changes which the voiced *m* of the original Indo-European language has undergone in the individual languages are known precisely: Indo-European *m* becomes Sanskrit *a*, Greek *a*, Latin *em*, Lithuanian *im*, Slavic *ē*, and, let us note for future reference, Germanic *um*. Having confirmed this from our table, the reader will understand the formulation *dekm — kmtom.*

Now, without going to the trouble of listing all the individual cases, we can simply say: The Indo-European word for 100 is derived from the word for 10: *dékm → kmtóm.*

Just as the second rank is intrinsically bound up with the first in the arrangement of the number sequence, the two are linguistically connected as well. But in which sense? All the "hundred" words contain the syllable -*to*-, which also occurs in the Slavic and Lithuanian for "ten." We have already mentioned this phenomenon (p. 81):

it turns the number word "ten" into the noun "ten-ness." The initial syllable *dé-* has dropped out of the *(d)-kṃtóm*; the stress accent has shifted to the last syllable. The Balto-Slavic "ten" form *desēti*, *dešimt* and the Sanskrit *daśatih* (p. 81) likewise mean the same as "hundred": a "ten-ness" of tens; the corresponding respective "hundred" forms have merely dropped the *de-* (see the table on p. 93).

Thus we arrive at the surprising conclusion that in the Indo-European languages a "hundred" is a "ten-ness" (of tens). This means that the second rank is not just somehow connected to the first, but that this relationship is the original ancient form, the grouping. This is the strongest evidence for the view stated previously (p. 46) that the first major gradations were built into the number sequence as groupings.

Now, what has happened with the German words *zehn* and *hundert*? These fail to fit the general pattern that is condensed in the table on p. 126, to a degree that no connection between them, such as can be seen in the other languages, would ever be suspected. This is the result of *phonetic shifts*.

The Two Germanic Phonetic Shifts. The *first* of these shifts, which took place sometime between 400 B.C. and A.D. 100, set the Germanic group of languages apart from the original Indo-European (and hence is called the *Germanic* phonetic shift); the *second*, which came to an end around A.D. 600, differentiated High German from the Low German, which preserves the old Gothic sounds. The line of separation between *Hochdeutsch* and *Plattdeutsch* runs through the center of Germany, from Aachen through Düsseldorf, Kassel, and Magdeburg to Frankfurt on the Oder (and has been called the *dat-das* or the *Benrath Line*, because it crosses the Rhine at the town of Benrath).

A phonetic shift, generally speaking, registers a regular change in a phoneme, for example, in the case with which we are already familiar, the alteration of the Indo-European laryngeal *k* into the sibilant *s* in the *Satem* languages. The most important of these phonetic laws apply to the so-called plosives, namely the hard consonants *k*, *p*, and *t* and the soft *g*, *b*, and *d*. The following tables present these phonetic shifts visually. Because the phonetic structure of Latin is quite

Table of Phonetic Shifts

	Indo-European	I Germanic	II High German initial	medial
1	k	*ch, h*		
2	p	*f*		
3	t	*þ*	*d*	
4	g	*k*		
5	b	*p*	*pf*	*f, ff*
6	d	*t*	*z*	*ss*
7	(Greek th)	*d*	*t*	

similar to that of Indo-European, I have chosen to start with the more familiar Latin words instead of their Indo-European equivalents. The material has been made as palatable as possible, so that the reader will not shy away from the effort of acquainting himself with these phonetic shifts. His reward will be a clear insight into the ways in which both German and English are thoroughly interconnected with the whole Indo-European linguistic family. These few phonetic equivalents are the most important threads by which German, for example, is drawn out from its isolation and woven into the great tapestry of Indo-European languages.

It will be seen that:

1. The Germanic phonetic shifts I — 1, 2, 3 in the table changed the three hard Indo-European plosives *k*, *p*, and *t* into the aspirated consonants *ch* (as in German "*ach*"), *f* and *þ* (*þorn*, the *th* in the English thorn), and the soft plosives *g*, *b*, and *d* into their hard equivalents (I — 4, 5, 6).

2. The second, or German, phonetic shift, labeled II, changed only some of the sounds, including the Germanic *d* into the German *t*, and in many cases differentiated between the initial and medial plosives (phonetic shifts II — 5, 6).

The next table gives examples in which the Germanic group is represented by Gothic and English, whose phonetic structures are closer to the original Germanic than those of other Germanic languages (the phonetic shifts referred to are those in the preceding table):

Indo-European >	Germanic	>	High German		
Latin	Gothic	English	Old	New	Phonetic Shift
cord-	*hairto*	*heart*	*herza*	*Herz*	1; 6
pater	*fadar*	*father*	*fater*	*Vater*	2; 7
turba	*þaurp*	*thorp*	*dorf*	*Dorf*	3; 5
ager	*akrs*	*acre*	*ackar*	*Acker*	4
papa	—	*pape*	*pfaffo*	*Pfaffe*	5
edere	*itan*	*eat*	*ezzan*	*essen*	6
Greek *thýra*	*daur*	*door*	*tor*	*Tor*	7
tincta	—	—	*tincta*	*Tinte*	No phonetic
praedicare	—	—	*predigon*	*predigen*	shift.

The last two examples shown are words that first entered into the German language after the second (German) phonetic shift and thus preserved their original sounds (*p-d* in the second) without change (to *f-t*). The German number words from *eins*, "one," to *neun*, "nine," which obey these phonetic laws, can be seen readily from the Table on p. 146.

The Germanic 10 and 100. Now let us change the Indo-European *dekm̥* and *km̥tom* into their Germanic forms according to the phonetic laws we have observed:

Indo-European *dekm̥* > Gothic *taíhun* (phonetic shift I — 6; 1),
Indo-European *km̥tom* > Gothic (c)*hund* (phonetic shift I — 1).

Taíhun is pronounced "tä-chun," the *ch* being voiced as in the German *acht*, "eight." Instead of the *-d* in *hund*, one would expect to see *-þ* from *t* (according to phonetic shift I — 3). This change holds true when the *t* follows an accented syllable (similarly with *k* and *p*). If however it precedes the stressed syllable as it does here in *km̥tom*, it is changed to the soft voiced *d* (as are *k* to *g* and *p* to *b*). Thus in front of a stressed syllable the hard plosives *k*, *p*, and *t* become the soft voiced *g*, *b*, and *d*, respectively (according to the rule known as Verner's Law). Still another example: Indo-European *fráter* > Gothic *broþar* (phonetic shift I — 3), but Indo-European *patér* > Gothic *fadar* and Greek *dekás* > Gothic *tigus* ("a ten, ten-ness"). Thus phonetic shifts have given us the Gothic rank levels *taíhun* and *hund*, the latter used only from 200, *twa hunda*, on up. These are the words used in the Gothic Bible (John 6:7; see Fig. 27):

twaim hundam skatte hlaibos ni gano (hai),
"Two hundred pennyworth of bread is not sufficient."

Fig. 27 "Two hundred," *twain hundam* (·) in the Gothic Bible. Lines are 14 cm long.

The Gothic word for "hundred," however, is not *hund* but is represented by two full word forms: *taíhun-taíhund*, "a ten of tens," in which the grouping "ten" is counted. Hence, it testifies to the primitive idea which clearly formed the "hundred" from the ten. This is a remarkable find of historical evidence, preserved for us by the Gothic language alone.

Besides this, however, there was also the (probably original) Gothic form *taíhunt-e-hund*, meaning something like a "ten-ness" of tens (not of hundreds!). This form affects the tens in the Gothic sequence all the way back to 70 (see table on p. 92). In the Biblical parable of the unfaithful steward (Luke 16:6 and 7), we see all three formations peculiar to the Gothic number sequence:

jah gasitands sprauto gamelai fim tiguns,
". . . sit down quickly, and write fifty;"

taíhuntaihund mitade kaurns,
". . . an hundred measures of wheat;"

nim þus bokos jah melei ahtaútehund,
"Take thy bill, and write fourscore."

Later we shall have more to say about the curious break in the Gothic formation of the tens between 50 and 80 (see p. I/163). Here we are concerned only with the formation of the word *ahtaút-e-hund*. The first part is the grouping "eight-ness"; the *-hund* is likewise quite clear. The connecting *-e-* is also found in the Greek tens, such as *hex-é-konta*, 60, for example. Because of this formation the Indo-European "ten" is also given with a final *-t*, as in *dekm̥(t)*, so that the

Gothic ten-form -*hund*, and above all the Greek -*konta* and the Latin -*ginta*, can be derived regularly from it.

This interpretation sheds a great deal of light on early ideas of numbers, for it shows not only the close connection between "ten" and "hundred," but far more — that the second rank level, 100, was formed verbally by a repetition, "ten-ten," of the already existing rank "ten." And the Indo-European *kṃtom*, 100, is thus also derived from "ten-ten":

dkṃt-dkṃt > dkṃt-(dk)om(t) > dkṃt-om > kṃtóm.

The German "zehn" and "hundert." Now we can arrive at these number words in German by applying the second, or German, phonetic shift, to the Gothic *taíhun*:

taíhun > Old High German *zehan* = New High German *zehn* (phon. shift II — 6)
hund > unchanged New High German *hund.*

Luther once wrote: *Die andern straffen und penen / solt man tzehenn ell tief begraben in die erden*, "... buried *ten* ells deep in the ground" — putting the Low German *t* and the High German *z* together in one word, as it were. Many German dialects preserve the palatal consonant in the word "ten," as in the Carinthian *zögn.*

But where does the German form *hund-ert* come from? The Gothic word "to count" is *raþjan*; *raþjo* means "reckoning, number" (see Latin *ratio*). In telling the story of the loaves and the fishes, the Gothic Bible (John 6:10) says (Fig. 28):

wairos raþjon swase fimf þusundjos
men number about five thousand
"So the men sat down, in number about five thousand."

The Gothic word was borrowed from the Latin *ratio*, "calculation," with the expanded meaning of "consideration, deliberation, reason," from *reor, rei, ratus sum*, "to reckon, consider, estimate, appreciate, believe." The numerical sense of the word is present in "rate" (< *rata pars*), the "calculated portion," and mathematicians use it in their rational — (that is, proportional) numbers. From Gothic, Old High German derived the word *redja* (Modern German *Rede*, "speech, discourse"); the expression *jemand Rede stehen*, "to answer to someone, give someone an account of ..." still has the connotation of setting down an account or reckoning.

Fig. 28 The words "number," *raþjon* (·) and "five thousand," *fimf þusundjos* (:) in the Gothic Bible. Lines are 14 cm long.
Figs. 27 and 28 from *Codex argenteus* Folio 48.

The Modern German *gerade*, in the sense of an "even" number, comes from the Gothic *ga-raþjan*, "to count."

Now if we follow the word "hundred" through other Germanic, Northern European, languages, we go back via the English *hund-red*,

Old Saxon *hunde-rod*, Old Norse *hund-rad* (*d* for voiced *þ*) to a Gothic word *hundra-raþ*, which means "hund(red) number" or "the number hundred."

The German word *hundert* first came into use in the 12th century, before which people said *zehan-zo* or *zehan-zug*, following the Gothic pattern of repeating "ten of tens" (*-zo, -zug*, see p. 96). Thus *zehanto-herosto* in Old High German means "leader of 100" or "captain," one who acquired his title after the model of the Roman *centurio*, commander of a "century," a military unit of 100 soldiers. The writer Notker of St. Gall (ca. A.D. 1000) once explained the name of the Greek goddess Hekate, who was among other things the goddess of growth or increase, by the number 100 (Greek *he-katón*):

Heizet si ouk Echate daz chit centum, uuanda der erduuuochr ofto chumit zen-ze-faltiger
"She is also called Hekate, that is *kentum* ["Hundred"], because through her the fruit of the earth often grows ten-ten-fold."

The simple Old High German *hunt*, which as in Gothic is used for the numbers from 200, *zwei hunt*, on up, was relatively late in replacing the word *zehanzo*.

The Image Latent in "Ten." Now that we have explained the origin and development of the first two rank levels, there is one more question to be answered: If "hundred" is actually a "ten of tens," what then does "ten" itself mean? What sort of image or idea does this number concept embody?

Unfortunately here we are not on such solid ground as before. The meaning of the Indo-European *de-km̥(t)* probably is "two hands." There are some quite substantial supports for this view. One is the role played by the hands in forming number words, as we have seen in various primitive cultures (see p. 35). In addition, the attached syllable *de-*, which in all Indo-European languages contributes to the formation of the number word "ten," could well be the shortened word *duo*, "two," whose *d-* is likewise very resistant to cnange in all languages. The dropping out of the semivowel *u* (= *w*, Gothic *t(w)ai-*) has analogues in the Greek and Latin two-forms *dis-*, respectively, *dia-* (see p. 175) in the Greek *dṓ-deka*, 12, and also in the pronunciation of the English word *two* as "too."

Further support is provided by the phonetic shift in Gothic:

de-km̥(t) > *taí-hun(d)*.

In Gothic, *handus* means "hand" and has virtually the same sound as the same word in modern German or English.

Thus it may not be entirely wrong to recognize *zwei-hand*, "two-hand," in the German word *zehn*, Old High German *ze-han*. And since the Gothic word *handus* is derived from *hinþan*, "to touch, grasp" (in many languages the word for "hand" means the "grasper," as in Old Norse *greiþ*), the word *de-km̥(t)* can be understood in its root meaning of "two grips" or "two handfuls." If this is so, moreover, then the English words *hunt* and *hound* and the German *Hund* "dog" — although at first this would have seemed ridiculous — may well be related, if very distantly, to the numbers "ten" and "hundred."

The Third Rank: Thousand

The Germanic "Thousand." For the number 1000, our tables on pages 92 and 93 show Greek *chílioi*, Latin *mille*, Germanic *þusundi*, Tocharian *wälts*, Sanskrit *sahasram*, Old Slavic *tysěšta*, Baltic *tukstantis* — a variety that makes it hard to believe that there could be any relationship between these words like that which was so strikingly revealed between the words for "hundred." And, in fact, with the "thousand" such a relationship no longer exists.

The linguistic bridge constituted by the first phonetic shift led us from the Latin *centum* without strain to the Germanic *hund*; yet in crossing the stream from the Latin *mille* we find no foothold on the Germanic shore: *þusundi*. Thus the Germanic tongues must have turned away from Indo-European sometime before the number sequence reached the 1000 level.

As our number sequences indicate, *tausend* or *thousand* is a Germanic word (see tables pp. 94–95). Although the word itself at first does not suggest any connection with *hundred*, "thousand" definitely resembles the Old Norse *þushundrad*. Hence "thousand" is a *þushundrad*. But what is a *þus*? If we trace the sound *þ* back from the Germanic languages, we arrive at an Indo-European root *tu*, which means "swelling" (phonetic shift I — 1, p. 127). From this root are derived, among others, the Sanskrit words *tavas*, "strong" and *tuvi*, "much"; Avestan *tuma*, "fat, plump"; Latin *tumere*, "to swell" (cf. "tumor"), as well as the Anglo-Saxon *þuma*, Old High German *thumo* and English *thumb*, which is literally the "stronger finger" and is thus distantly related to "thousand." Hence our "thousand," which can be traced through the intermediate steps of Modern German *tausend* to Middle High German *tusend* < Old High German *dusunt* and *thusunt*, Gothic *þusundi* > English *thousand*, is nothing more than "many hundred," a "strong" or a "great" hundred.

If we go back to the root meaning of "hundred," the first three rank levels in the Germanic languages can be arranged in ascending order semantically as well as grammatically:

German: *zehn > Gezehnt   > starkes Gezehnt*
English: ten > ten of tens > strong ten of tens.

Even if we disregard the perfectly plausible meaning of "ten" as "two-hands," we can still observe here a beautiful example of how the perfected number sequence grows slowly and gradually out of the simplest and most basic conceptions of ordinary people: no conscious planning, no insight into the laws of gradation, no knowledge of infinite progression — in other words not a deliberate intellectual act but a gradual advance of the number sequence, an unconscious accretion from one level to the next, induced by nothing more than expanding necessity. Languages provide remarkable testimony to this slow and tedious growth, for beyond "ten" they do not devise an essentially new number word but follow along in the same tracks and derive the new "hundred" from the old "ten," which has merely been grouped into a "ten of tens."

"Many hundred" at first is "many" or "countless many," because it proceeds through that stage to the numerical hundred. Then the limit of counting is gradually pushed farther outward so that it increasingly displaces the wavering shadow of the indefinite "many." The next step brings us to a knowledge of the — but now there is no word for it! As the fluctuating, indeterminate "many hundred" was absorbed into the established number sequence, it was condensed to the *specific* number word "thousand."

Semantic Specification. We must now deviate from our main course for a moment and say something about this important piece of evidence for the step-by-step growth of the number sequence. The phenomenon of semantic specification occurs in many languages. A word for "many" or "numberless," such as the Greek *myríos*, whose semantic content is unspecified or variable, becomes a number word with a clear and fixed meaning like the Greek *mýrioi*, 10,000. The intangible and tenuous idea is *condensed* or *specified* into a fixed, solid numerical concept.

Homer was familiar with *chérados myríon*, "pebbles without number" (*Iliad* XXI, 319), but not with the specific meaning of this word, for he wrote *déka chílioi*, 10'1000. The word *mýrioi* first appears in Hesiod and Herodotus. We mentioned the probable derivation of this word from *myrmex*, "ant," when we spoke of the "number beasts" whose names or pictorial symbols have been condensed into number words or numerals; the reader may refer to these again on p. 122. Here we shall cite a few more illustrative examples, beginning with the following one from Hebrew. In the Old Testament (I Samuel 18:7,8) it is said that

"Saul hath slain his thousands, and David his ten thousands. And Saul was very wroth... and he said 'They have ascribed unto David ten thousands, and to me they have ascribed but thousands'"—

    nathnu   *l-Dawid*   *rebaboth*   *w-li*
"they have ascribed unto David ten thousands and to me

    nathnu   *ha-alafim,*
they have ascribed thousands."

The sense of this passage is an augmentation: If Saul killed a thousand, David slew many, many more, "thousands without number." In fact the word *rebaboth* was condensed from this indefinite original meaning to the fixed 10,000, as we have already mentioned (see p. 116). Luther translated the number in Psalm 68:17 into the old indefinite meaning of "many": "The chariots of God are many thousand times a thousand":

    recheb   *elohim*   *ribbothayim*   *alfe*   *šin'an,*
literally: the chariots of God two-ten-thousand thousands repeat. (King James version: "The chariots of God are twenty thousand.")

Here the grammatical dual very clearly expresses the plural "many," and semantic specification thereby takes a further step away from antiquity (see p. 14).

The original meaning of the Tocharian *tmam*, 10,000, is also "innumerable"; this is related to the Old Slavic word *tuma* (Russian *tma*), which means "dark." The Tocharian word *wälts*, 1000, is likewise contained in the Old Slavic *velije*, "great, mighty."

That the indefinite plural of a number word can be specified to a fixed meaning, something like the manner in which "many tens" became "twenty," is shown by excellent examples from both Hebrew and Danish, which we have already seen in a different connection (p. 14). Moreover, the Akkadians took their number word *meu*, 100, from the Sumerians, in whose language it served, in the form of *me*, to indicate the plural of any word. Thus the number sequence of the Akkadian invaders had not yet advanced to 100, whereas the Sumerians were already counting up to 60^3 — a striking example of the way cultural events in the history of two peoples can be recognized in their number sequences.

An uncommonly fascinating instance of such specification is presented by the Japanese word for 10,000, *yorozu*. In the Japanese number sequence *yo-tsu* stands for 4, the *-tsu* being merely a suffix that identifies it as a number word (see table on p. 450 and pp. 452 ff.). *Yo* originally meant "many" or "numberless" and was condensed from this to the specific "four." But then the original meaning was revived, so to speak, as the number sequence advanced, and higher number words had to be found: *Yo* was joined to the syllable *ro*, which indicates that the objects exist in an unspecified number. Thus *yo-ro* literally has the tautological meaning of "indefinitely many" and then, specified by the number suffix *-tsu*, became the fixed number word for 10,000.

There is an instructive analogue to our "thousand" in the language of the gypsies. Here *deš* is 10 and *šel* is 100 (see p. 191). For 100 the gypsies in Wales say *baro deš*, "great ten," and for 1000 those in Russia say *baro šel*, "great hundred" — exact parallels of the expression "strong hundred."

Although this extension of an existing number word and its subsequent specification took place in prehistoric times in the case of our "thousand," the same process in the case of a million is much closer to our own time: what is a "million?" nothing but a blown-up Latin *mille*!

Compare the Italian *padre–padrone, sala–salone*; the ending *-one* has an augmentative effect. *Padrone* is a patron or protector, and thus a heightening of "father"; *salone* is a "great" hall. Similarly *milli-one* is a "many-thousand," the exact counterpart of the "many hundred" in our "thousand." Only one thing is different: "million" is a synthetic word. It may be natural to an Italian, but not to an Englishman. The further steps "billion," "trillion," and so on, are just as artificial in Italian.

Examples of such specification can be found not only in the history of language, but also in that of writing. To the "number beasts," which should be recalled here (see p. 122), we must add the Babylonian numeral for *šar*, $3600 = 60^2$, and thus for the second rank in the sexagesimal system (see Fig. 35). This is a large circle, formed by four cuneiform wedges, and originally represented "universe" or "cosmos" and thus stood for an indefinite, limitless abundance. Another analogue is the Egyptiah *hh*, "infinitely many" (see Fig.

29), represented by the god of space *Hh*, whose pictorial symbol acquired the fixed meaning of million, but later, when the culture of the ancient Egyptian empire declined as a result of foreign invasions and its range of numbers shrunk, it was again relegated to the meaning of "infinity."

Fig. 29 The Egyptian god of air space — the symbol for million.

Other Forms for 1000. Now let us return once more to the linguistic history of the third numerical rank. We shall look for explanations of "thousand" in the other Indo-European languages.

The only other forms belonging to the same class as the Gothic *þusundi* are the Balto-Slavic forms: the Lithuanian *tukstantis* and Old Slavic *tysēšta*; these may have been borrowed from the Germanic languages. Number sequences have advanced by taking over foreign number words just as often as by semantic specification of native terms. Besides specification, the *borrowing* of a rank is another important piece of evidence for the gradual advancement of the limit of counting; we shall have more to say about this in its proper time (p. 190).

But what is the situation with the other 1000 forms (Sanskrit, Greek, Latin, Tocharian, and Celtic), which have nothing whatever in common with the Germanic "thousand?" Those of the first three languages listed, *sa-hasram, chílioi,* and *mille,* are all interrelated and are derived from a single Indo-European ancestor, *sm̩-gheslo-m,* in which the element *sm̩* means "one" (see p. 93). From this derives the ancient Sanskrit *sa-hasra-m,* in which the *l* has changed to *r* (and also the Avestan *ha-zanra* > Persian *ha-zar*). The prefix "one" occurs again in the Greek *he-katón,* "one hundred" (Indo-European *s-* > Greek *h-*; cf. p. 146). Then from the Indo-European root *-gheslo-* came the Greek *chésloi,* with the dialectic variants: Lesbian *chéllioi,* Doric *chélioi,* and Attic *chílioi.* The Indo-European feminine form *(s)mi-ghsl-i* gave birth to the Latin *mixli,* which later was smoothed out to *mille.* We know of similar verbal polishing in such words as Latin *seni* < *sex-ni,* "every six," and *ala,* "wing" < *axilla,* "axle" < Indo-European *aksla.*

Thus the Sanskrit, Greek and Latin, "thousand" actually mean "one *gheslom.*" But what is a *gheslom*? That remains obscure. Attempts have been made to connect the Old-Sanskrit *sahasra* with the Sanskrit *saha-,* "power, strength, victory" (Gothic *sigis,* "Sieg"-fried, "Vic"-tor?, from an Indo-European root *segh,* meaning "victory"), which would render the "thousand" a "force"-hundred and thus to draw it into the same semantic area as the Germanic *þus-hundi.*

The Tocharian *wälts,* 1000, seems to point in the same direction if it is connected with the Tocharian word *wäl,* "prince."

If we add that the Celtic *mil* is borrowed from the Roman *mille,* we then have traced all the Indo-European words for the rank level of 1000 back to their root meanings.

The Higher Ranks

The Number Towers of India. If the words for the third numerical rank show little relationship among the Indo-European languages, there is none at all in the case of the higher ranks. Some individual languages still form words for the fourth rank, 10,000: e.g., Greek *mýrioi*, Tocharian *tmam*; in the Germanic group, however, "thousand" remained the highest indigenous number-word, for "million" and the subsequent ranks are alien and artificial creations that did not grow out of the vernacular.

Only a few peoples in the world have devised words for the higher ranks of numbers out of their own vocabularies. We have seen the Egyptian example (p. 134), but we also saw how this likewise lost the impulse to climb higher up the ladder of numbers. The ancient Egyptians went no higher than a million, and they later retreated from this last rung. But one Indo-European people, in contrast to all others, uniquely erected the boldest tower of numbers in existence: the Hindus. It is curious that one people is inclined in a particular direction while another follows a quite different tendency. The Indians have shown a unique gift for abstract numbers. They knew of no better way to honor sacred divinity than counting, which they esteemed far above all merely human endeavors. Here is the "manifestation" of Gautama as the Buddha (the Enlightened One):

> The Buddha will be recognized by 32 principal and 80 secondary symbols, his mother by 32, and the house in which he will be born by 8. His mother, the Queen Maya-Devi, will be attended by 10 million ladies-in-waiting. Hundreds of thousands of holy men and hundreds of thousands of millions of the enlightened will pay homage to the Buddha. His throne is constituted of good works performed during hundreds of thousands of millions of *kalpas* [i.e., mythical periods of 4320 million years]. But the great lotus which blooms during the night of the Buddha's conception opens its blossom to an extent of 68 million miles."

These sacred numbers, except perhaps for the lowest ones at the beginning of the passage just quoted, are not mythical or mystical numbers like 7 or 13, but they are sanctified by their very magnitude, which leaves all mortal things far behind. There is no more striking example of the Indians' intrinsic affinity for numbers than Buddha's Temptation. This legend is contained in the book of the Lalitavistara, which tells the story of the Buddha's life. It is the strongest proof of the Indians' knowledge of the infinite progression of the number sequence, a knowledge which they did not obtain from experience and which was not forced on them, like the first three ranks, by economic demands.

When Buddha reached the age of manhood he courted Gopa, the daughter of the Prince Dandapani. But he would be rejected unless he could show public proof of his abilities. And so, together with five other suitors, he was put through trials in writing, wrestling, archery, running, swimming, and number skills. In all these contests he brilliantly defeated his rivals. After the competition, Gopa's father commanded him to pit himself against the great mathematician Arjuna, who was to be the measure of Buddha's knowledge. Arjuna instructed him to list all the numbers (that is, the numerical ranks) above 100 *kotis*. *Koti* was the name of the seventh rank, meaning 10^7 or 10 millions. Beyond *sahasra* (10^3), India at an early date had *ayuta* (10^4), *niyuta*, also called *lakša* (10^5), and *prayuta* (10^6 = a million).

Buddha answered:

koti (10⁷, abbreviated 7)	*hetuhila* (31)
ayuta = 100 *kotis* (9)	*karahu* (33)
niyuta (11)	*hetvindriya* (35)
kangkara (13)	*samaptalambha* (37)
vivara (15)	*gananagati* (39)
akšobhya (17)	*niravadya* (41)
vivaha (19)	*mudrabala* (43)
utsanga (21)	*sarvabala* (45)
bahula (23)	*visandjnagati* (47)
nagabala (25)	*sarvasandjna* (49)
titilambha (27)	*vibhutangama* (51)
vyavasthanapradjnapti (29)	*tallakšana* (53).

In Buddha's numerical ladder *ayuta* and *niyuta* refer to 10^9 and 10^{11}, respectively, not 10^4 and 10^5 as elsewhere. Indian religious poetry contains various number ladders of this kind, modeled after Buddha's, but the number words and their meanings show little consistency from one scale to another (see p. 47 for the formations with *maha-*). Their importance lies not in any clear conception of numbers (for who can visualize a number like 10^{53}?) but in their under-lying recognition that a tower can be built up, level upon level, and that the tower of numbers rises high above the world of mortals to the realm of the gods and the "Enlightened." For us, however, their greatest significance is the fact that their gradations clearly build up to invisible, superhuman heights, like the successive levels of an Indian temple (Fig. 30); and — most important of all — the ranks are named! Up to *koti*, 10^7, they are true decimal gradations; from there the numbers rise by levels of 100. The ranks up to *koti* must have been more useful and practical than the subsequent, artificially invented ranks. In other numerical ladders the decimal gradation is used throughout. (In this connection we may recall the Mayan "sacred" tower of numbers; see p. 62).

But Buddha is far from finished: this number tower is only the first, the *tallakšana* count; superimposed on this is the *dvadjadravati* count, above that there is the *dvajagranisamani* and above the latter there are still six more number sequences! What is the ultimate number of the last of these sequences?

First count:

$$10^7 \cdot \underbrace{10^2 \cdot 10^2 \cdots 10^2}_{23 \text{ times}} = 10^7 \times (10^2)^{23} = 10^{7+2 \times 23} = 10^{7+46} = 10^{53}.$$

Second count:

$$10^{53} \times 10^2 \times 10^2 \cdots \times 10^2 = 10^{53} \times 10^{46} = 10^{7+2 \times 46} \text{ etc.},$$

to the ninth count which in the end leads to $10^{7+9 \times 46} = 10^{421}$, and thus to the monstrous number 1 followed by 421 zeros.

Yet Arjuna was still not satisfied, and he gave Buddha a further

Fig. 30 Indian temple: clearly super-imposed, the structure rises up level by level.

test: to name all the divisions or atoms of a *yoyana* (a mile). Buddha began:

7 atoms make a very minute particle,	
7 very minute particles make 1 minute particle, and hence	(7^2) atoms
7 minute particles make one that the wind will still carry	(7^3)
7 such particles make 1 rabbit track	(7^4)
7 rabbit tracks make 1 ram's track	(7^5)
7 rabbit tracks make 1 ram's track	(7^5)
7 rams' tracks make 1 ox track	(7^6)
7 ox tracks make 1 poppy seed	(7^7)
7 poppy seeds make 1 mustard seed	(7^8)
7 mustard seeds make 1 barleycorn,	(7^9)
7 barleycorns make 1 knuckle	(7^{10})
12 knuckles make 1 handbreadth	$(12 \cdot 7^{10})$
2 handbreadths make 1 ell	$(2 \cdot 12 \cdot 7^{10})$
4 ells make 1 bow	$(4 \cdot 2 \cdot 12 \cdot 7^{10})$
1000 bows make 1 *krosa* (a measure: 1 "cry")	$(10^3 \cdot 4 \cdot 2 \cdot 12 \cdot 7^{10})$
4 *krosa* make 1 *yoyana* (mile)	

One mile therefore contains $4 \times 10^3 \times 4 \times 2 \times 12 \times 7^{10} = 384{,}000 \times 7^{10}$ atoms, a number I shall gladly let the reader work out for himself. To this Buddha added that just as he had numbered the atoms in a mile, one could also number all the atoms in all the real and mythical lands in this world and even in the 3000 thousand worlds contained in the universe.

This enumeration of particles is very likely the oldest example of calculation with vast numbers; it was known long before the Lalita-vistara was written (300 B.C.) and was perhaps included in one of the sacred books of India precisely because of its popularity. The Indians' adeptness and familiarity with such vast magnitudes is evidenced by the large numbers of such numerical towers in southern and northern India. We shall content ourselves with Buddha's tower of numbers.

After stripping away all the legendary elements, we are left with the Indians' successful attempt to devise a tool with which they could grasp such huge numbers verbally and thus also mentally. But conceiving such numbers verbally does not mean visualizing them; it means mastering the undefined, the innumerable, the obscure "many" by using the clear, verbally defined concept of the counted, of the number which falls within the scope of the short number sequence created by the requirements of ordinary life. The significance of such an intellectual attitude for the entire culture of a people, and the difference between the Hindu approaches to numbers and those of other peoples can be seen here very distinctly. The contrasting attitude of the Romans, for example, is revealed by Cicero, who with surprising perceptiveness compared his fellow countrymen with the Greeks:

In summo honore apud illos geometria fuit, itaque nihil mathematicis illustrius. At nos metiendi ratiocinandique utilitate huius artis terminavimus modum.

"By them [the Greeks] geometry was held in the very highest honor, and none were more illustrious than mathematicians. But we [the Romans] have limited the practice of this art to its usefulness in measurement and calculation."

We have already touched on the importance of having a name for each and every numerical rank in developing the perfected number sequence, but it may be stated here again:

Sanskrit 5 *koti* 7 *prayuta* 3 *lakša* 2 *ayuta* 6 *sahasra* 4 *šata* 3 *daša* 2

expresses the identification of the numerical ranks in exactly the same manner as the Chinese (see p. 54). If we eliminate the names of the ranks, the units alone will suffice to express this number according to the positional principle: 57,326,432.

Archimedes' Counting of the Sands. The European peoples did not generally have the Indian propensity for the poetry of numbers. The Greeks were more attracted by things that can be visualized and sensed than by abstract numbers. Yet it was a Greek who once "counted the sands," almost certainly after the Indian model: Archimedes, who lived in Syracuse in the 3rd century B.C. He was the most original and creative of all the Greek mathematicians. Foremost among his achievements was the "infinitesimal" analysis, related to the concept of infinity, by which he arrived at a large number of new insights into the measurement of the circle, the volumes of bodies bounded by curved surfaces, the areas of plane figures with curvilinear edges, and many others. Some two thousand years later the ideas with which he broke new ground became a logical part of the integral calculus developed in western Europe. The historian Plutarch relates fanciful stories of Archimedes' skill in engineering and of his mechanical genius (it was he who, among other things, discovered the principle of the lever and the law of fluid displacement). During the siege of Syracuse by the Romans he lifted ships into the air and set them afire with huge magnifying lenses. The story of Archimedes' death is famous. After the fall of Syracuse, as he was struck down by one of the Roman invaders while meditating on the diagrams he had drawn in the sand, he cried out: "Do not disturb my circles!"

In his "Sand-reckoner" (*Psammítes* < *psámmos*, "sand") Archimedes proposed to count the seemingly uncountable by creating a number sequence similar to that of the Indians, which could count all the grains of sand contained in a sphere the size of the universe:

Many people believe, King Gelon, that the grains of sand are without number. Others think that although their number is not without limit, no number can ever be named which will be greater than the number of grains of sand. But I shall try to prove to you that among the numbers which I have named there are those which exceed the number of grains in a heap of sand the size not only of the earth but even of the universe.

With these words, in which he clearly stated his intention, Archimedes began his treatise. Despite its resemblance to the Indian example, this was essentially a different undertaking: He did not wish to demonstrate any superhuman skill but merely to show by counting the sands that the ordinary Greek number sequence, which went only as far as *mýrioi*, could in principle be extended without limit, and that any imaginable number could thereby be expressed verbally. Archimedes did not create poetic fantasies out of his feeling for numbers but cogitated soberly about a number sequence that was to do nothing but count. This was the essential difference. It was also the difference between the solitary thinker contributing to the advancement of science and the religious tradition of the Indians with its powerful ties to the people.

In presenting the arrangement and structure of the Archimedean ladder of numbers, we leave it to the reader to find the similarities and differences between it and the "Testing of Buddha." The numbers from 1 to $10^8 = 10^4 \times 10^4 = a$, the product of *mýrioi × mýrioi*, was called by Archimedes the *first eight-power series*; *mýrioi* is the highest numerical rank in Greek. The final $a = 10^8$ becomes the unit of the second eight-power series, which in turn results in still another eight-power series:

$$a, a + 1, a + 2, \ldots, a + a = 2a, 3a, \ldots, 10a, \ldots, 10^8 a$$
$$= a \times a = a^2,$$

and so forth. Thus eight-power series succeed each other up to $10^8 = a$ powers:

$$a^{a-1}, a^{a-1} \times 2, \ldots, a^{a-1} \cdot a = a^a = 10^{8 \times 10^8} = p,$$

in which the largest number p is 1 followed by 800 million zeros.

According to Archimedes, all the numbers from 1 to p belong to the first period. This is followed by the second period, in which p is the lowest number:

1st eight-power series $p \times 1, \ldots, p \times 10^8 = p \cdot a$,
2nd eight-power series $p \times a, \ldots, p \times a^2$, and so forth, up to the
ath eight-power series $p \times a^{a-1}, \ldots, p \times a^a = p^2$.

These in turn are followed in the same manner by the periods p^3, p^4 and so on, to the $10^8 =$ the ath period:

1st eight-power series $p^{a-1}, \ldots, p^{a-1} \cdot a$, etc. up to the
ath eight-power series $p^{a-1} \times a^{a-1}, \ldots, p^{a-1} \times a^a = p^{a-1} \times p = p^a$.

This last number, the pinnacle of this vast tower, is the ath-number of the ath eight-power series of the ath period or, as Archimedes himself calls it,

hai myriakis-myriostâs periódu myriakis-myristôn arithmôn [eight-power series] *myríai myriádes*,

"the myriad of myriads of the myriad-myriadth eight-power series of the myriad-myriadth period."

Whereas the final number p of the first period is 1 followed by 800 million zeros, the largest number of all, p^a, has 10^8 times more, 8×10^{16} zeros.

Now Archimedes can begin his actual counting of the sands: how many grains of sand can be contained in the celestial sphere, at the center of which stands the earth and over which the sun and stars trace their paths? The problem is to compute how many times a particle of sand S with a diameter s can be divided into the space H of the heavenly sphere whose diameter is h, or $H \div S$. Since the volume of a sphere is a function of the cube of its diameter, $H \div S = h^3 \div s^3$.

Then Archimedes expressed the diameter of the celestial sphere in terms of the diameters of the grains of sand s:

$h = 10^4$ diameters of the earth of 10^6 stadia, each of 10^4 finger-breadths, each of 40 poppyseeds, each of 10^4 sand-grain diameters s, or

$h = 10^4 \times 10^6 \times 10^4 \times 40 \times 10^4 s = 4 \times 10^{19} s.$

Hence the number of grains of sand contained in the sphere of the heavens

$H : S = (4 \cdot 10^{19})^3 = 64 \times 10^{57} < 10^{59} = 1000 \times 10^{8 \times 7};$

that is, it is smaller than the 1000th number in the 7th eight-power series of the first period, or 1 followed by 59 zeros, and thus much smaller than even a single period!

But even this is not enough. Archimedes continues: The astronomer Aristarchos of Samos taught that it is not the earth but the sun which stands at the center of the universe, that the earth rotates around the sun, and that everything is contained in a hollow sphere upon which the stars move and which is much larger than the earth-centered sphere. Yet even for this sphere the grains of sand can be counted; their number is $10^{8 \times 7 + 7} = 10^{63}$, that is, 1 followed by 63 zeros, and thus smaller than the 8th eight-power series 10^{64}.

"I suppose, King Gelon," Archimedes concludes, "that all this seems unbelievable to one who is not skilled in mathematics, but to the mathematician it is quite clear. Therefore I have thought it worth your while to learn of it."

Great as the last number is, it is still far smaller than the last number of Archimedes' sequence. The whole point of Archimedes' discussion is that even so inconceivably large a number as that of the grains of sand contained in the universe can not only be clearly understood and verbally expressed, it can even be easily exceeded: the limitless progression of the number sequence has finally been recognized!

Scientifically recognized, we should add, looking back on the Indian tower of numbers, for the latter merely gives colorful names to the individual stories or ranks: *koti*, "peak"; *vardha*, "sea"; *padma*, "lotus-blossom." Archimedes, on the other hand, expresses only the gradation, without the colors and textures that are inherent in the Indian names; here, finally, we have the concept of abstract numbers put into words.

This number sequence, which was thought out in isolation by one individual, bears no traces of a slow, unsystematic growth, as it would if it had evolved gradually and anonymously out of the culture of a whole people; it is an artificial creation with an astonishingly logical and coherent structure.

The great mathematician Gauss had so much admiration for Archimedes that he said regretfully: "How could he have missed [the discovery of our present positional system of writing numerals?] To what heights science would have risen by now, if only he had made that discovery!" For Archimedes made his computations with the peculiar Greek letter numerals, about which we shall speak later.

But this, too, shows how even Archimedes, who in one respect was fortunate enough to rise to a level of thought far above that of his century, in other respects remained enmeshed in the intellectual fabric of his own time.

The Higher Ranks of Numbers in our Own Culture. Equally artificial is the continuation of the number sequences in western Europe through the higher ranks, the million, billion, trillion and so forth. Its genesis is one more striking documentation of the essential difference between the individually created and the evolved number sequence.

By the time the Germanic peoples first appeared on the scene of recorded history, they had already formulated the first three ranks, the tens, hundreds, and thousands, of their number sequence. The broadening of their commercial activity was naturally followed by an expansion of the domain of numbers; numbers above a thousand now cropped up everywhere. But popular language no longer kept up: the thousand was the last and highest named rank, and so it remained throughout the whole Middle Ages. "Ten thousand, hundred thousand, and thousand thousand" wrote the calculators when employing their vernacular scholars, who spoke Latin, followed the Roman fashion: *decem milia, centum* or *centena milia, mille milia* or *decies centena milia* (see pp. 28–29).

Adam Riese's book on arithmetic contains the number

86789325178

"This is six and eighty thousand thousand times a thousand / seven hundred thousand times a thousand / nine and eighty thousand times a thousand / three hundred thousand / five and twenty thousand / one hundred eight and seventy."

Symbolically: $86T(T \times T)$; $7H(T \times T)$; $89(T \times T)$; $3 H T$; $25 T$; 178. The most curious aspect in all this is the isolation of the hundreds, perhaps to set them off from the units and the tens in their inverted word order.

We see very clearly that 1000 is the highest level of the number tower in both our own and the Latin languages; if a higher number is required, another, identical, tower is imposed all over again upon the first. The Indians, however, kept on building the same tower higher and higher, level after level:

8 *mahapadama* 6 *padma* 7 *vyarbuda* 8 *koti* 9 *prayuta* 3 *lakša* 2 *ayuta* 5 *sahasram* 1 *śatam* 7 *dašan* 8.

Imagine what the number of particles contained in the Indian mile would have looked like in medieval German (see p. 138)!

"Numeration" — the reading and writing of numbers — was therefore a separate branch of the art of computation in the Middle Ages. The number was broken up into "triads" by the use of periods, strokes, or commas. Adam Riese recommends:

Seind aber mehr dann vier ziffer vorhanden / so setze auff die vierdte ein pünctlein / als auffs tausendt / Vnd heb gleich allda widerumb an zu zehlen / eins / zehen / etc. biss zum ende. Als dann sprich auss / souiel punct vorhanden / so manchs tausendt nenne. Das hundert ist die dritte figur nimb allein in benennung / Als dann die erste vnd ander mit einander,

"But if there are more than four digits / then above the fourth place a small dot / as being over the thousand. / And there take up counting again / one / ten / and so forth, all the way to the end. Then say / how many dots there are / so many thousands are to be named. The hundred is the third figure; name this separately. / Then take the first and the others together,"

as in the example we have just seen; compare this with the corresponding instructions for the medieval counting board (see pp. 319 ff. and 332 ff.). Others, unlike Riese, placed not one dot but, going from right to left, first one, then two, and then three dots, and so on, over the respective 1000 digits. This division of the number into groups of three when reading the number, a custom we still follow, originated at a time when the language had formed only three numerical ranks and is thus a feature rooted deep in the history of our number sequence. The Greeks, whose language has a fourth rank (M) in the word *mýrioi*, break up long numbers into "tetrads" or groups of four each:

867 MM 8932 M 5178.

Quite understandably, men who concern themselves scientifically with the study of numbers are apt to need and find names for the higher ranks; for people in general the "thousand" is more than enough.

The word "million" first made its appearance in Italy, probably in the 14th century, as the word itself testifies (see p. 134). Its invention has been ascribed to the Venetian merchant Marco Polo, who had been in China and described his travels among its people, saying that he saw "many thousand" (literally: *milli-one*) persons there. As a specific number word, *millione* was first used by Italian merchants. By the end of the 15th century it appeared in printed texts on arithmetic. In his great treatise on mathematics, published in 1494, Luca Pacioli says:

mille migliara che fa secondo el volgo el millione,
"a thousand thousands, which in popular speech is called a million."

The word first appeared in German textbooks on arithmetic a generation later (around 1530), and was occasionally used by Adam Riese, but took a long time to become established. Riese himself, as we have seen, used the old 1000 × 1000 formation, which lasted well into the 18th century. When the new word was mentioned, it was distinctly pointed out: "And when a million is written in numerals as 1000000, this is ten times a hundred thousand."

The "million" entered commercial parlance from the Dutch language as a large unit of money: 10 tons of gold = 1 million guilders. The "ton of gold" was worth 100,000 units of the national coin in use at the time. Hans von Schweinichen, the Duke of Liegnitz's steward, tells of being invited to visit the wealthy Fugger family at Augsburg in 1575: "There was a sideboard extending along the entire hall, loaded with drinking vessels and beautiful Venetian glass; this was said to be worth far more than a ton of gold." Thus a million was not the fourth rank, the one following a thousand, but the sixth.

In a Spanish text on computations of 1505, whose author was Philip II's teacher in arithmetic, the word *million* is used even for the twelfth numerical rank, while a special word, *cuento*, was adopted for the sixth (literally: "counting" < *contar*, "to count, reckon"); in Portuguese *un conto de reis* is "one million reis" (a unit of coinage); today both words have changed their meanings in Spanish.

These examples show that this term arose less from the scientific need to extend the number sequence by decimal gradations than from the desire for verbal abbreviation and the need to visualize higher numbers. In contrast to the Indian method, such a verbal projection of the number sequence, in which whole gradations are omitted, hardly resembles the positional system of writing numerals.

However convenient the shortened expression "million" may be, Italy, where the word was invented, was very slow and very late in adopting the word "billion" and always regarded it as a curiosity. Verbally this is a contraction of *bis-million*, "two times a million," but with the meaning of "a million times a million." In America it was used into the 18th century for the ninth, not the twelfth rank, and in France it still has this significance: 10^9 is called either *billion* or *milliard*. What better proof can there be of the synthetic character of this formation?

As far as we know from documents still preserved today, Nicolas Chuquet near the end of the 15th century was the first to use a whole series of such words to extend the number sequence:

745324˙804300˙700023˙654321 ... *le premier point peult signifier million | le second point byllion, ... tryllion, quadrillion, quyllion, sixlion ... non- yllion et ainsi des aultres se plus voultre on vouloit proceder.*

"The first dot indicates a million, the second a billion, ... trillion ... nonillion and so on, as far as one may wish to go."

Later the series was continued by others in the same or similar fashion, either independently or following in Chuquet's footsteps; some writers used the forms *bi-million*, *tri-million*, etc., instead of the contraction "billion." In general, all these sequences proceed upward by steps at every sixth rank: 10^6, 10^{12}, 10^{18}, and so forth

Milliard is a French word which appeared for the first time in the 16th century. Originally it designated the 12th and later came to stand for the 9th rank; it did not come into common use in France until the 19th century, and in Germany not until after 1871, when it was made popular by the "5 milliard" of war reparations imposed on France after the Franco-Prussian War.

The history of written numerals has an exact parallel of this artificial expression, which in one stroke removed the then existing limits but provided only a colorless extension of the sequence of numerals. For their first three numerical ranks, the Romans initially used the symbols X, C, and CIↃ; these were all different and had different origins. But for the subsequent 10^4, 10^5, ..., etc., the last symbol was artificially expanded, as in ((I)) and (((I))), by adding more of the curves which initially, in the 10^3 symbol CIↃ, did not have any numerical meaning but were merely strokes of the pen (see Fig. 8). This corresponds exactly to the artificial extension of the number words *bi-*, *tri-llion*, etc. from *mille*. Because of the rarity of the higher Roman number symbols, I shall illustrate three Roman denarii, which bear numbers indicating expenditures and not monetary values (see Fig. 31; note the "half" symbols on the first and third).

The artificiality of this extension of our old, naturally evolved number sequence is perfectly obvious. It is shown by its origin in the scholar's study, by its omission and skipping over some of the ranks of numbers, by its ambiguity, by its word formations and, finally, by the verbal "statelessness" which reveals itself at the end of all such synthetic devices: the word "thousand" is Germanic, but what is "billion"?

Along with their native homes, words and things also lose their character. They become pale and barren and featureless, they are shrunk down to gray abstract concepts which now have unique, specific meanings but no longer the resonance of history and culture. *Katzengold*, "fool's gold," has been boiled down to FeS_2. Whereas such concepts, because of their colorlessness and lack of ambiguity, enter immediately into the language of science, "billion," "trillion," and their fellow orphans were not so fortunate. We have become accustomed to billions in financial statements, and the milliard still plays a role here and there in France and Germany, but the trillions and their successors have completely disappeared from scientific communication. "One cubic centimeter of gas at $0°$ C and 760 mm pressure contains 28×10^{18} atoms," says the scientist; not twenty-eight trillion atoms. He needs no number words other than the commonest and most ancient ones; the higher ranks of numbers he counts just as Archimedes once did.

With this we have once and for all finished with the history of the higher numerical ranks. Now let us go back to our starting point once more.

Fig. 31 Three Roman denarii of C. Calpurnius Piso (64 B.C.) with the symbols for 5000, 10,000, and 50,000 shown behind the head of Apollo. The symbol for 100,000 is inscribed on the *Columna rostrata* (see Fig. 10). British Museum, London.

Meanings of the Units and the Tens

The Units. Although we shall find some very intriguing nuggets of information in this area, the total yield will be quite meager. Many hypotheses about the original meanings of "one," "two," or "three" have been advanced: it has been suggested that the personal pronouns "I," "thou," and "he" were incorporated into the first few number words; but no matter how much philological acumen has been lavished on this subject, it has thus far been fruitless and will probably always remain so. The few suppositions that have any merit at all are stated below. We shall begin by reviewing the connection between our Germanic words for the units and the Indo-European words; the latter, along with the Latin equivalents, reveal their relationship to our number words through the first and second phonetic shifts. We shall draw on our tabulation of these shifts (p. 127), referring especially to the last column. An asterisk (*) in the following table refers to the discussion of the units. There we shall also see what can be plausibly said at present about the meanings of the number words themselves in the other Indo-European languages.

Relationship of the Germanic to the Indo-European Units

	Indo-European and Latin >		Gothic and English >		German	Phonetic Shift (p. 127)
1	*oi-nos, oi-qos*	*unus*	*ains*	*one*	*eins*	—
	sems, smiə, sem					
2	*duųo, dųo*	*duo*	*twai*	*two*	*zwei*	6
3	*trejes, trie*	*tres*	*þreis*	*three*	*drei*	3
4	*qųetųōr*	*quattuor*	*fidwōr*	*four*	*vier*	*
5	*penque*	*quinque* (Greek *pénte, pémpe*)	*fimf*	*five*	*fünf*	2
6	*sųeks, seks*	*sex*	*saíhs*	*six*	*sechs*	1
7	*septṃ*	*septem*	*sibun*	*seven*	*sieben*	*
8	*oktou*	*octo*	*ahtaú*	*eight*	*acht*	1
9	*neųn*	*novem*	*niun*	*nine*	*neun*	—

One. For "one" there are two primeval Indo-European forms:

1. *oi-nos*, which leads readily to the Latin, Celtic, and Germanic "one." In the word *oiné* the Greek language recognizes "the One on the die," which also exists in the dialect form of *oinós*, "wine"; the Lithuanian *w-ienas* belongs to the same category. The Slavic *jed-in* > Russian *od-in* may perhaps go back to an expanded Indo-European form *e-dhi-no*, the linguistic explanation of which would lead us too far away from our proper course. The Indo-European *oi-qos* > Greek *oíos*, "solitary, alone" is also related to the Sanskrit *e-ka*, "one." Other cognate words probably mean "the only, the very."

2. In contrast to this specific meaning, there is also the "collective" meaning of "one": Indo-European *sem*, from which the Sanskrit *sam* and the German *zu-sam-men*, "together," *sam-meln*, "to collect," and *säm-tlich*, "jointly, collectively," are derived. From the same root have come the Tocharian word *sa* and the Greek *heîs*, *(s)mía*, *hén*, "one," which is already familiar to us in *he-katón*, "100" and *há-pax*, "once " (see p. 25). If we recall that the Indo-European *s*-before a vowel becomes Greek *h*- (as in *septṃ* > *heptá*, "seven"; Indo-European *su*, Latin *sus* > Greek *hys*, "pig, sow"), we shall arrive at Greek *háma*, "at once, at the same time," *hamâsthai*, "to collect," *hómalla*, "sheaf, pile," and also *homos*, "the same, common, mutual," and *homoîos*, "like, similar" — that which is "one" with something else. In Latin we find *sim-ilis*, "similar," *sim-plex*, "simple" (cf. German *ein-fach*), *sin-guli*, "unique," *sem-el*, "once, one time," and *sem-per*, "once and for all, always"; examples in the Germanic group are Gothic *simle*, "one," Gothic *sama*, and Old High German *samo*, "the same," (cf. Russian *sam*, reflexive pronoun with the same meaning) Gothic, *samana*, Old High German *saman* and *zi-samane*, Modern German *zu-sam-men*, "together," as well as Gothic *sums*, "any," which reappeared in the English word *some*.

We have already discussed the connection between *sem* and the Latin *mille* (see p. 135); the Latin word *semester* has no relation to this (see p. 181).

Two. This word is clearly discernible through all Indo-European languages. We have already discussed at length the early, primitive ideas reflected in this word (see pp. 12 ff.).

Three. The possible connection of "three" with the notion of "beyond, trans-, on the other side" has also been mentioned (see p. 17). One curious feature worth noting here is that only Sanskrit and the Celtic languages have preserved an ancient feminine form for three and four: Indo-European *t(r)i-sores* and *quete-sores* > Sanskrit *tis-rah* and *čatasrah*, Celtic *teoir* and *cetheoira*, which are perhaps related to *sor*, "wife, woman." This 3 form may have taken its name from the "third wife," who followed the two principal wives; the feminine 4 form may have resulted from the remote influence of the feminine "three," if it did not itself originate in the same way. Analogously the "husband, man" appears to have entered into the masculine 4 form *čat-verah* (Sanskrit *virah*, Latin *vir*, Gothic *wair*, "man"; cf. Germanic *wer* as in *Wergold*, the fine imposed for killing a man).

Four. Derivation: the Gothic *fidwōr* calls for an Indo-European form *petwor* in place of *quetuor*, beginning with *p-* instead of *q-* (see Phonetic Shift I—2, table on p. 127). The initial consonant *f-* did not originate in a phonetic shift, however, but resulted from the backward influence of the initial *f-* in *fimf*, "five." Wherever, as in the number sequence, words commonly stand directly adjacent to one another and are eroded by constant repetition, such adaptation or assimilation of the sound or rhythm of the word often takes place. We have seen a beautiful example of this in the Celtic number sequence (p. 125); to mention still another case of phonetic assimilation: Lithuanian *devyni* and Slavic *dēvetj*, "nine," exchanged their initial consonant *n-* for *d-* as a result of the influence of the subsequent number words *dešimt* and *desetj*, "ten" (see p. 93).

In Greek, the Indo-European *qu* may become *k*, *p*, or, as in this case, *t*: *tettares*. But what is the underlying meaning of the Indo-European *quetuor*? Here we can only offer a conjecture. I shall suggest two new explanations, whose validity the reader must judge for himself.

The first of these bases the number word "four" on the "finger-tip sequence" of one hand (omitting the thumb). This hypothesis derives the Indo-European *quetuor* from a conjectural word *oke-to-uoro*. The latter contains the elements *ok-* (also *ak-*), "tip, peak" (as in the Greek *akris*, "mountain peak"; cf. Acro-polis, literally "city on the heights," and Latin *acer*, "sharp"); *-to-*, the same syllable that occurs in *km-to(m)*, "hundred," with the sense of "sequence, series" as in *oketom*, "row of tips"; and the final syllable, which became the Sanskrit *vara*, "row," intensifies the conception of a sequence. In the case of "four," *quetuor*, the initial syllable *o-* was dropped. In "eight" it was retained, and at the same time the word took the dual form, *oktou*, to express the idea of "two times four finger tips."

It is hard to say how far this linguistic hypothesis can be pushed. Internal evidence indicates that it may well be true, especially since it provides a plausible explanation of the Indo-European dual form *oktou*; the fingers, then the hand without the thumb, which have caused the sharp break after "four," and, finally, the two hands together have, as we have shown, played a decisive role in developing the number sequence. The hypothesis also finds some support in the Indian Kharosthi numerals, of which the 8 has the form of a doubled 4 (see Fig. 12); the Chinese character for the number 8 may also have originated as an ideogram representing two 4-fingered hands (see Fig. 13).

The other interpretation proceeds from the cross with four arms or "tips," probably the most descriptive symbol for the number 4. Latin *tri-quetrus* means "three-cornered," the idea of "peak" or "corner" being embodied in the *quetrum* and derived ultimately from the Indo-European *quetuor*. From *quetrum* (not from *quattuor*) came the Latin word *quadra*, "square, cross"; in this word the old meaning again comes to the surface. "Eight," which is grammatically a dual, seems to have originated in a "doubling" of the cross, in that another arm or beam was inserted perpendicular to each of the four arms, thus forming the hooked cross or swastika that has been used as a primitive symbol by many unrelated cultures.

Now we do sense that for primitive man, when he wanted to count to four, such a symbol as a model for a number word was far too remote, even if it did exist in his environment. Quite aside from this, the swastika is never felt as a "doubling" (dual) of the simple, four-armed cross.

Both these hypotheses connect the number word "four" with "point," or "tip." The German language contains a surprising expression of this connection between "four" and "point": the Old High German word *ort* did not mean "place," as it does in Modern German, but "point" or "corner"; in many regions of Germany the sharp point of a cobbler's awl is still called *Ort*. *Ort wider orte*, "tip against tip," "point against point," says Hildebrand's Song, and Albrecht Dürer once remarked that an artist should draw in "*die allerkleinsten Rünzelein und Ertlein*," the most minute wrinkles and corners." But how is this word related to "four"? Numismatics designates a coin divided by two perpendicular snips as being in *vier Orte*, "four sharp wedges" ⊕, or four "corners"; hence, the quarter-thaler was also called the *Ortstaler*, and the quarter-guilder also an *Örtel* or an *Eckele*. From the domain of coinage, the *Ort* came to be used for a quarter unit in measures and weights; the Dutch word *oord* is a "quarter measure." It is also quite significant that the odd, or "uneven," number was felt to be a "point": *En Stier hab i örti*, says the Bavarian peasant, "I have one 'point' ox," meaning that he has one ox in addition to the even-numbered yoke.

Five. The Germanic word "five" should, since it derives from the Indo-European *penque*, have once had a laryngeal as a final consonant; this is still actually present in the Swabian dialect form *fuch-zk*. Today it has assimilated the initial *f-*, just as, to cite a parallel example, the medial *k*-sound in the Latin *quinque* has transformed the initial *p-* in its Indo-European ancestor to a *k-* (*qu-*) in Latin.

From our previous discussion we know the role which the "5 fingers" or "hand" played in the arrangement of the number sequence. Thus the number word "five" has always been associated with these two images. The Gothic word *fimf* is probably connected with the Gothic *figgrs* (pronounced "fingrs") "finger," just as the Slavic *pētj* is related to the Slavic *pěsti*, "fist." It is also certain that the Egyptian "five" is the same as their word for "hand"; the same thing is true of Austronesian and many other languages of primitive peoples (see p. 52).

Now we come to something quite remarkable. If we compare the numbers 5 and 4 in the *Kentum* and *Satem* languages (see tables on pp. 92–93) — Sanskrit *panča* — *čatvaras*, Greek *pénte* — *téttares*, Latin *quinque* — *quattuor*, Lithuanian *penki* — *keturi*, Irish *coic* — *cethir*, Welsh *pimp* — *petwar* we see that the word for 4 invariably begins with the final syllable of 5. Thus it seems that some root *R* attached itself (like the Sanskrit -*ča*-) in front of a syllable *s* in the word for 5 and after a syllable *s'* in the word for 4; thus, diagrammatically, $5 = s + R$ and $4 = R + s'$. Dare we see more than mere coincidence in this pattern, which extends through virtually all the Indo-European languages? Just what does the root *R* mean? Perhaps "hand," so that 5 could mean something like "the whole hand" and 4 a "hand less 1"; or else 5 could be "1 more than a (four-fingered) hand" and 4 a "hand less the thumb." Such formations appear elsewhere in the number sequence (cf. eleven, *ein-lif*: pp. 83 and 157). At any rate, one could hardly find a better explanation for this peculiarity, any more than for the interpretation of the Indo-European *penque*, which must likewise have been some actual thing, as "hand and 1," taking the postpositive -*que* to mean "and," as it does in Latin.

Six and Seven. On the derivation of "seven": According to the rules of Indo-European linguistics, the Latin *septem* should have led to *seftem* or, by dropping the *t* between *f* and *n*, to *sifun* instead of *sibun*, And, indeed, we do have the Anglo-Saxon words *seofon*, English *seven*, and Dutch *zeven*.

The resemblance of these two number words to the corresponding words in the Semitic family is striking: original Semitic *šeššu* > Assyrian *šišša*, Hebrew *šeš*; cf. Sanskrit *šaš*, 6; and Semitic *sabu* > Hebrew *šeba'*, Assyrian *sibi*, Arabic *sabun* (cf. Sanskrit *sapta*, 7). Whether this similarity is purely accidental or the result of borrowing (in which case the words must have gone from the Indo-European into the Semitic languages) is hard to tell.

Eight. This number has already been discussed in connection with "four."

Nine. We have mentioned the connection between this word and "new," as well as the interpretation that one "new" number has been added to the two 4 groups of both hands (not counting thumbs) (see p. 23). Ancient Egypt provides some support for this postulated connection between "nine" and "new": the root of the Egyptian word for "nine" is also used to indicate the rising of the sun in the east and the first appearance of the new moon. See p. 170 for a different interpretation of "seven," "eight," and "nine," which may explain the break that exists in the Indo-European sequence of tens.

The Tens

Greek *-konta* and Latin *-ginta.* The Indo-European languages form the tens by coupling the units with the first numerical rank, "ten," usually not in the multiplicatory form "three times ten" which we might expect but rather according to the old grouping, "three tens": Greek *triá-konta,* Latin *tri-ginta,* Slavic *tri desēti* (see p. 126).

The Greek *-konta,* "ten" or "ten-ness." The verbal device used to form the tens (except for twenty), is derived from the Indo-European *-koṃta,* an expansion of the *dekṃ(t),* "ten" which in turn comes from *(d)kṃt.* This contraction was still present in the Doric word *Fi-kat-i,* 20 (Indo-European *ṃ* > Greek *a*; see p. 126). The final *-i* indicates the grammatical dual; the first syllable prefix *Fi-,* Latin *vi-* < Indo-European *dui, ui* is a dual formation of "two," *duo.* The number word for 20, corresponding to the Latin *vi-gint-i,* is thus picturesquely built up from a threefold two form, testifying to the special status and great age of the word for 20. The archaic Greek letter *F*, pronounced "w," is called di-gamma, or double *Γ*, because of its shape. We also find the prefix *ui-* in the Attic dialect word for 20, *(w)ei-kosi,* cf. Sanskrit *vim-satih* and Irish *fi-che,* as well as in the Gothic dual *wit,* "we two, both of us" (see p. 13); in the Latin word *di-vi-dere* it appears in the divisive sense of two. For the erosion and elision of the tens in modern Greek, see p. 125.

A curious aspect of linguistic history is the derivation of the Latin *-ginta* from the Latin *konta*: The Etruscans used only the hard final consonants *k*, *p*, and *t*. But after they adopted the Greek alphabet for writing their language, the soft gamma *Γ* became the hard sound *k* in Etruscan. And the Romans, who in turn took their letters from the Etruscans, changed the *Γ* to ⟨ and finally to C, which stood for both the hard *k* and the soft *g* sound. The abbreviations C., for Gaius and Cn., for Gnaeus, and also the interchangeable forms *vi-c-esimus* and *vi-g-esimus* for "twentieth," all document this phonetic ambiguity, which was finally ended in the 3rd century B.C., by the addition of one line, turning the C into G for the representation of the *g*-sound. As a result of this early double meaning of C, however, the 20 form *-ginti* arose from *-konti.* In addition, the *-o-* bracketed by *i* on both sides in *vi-konti* was changed to *i*, so that the form *viginti* finally emerged. All subsequent tens from 20 on have the ending *-ginta,* resulting from the influence of the *-ginti* (the *-a-* sandwiched into the words for 50, 60, and 70 is introduced by analogy with *quadra-ginta*).

German *-zig.* What is the explanation of the ten forms *vier-zig,* *fünf-zig,* etc. in German? The suffix *-zig* has the sense and form of a "ten" ending just like those discussed in the preceding section.

In the tabulation of Germanic and German number words (see p. 96) this ending can be traced back as follows:

German *vier-zig* < Old High German *fior-zug* < Gothic *fidwor-tigjus.* The Gothic *tigjus* is the plural form of *tigus,* "ten, ten-ness," an auxiliary form of *taihun* (like the Greek *dekás* and *déka,* "decade" and "ten") which is encountered in all the Germanic languages: Old Norse *-tiger,* Old Saxon and Anglo-Saxon *-tig,* English *-ty,* Icelandic

-tiu, Swedish *-tio* and Danish *-tyve*. The German phonetic shift altered *-tig* to *-zug* and later to Middle High German *-zec* — the Nibelungenlied says *drizec tusend* — and finally to *-zeg* and *-zig*.

The force of the noun "ten" can be seen in Old High German from the genitive case that follows it: *feorzug wehhono*, "40 of the weeks" (*wehha*, "week") and *zehenzog scafo*, "ten-ty (a hundred) of the sheep" (*scaf* = sheep).

The Germanic and the German "twenty" and the Balto-Slavic forms of this word are remarkable among the Indo-European languages for their regularity (see tables on pp. 92–96)! The Germanic form is simply *twai tigjus*, "2 tens" and thus not a grammatical dual, to say nothing of a doubled word form like *viginti*; the only exception is the Old Norse *tuttugu*.

The Romance languages derived their words for 20 from Latin, although Rumanian did not, probably influenced by Slavic proximity: "twenty" in Rumanian is *doau zeci* (see table on p. 99).

In looking through the old Germanic forms of the tens, we discover a striking break after 60, as in the Gothic *saihs-tigjus* but 70 *sibunt-ehund* (see p. 94). The reader may be surprised to find that this break after 60 occurs in many of the *Kentum* and *Satem* languages (tables on pp. 92–93). The explanation cannot be found within the Indo-European family alone. We shall learn far more if we extend our research from the narrow sphere of historical linguistics into other domains that appear relevant to this problem — to the "great hundred" and to the special meaning of 12 and 60 in cultural history generally.

First, however, let us pause for a moment, for we have reached the end of the road in our search for enlightenment about number words. What some readers may have hoped for, namely a revelation of the original formations of the number words, has not been realized. But we did uncover, sometimes to our amazement, the pulsating evolution of the number sequence, which sometimes forged ahead vigorously and at other times remained stagnated at a conventional boundary. We have seen a constant shuttling to and fro as the loom of cultural history wove the tapestry of the number sequence. And whenever we carefully pulled one thread or another out of the fabric, in order to trace its convolutions back to the very beginning, we were rewarded with profound glimpses into the early condition of the human mind.

Now, however, let us once again emphasize the two most important conclusions to be drawn from our discussion:

1) The Indo-European number sequence built up around the number 10.

2) The extraordinary interweaving of our own number sequence with those of the other Indo-European languages.

Was There a Babylonian Influence on Our Number Sequence?

THE BREAK AFTER SIXTY

This is too obvious to be overlooked. Indeed, our comparative tables show that it is present in all Germanic number sequences (pp. 94–96):

	Anglo-Saxon	Old Saxon	Gothic	Old High German
50	*fíf-tig*	*fíf-tig*	*fimf-tigjus*	*finf-zug*
60	*six-*	*sehs-*	*saihs-*	*sehs-*
70	*hund-seofontíg*	*ant-sibunda*	*sibunt-e-hund*	*sibon-zo*
80	*-ahtatig*	*-ahtoda*	*ahtaút-*	*ahto-zo*

It is also found in other Indo-European number sequences (pp. 92–93):

	Greek	Latin	Celtic	Tocharian	Sanskrit
50	*penté-konta*	*quinqua-ginta*	*coi-ca*	*pñ-ak*	*pança-śat*
60	*hexé-*	*sexa-*	*ses-*	*säks-äk*	*śaš-tih*
70	*hebdomé-*	*septua-*	*secht-mogo*	*sapt-uk*	*sapta-*
80	*ogdoé-*	*octo-*	*ocht-moga*	*okt-*	*aši-*

In Gothic, 60 is called "six tens," but then one finds "the sevenfold, (or 'seven-ness'), the eightfold, the ninefold, and the tenfold of tens" (cf. pp. 150 and 129). The Old High German *-zo* was probably derived from the Gothic *-te-(hund)* after the *-hund* had been dropped. Still more peculiar are the Germanic formations, in which *hund-* or *ant-*, with the original Indo-European meaning of "ten," is placed in front of the number word. Thus *hund-seofon-tig*, for example, means 10-7-10! This is also the explanation for the hitherto incomprehensible Dutch word for 80: *tachtig < (an)tachtig*.

The break disappears in the more modern languages, in which the simpler word forms for the tens up to 60 have displaced those for higher numbers (another example of remote influence). Old Norse, to be sure, also lacks this break (or no longer has it? see p. 94) in its number sequence, but the expression "so and so many decades old," is counted in this manner: first from 20 *tvi-tögr . . .* up to 60 *six-tögr*, but then 70 *sjau-tögr* as well as *sjau-raedr*, 80 *atta-raedr*, . . . up to 120 *tolfraedr* (for *raedr*, Gothic *raþjo*, see p. 130). Here the break is immediately apparent.

In Greek, Latin, and Celtic the units in the ten forms from 60 on occur in their ordinal and not their cardinal forms: instead of *hepté-konta*, "7 tens," we have *hebdomé-konta*, "the 7th tens," so that the plural form *-konta* is nonsensical. Its presence can be explained only by analogy with the earlier tens, for the logical form in Greek would be *hebdomé-kas*, "the 7th ten." Even if, as philologists have recently come to believe, the Greek forms from 70 on were not derived from the ordinal forms of the units but from some other

[152]

phonetic rule, the break after 60 — which is what concerns us here — still remains.

Latin *septua* < Old Latin *septuma* > later *septima*, "seventh," is formed like *decimus* < *decumus*, "tenth"; the *porta decumana*, the main gate of a Roman army encampment, was the point at which the tenth cohort of the legion was always stationed. Similarly *octua* and *nona* < *novena*.

Tocharian also shows a change in the ten forms, *-ak* > *-uk*, in going from 60 to 70.

It is interesting that in Sanskrit the break occurs after 50. Perhaps this arose in going from one hand to the other, if the tens were counted off on the fingers.

In Slavic and Lithuanian, however, we find no change after 60; the break after 40 in these languages has already been mentioned (see p. 23). This, too, is an indication that the break after 60 dates from a very early time, for the Slavic number words are relatively young, as we gathered from borrowings and from the Lithuanian *lika-* formations (see p. 84). Since there can be no doubt at all about the existence of the break at 60 in many or most of the Indo-European languages, is there an explanation for it?

THE ROLE PLAYED BY THE NUMBER SIXTY IN CULTURAL HISTORY

In Greece both 60 and its multiples such as 360 were frequently used as "round numbers" — that is, numbers whose specific meanings are inflated into the indefinite "many," as in the expression, "If I've told you once, I've told you 'a hundred times' that . . . ." The inflation or semantic fading of a number word is the opposite of the specification of its meaning and usually involves the three ranks 10, 100, and 1000, but it can also happen to any other number which for one reason or another has acquired special significance (as in *seine seiben Sachen packen*, "packing up his seven things").

The Greek historian Herodotus says that Darius, the Persian King, had 60 knots tied in a thong and then ordered the Ionian princes to untie one knot each day (*History* IV, 98); they were to guard the bridge of ships over the Istros River in Darius' absence until the last knot had been untied. Xerxes in fury commanded the Hellespont to be flogged with 300 lashes, because of the destruction the waters had caused in a storm. The oldest known Greek example of the significance of this number is provided by Eumaeus' herd of swine (*Odyssey* 14, 20):

"The swineherd always sent them (Penelope's suitors) the best one of the fatted pigs for them to feast on; and the number of swine remaining was only three hundred and sixty."

Among the Romans, *sexaginta* and especially its tenth multiple *sescenti*, 600, were commonly used as round numbers meaning "many."

The Old High German Song of Hildebrand also once uses 60 as a round number:

Ih wallota sumaro enti wintro sehstic ur lante,

"I spent sixty summers and winters (= many years) outside the land."

In the *Annolied*, which dates from the beginning of the 12th century, wo find the round number 300:

Romere scrivin cisamin	"The Romans wrote down together
in einer guldiner tavelin	upon a golden tablet
driu hunderit altheirin	three hundred old men
die dir plegin zuht und erin.	who render thee homage and honor."

Even to the present day, however, sixty is sometimes used exactly and at other times to mean "a heap" or some indefinite large number. In this latter sense we can make out the original use of *Schock* (a German word meaning a large lot of sixty units) exclusively for "60 sheaves" of grain and hence for a pile of sheaves (cf. the archaic English word *shock* = a pile of sixty sheaves of grain): *tein scok gavarno* reads an Old High German passage dating from the 9th century. In the Swabian dialect *schocken* still means "to pile up grain"; in the dialect of Hesse a building's walls *schucken*, "are made up of," bricks. As a measure of quantity,

1 *Schock* = 60 pices = 3 score = 4 *Mandel* = 5 dozen;

1 *Großschock* ("great shock") = 64 pieces (a shock with an "excess").

In northern Europe the doubled sixty was even more frequently used, and was known as the great hundred.

GREAT HUNDRED
In his novel "The Vagabond," the Norwegian writer Knut Hamson says: "They counted a so-called great hundred for every 120 fishes." This term still measures quantities of fish, even in Germany, where it is used for other things as well. In Lübeck there was a *Hundert Bretter* = 120 items = 10 *Zwölfter* ("twelves") or, as they said in Mecklenburg, 10 *Tult*. English differentiates between the *long hundred* of 120 units and the *short hundred* which has only 100. Today, however, we recognize "hundred" only in its meaning of 100.

The northern Germanic region, primarily Iceland, was the home of the great hundred; there *hundraþ* meant 120 in all monetary calculations and in designating military units, until the introduction of Christianity around the year 1000. Thereafter it came to stand for the small hundred of 100 in ecclesiastical and learned writings (see p. 79), and the two hundreds were sometimes, but not always, distinguished as the 10-(*ti-roed*) or the 12-(*tolf-roed*). A writer around the year 1250 once designated the 360 days of the year in the old manner:

III^c daga tolfroed — "three hundred days by the 12-count,"

and commented: "Nevertheless in the book language (Latin) all the hundreds are reckoned by the ti-count (*oll hundraþ tiroed*), which according to the proper count is *III^c tiroed ok LX daga*."

We may well ask why the writer did not simply use the 100 form corresponding to *tio tiger*, "10 tens" — in this case, "36 tens" — which our number sequence provides (see p. 94)? Because it means ten tens in a 12-count great hundred enumeration, but not 100 as such. We can understand this better if we read from an Old Norse tax roll:

"Whoever has property worth 1 ten shall pay 1 ell of *wadmal*;
"Whoever has property worth 2 tens ... etc.; then
"Whoever has property worth *halft-hundrap* (6 tens) shall pay 4 ells...
"Whoever has property worth *tio-tiger* (= 10 tens) shall pay 6 ells."

Thus *tio tiger* meant 10 tens in the 12-count hundred system of counting, not an "independent" 100; *tio tiger ok priu hunderap* is $460 = 10'10 + 3'120$.

The *wadmal* in the foregoing list is baize or frieze cloth, which was also used as a medium of exchange or "money," in addition to cattle and, later, silver, because the poorer man who had no cattle or silver could weave the cloth himself, and it was easy to divide. The standard of values was originally

1 *hundrap silfrs* = 120 ounces of minted silver = 2400 ells of frieze-cloth,

whence 1 ounce of minted silver = 20 ells of frieze-cloth,

an equivalence in which we see both the great hundred and the old vigesimal grouping.

The Gothic Bible, however, has only the 10-count *hund* for multiples of 100, perhaps because of the Greek from which it was translated; nevertheless, it must have been in use among the people. Luke 7:41 reads:

There was a certain creditor which had two debtors: the one owed five hundred pence, and the other fifty.

Here the 500 is translated as *fimf hunda*, but the very same number in I Corinthians 15:6 has:

After that, he was seen of above five hundred brethren at once,

expressed in Gothic as *fimf hundam taihun-tewjan bropre* — that is, "5 hundred by the ten-count"; the Gothic word *tewa* means "series, order." Thus the great hundred was apparently commonly used in popular speech, and was later displaced by the "Christian" 10-count hundred.

Next we come to the great thousand, which contains 1200 units; it has "10 hundreds, but the hundred contains 6 score or 2 shocks," says an arithmetic text of the year 1706. As a "tun," a cask measure amounting to 1200 liters, it is still in use today.

Now we can ask the question that we have wanted to voice ever since we first observed all these bits of evidence: How did the great hundred come into being? Is it made up of 2 shocks ($120 = 2 \times 60$), or does the ancient vigesimal grouping play some part in it ($120 = 6 \times 20$), or was it originally made up of twelves ($120 = 10 \times 12$)? The evidence we have clearly points to the latter.

THE NUMBER TWELVE AS THE BASIC UNIT OF THE GREAT HUNDRED

As we just noted, Old Norse distinguished between hundreds made up of ten tens and those of ten twelves; in Anglo-Saxon 120 was expressed as *hundtwelftig*. The word *tylft*, the "twelft," to which the Mecklenburger term *Tult* is related, is the most widespread small measure of quantity: *tvitylft* = 24, *þrennartylft* = 36 items. "Fifteen years old" in Norse poetry (the 12th-century *Heimskringla*) is:

gammal vetra tolf ok þriggja, "12 and 3 winters old"

Charlemagne's monetary standard of the year 780, which had a lasting influence on medieval European coinage, clearly embodied the basic 12-unit:

(Latin)   1 *libra* or *talentum* = 20 *solidus* = 12 *denarius*,
          1 pound of 20 shillings each of 12 pennies = 240 pennies,
or   1 pound = 8 long (Bavarian) shillings of 30 pennies each
          = 240 pennies,
but also: 1 pound = 12 ounces of 20 pennies each = 240 pennies.

In France the table of equivalents was:

1 *livre* of 20 *sou* (< *solidus*) of 12 *denier* (< *denarius*) each = 240 *denier*, whence the *sou* also came to be called a *douzain*, a "twelver."

This old standard of monetary units is still in force in English currency. The shilling as an old measure of 12 units also derives from it. Luther said in one of his *Tischreden*,

Wenn gott sihet, dass wir undanckbar sind, so lässet er uns durch den teuffel einen guten schilling geben,

"If God sees that we are not thankful, he lets the Devil give us a good shilling,"

a colloquial expression for a drubbing of 12 or even 30 blows. In the dialect of Silesia the word *Schilling* has been shortened to *Schilg*, as in *ein Schilg Eier*, "a dozen eggs," and has been inflated to the indefinite meaning of "many" in the expression *schilgemal* (literally: *Schilg* times means a dozen times).

The various fines and penalties imposed under old Germanic law refer to a basic number: in the Alemannian, Bavarian, Friesian, Saxon, and Burgundian tribes this was 12, in Frankish law it was usually 10, and among the Lombards it was 12 for inflicting wounds or injuries and 10 for other infractions. In the *Lex ripuaria*, the code of the Riparian Franks, a stallion, a coat of mail, and a hunting falcon were valued at 12 *solidi*; a helmet at half this amount, or 6 *solidi*; a sword with its scabbard at one third, or 4 *solidi*; a cow, a mare, and a sword without scabbard at one fourth, 3 *solidi*; an ox, and a shield and lance at one sixth, or 2 *solidi*.

In the *Lex salica*, the law code of the Salian Franks, a fine was once specified thus:

unum tualepti | sunt denari CXX | culpabilis iudicetur,

"the guilty one is sentenced to pay a Twelve, that is 120 pennies"

— a document which again bears witness to the number 12 as the basic unit of the great hundred.

But why precisely 12? The importance of this number in the daily lives of common people, in commercial transactions, and in legal affairs is probably due to its easy divisibility in so many ways (see the schedule of fines of the *Lex ripuaria* on the preceding page). The commonly used fractions of the *tylft* or the shilling could all be expressed in terms of whole numbers of pennies:

1	$\frac{1}{2}$	$\frac{1}{3}$	$\frac{1}{4}$	$\frac{3}{4}$	$\frac{2}{3}$	of a shilling
12	6	4	3	9	8	pennies.

For this reason the north European *tylft* is an original, native measure, and not one that was first brought in by way of the Carolingian coinage system. For, quite apart from its ready divisibility, it was also consistent with the Roman pattern in the table of ounces.

As a measure of quantity we have retained the Twelve in our "dozen" (< French *douzaine* < *douze*, 12) and in our "gross" of 12 dozens (< French *grosse douzaine*, "strong dozen"). The *douzain* was actually another term for the French shilling, the *sou*, and thus came into common usage (see p. 156). In German commercial terminology, to be sure, these words are fairly recent acquisitions. The former made its first appearance in Alsace as *totzen* in the 14th century, while the gross migrated from France through the Low Countries into Germany in the 16th century.

In addition to being readily divisible into so many different fractions, we must not overlook another reason for the universality of the number 12 as a measure of quantity — namely the "excess," which was used by many tribes in the north, primarily in legal writings, in speaking of time limits set by law. It is still used today in commercial dealings. The 1001 "Arabian Nights" give a very clear idea of the "excess," as do such common expressions as the German *acht Tage* for a week, the French *quinze jours* for two weeks, and the English "baker's dozen" for 13.

A term set by law is considered to have elapsed when the period of time following it has been entered upon; thus a portion of the latter was added to fill out the legal term. The Germanic peoples counted time by nights, and therefore defined the legal term of a week as "7 nights and a day," from which comes the week's reckoning as eight days. The expression "after a year and a day" also gives the day as an excess. "For one hundred years and a day" was the old legal formula for eternal banishment.

In complimentary salutations an "excess" is always added on (20 + 1 or 100 + 1), and a "baker's dozen" is actually 13. Examples of this kind are easy to find. But we already have enough to illustrate the customary use of the excess, as a result of which not ten, but twelve as "ten with an excess" may have come into common use. Did this perhaps even give rise to the true Germanic word-forms *ein-lif* for 11 and *zwei-lif* for 12, in the sense of "1 (or 2) left" or "left over"?). When we find that the Icelandic legislature, as we know from documents dating from around the year 1000, was made up of

"12 men from each district" — *tylptir manna* —

the number 12 here does not indicate a pointless choice of a readily divisible number, but reflects the customary legal usage of the excess. It is certainly not the "mystical" number 12 (standing for the

Apostles), which could hardly have come into use in Iceland before the advent of Christianity. Now we can also understand the great hundred in terms of 10 such twelves, wherein the excess of 20 over the 100, as the great hundred can also be interpreted, was fully consonant with the sensitivities of the people.

And the *shock* — is it half a great hundred? If we interpret it thus we shall never make sense of it, and it will appear inconsistent with northern European habits of thought; it is more likely to have been a concept of five twelves (or 50 sheaves with an excess of 10).

Is the *shock* borrowed from the Graeco-Babylonian *sossos*, 60? The similarity in both sound and meaning is surprising. To form an opinion on this, we must examine the Babylonian sexagesimal system, one of the most peculiar systems of numbers in history. But before we set out on this path, let us complete our study of the number 12 with the following discussion.

THE ROMAN DUODECIMAL FRACTIONS

A few pages back these offered us some fascinating insights into the primitive state of mind in which men incorporated numbers verbally and mentally into their ordinary lives.

The Roman fractions were originally based on a system of equivalent weights: 1 *as* (or pound) = 12 *unciae* (ounces). Linguistically, *as* is the same word as the Greek *heis*, with the dialect pronunciation *'hâs*, *âs*, "one"; the Latin *uncia* is "unity" < *unus*, "one." Thus a large unit was divided into smaller ones, as in any system of measures. Since money was originally weighed out for transactions, the *as* was an amount of copper in the Roman system of coinage. Its subdivisions were the *unciae* or "ounces." Our illustration (see Fig. 32) shows a quarter *as*, a *quadrans*, valued at 3 ounces. Verbally it has the value of a fraction, but mathematically it is a whole number, as expressed in the three embossed circles at the bottom of the coin. The underlying tendency of all subdivisions of measures is to avoid fractions and to express them instead as whole numbers of smaller units. These fractional measures gradually become separated from their content of weight or quantity and, in terms of the abstract number sequence, become "pure" fractions: a *heres ex quadrante*, for example, is an heir who inherits one quarter of the estate.

In their computations the Romans used no fractions other than these duodecimal fractions (plus a few fractional subdivisions of the *uncia*). If a fraction like $\frac{5}{9}$, for example, turned up in their computations (or in those of the Middle Ages, which adopted the Roman fractions), they either made do with the nearest approximation in terms of a duodecimal fraction:

$$\frac{5}{9} = \frac{20}{36} = \frac{20}{3} \times \frac{1}{12} = \frac{20}{3} \text{ unciae} = 6\frac{2}{3} \text{ unciae} \sim 7 \text{ unciae (septunx),}$$

or else they laboriously translated it into its exact duodecimal fractional equivalent:

$$(6 + \tfrac{2}{3}) \text{ unciae} = \tfrac{1}{2} \text{ as} + 16 \text{ scruples} = 150 \text{ scruples.}$$

Now and then they also expressed a fraction in the Egyptian fashion, as the sum of certain standard fractions. Pliny the Elder, for example,

Fig. 32 A *quadrans* issued by a city in Italy: $\frac{1}{4} = \frac{3}{12}$ *as* = 3 *unciae*. The *quadrans*, or "quarter *as*," was verbally a fraction but computationally in terms of its monetary value a whole number (3 ounces). Diameter 4.5 cm. Municipal Numismatic Collection, Munich.

in estimating that Europe's area was something more than $\frac{11}{24} =$ $\frac{1}{3} + \frac{1}{8}$ that of the whole world, said (*Natural History* VI, 12):

totius terrae tertiam esse partem et octavam paulo amplius —
"somewhat more than the third and the eighth part of the whole earth."

Thus out of the infinite realm of fractions with all possible denominators, these duodecimal fractions that were tied to measures made up only a very scanty sequence, thus constricting all abstract fractional calculations and causing the computer a great deal of misery whenever he "came up against fractions." From these examples we can readily understand how the introduction of Indian numerals and their method of writing fractions really rescued medieval arithmeticians from a nightmare. But for our present purposes, these peculiar Roman computations with fractions are merely one more instance of the difficulties and complications confronting man throughout cultural history in his search for the key to the proper understanding of numbers.

In the table immediately following, the fractions are shown in their simplest form in column 1, the Roman duodecimal fractions with their denominator in ounces in column 2, their Latin names in column 3, translations of the latter in column 4, the Roman symbols for the various fractions in column 5, and the numerical meanings of these symbols in column 6.

The Roman Duodecimal Fractions

1	2	3	4	5	6
	$\frac{12}{12}$	*as*		I	
	$\frac{11}{12}$	*deunx* (< *de uncia*)	—1 ounce	S ∴	$\frac{1}{2} + 5$
$\frac{5}{6}$	$\frac{10}{12}$	*dextans* (< *de sextans*)	—$\frac{1}{6}$ *as*	S....	$\frac{1}{2} + 4$
		decunx (< *decem unciae*)	10 ounces		
		semis et triens	$\frac{1}{2} + \frac{1}{3}$		
$\frac{3}{4}$	$\frac{9}{12}$	*dodrans* (< *de quadrans*)	—$\frac{1}{4}$ (*as*)	S...	$\frac{1}{2} + 3$
$\frac{2}{3}$	$\frac{8}{12}$	*bes* (< *duo partes assis*)	2 partial *as*	S..	$\frac{1}{2} + 2$
	$\frac{7}{12}$	*septunx*	7 ounces	S.	$\frac{1}{2} + 1$
$\frac{1}{2}$	$\frac{6}{12}$	*semis*	$\frac{1}{2}$ (*as*)	S	$\frac{1}{2}$
	$\frac{5}{12}$	*quincunx*	5 ounces	.....	5
$\frac{1}{3}$	$\frac{4}{12}$	*triens*		....	4
$\frac{1}{4}$	$\frac{3}{12}$	*quadrans* (B 169)		...	3
	$\frac{2}{12}$	*sextans*		..	2
	$\frac{1}{12}$	*uncia*		.	1
			$\frac{1}{3}$ (*as*)		
$\frac{1}{4}$	$\frac{3}{12}$	*quadrans* (see Fig. 32, p. 158)	$\frac{1}{4}$ (*as*)		
			$\frac{1}{6}$ (*as*)		
			1 ounce		

Later on the *uncia* or "ounce" was subdivided in common popular usage into $\frac{1}{2}$, $\frac{1}{3}$, ... $\frac{1}{24}$ ounces = 1 scruple, and the latter likewise into $\frac{1}{2}, \frac{1}{4}, \frac{1}{6},$ and $\frac{1}{8}$ scruple $= 1$ *calcus,* resulting finally in the following table of fractions:

As	1	$1\frac{1}{12}$	...	$\frac{1}{12}$					...	$\frac{1}{288}$	...	$\frac{1}{2304}$
Ounce	12	11	...	1	$\frac{1}{2}$	$\frac{1}{4}$	$\frac{1}{6}$	$\frac{1}{8}$	$\frac{1}{24}$			
	Scruple	24	12	8	6	4	3	1	$\frac{1}{2}$	$\frac{1}{4}$	$\frac{1}{6}$	$\frac{1}{8}$
						Calcus	8	4	2	x	1	

The peculiarity that $x = \frac{1}{6}$ scruple can be explained as $\frac{1}{3}$ of $\frac{1}{2}$ of a scruple $(= \frac{1}{3}$ *obolus*). From the table of fractional values above we see, for example, that

1 scruple $= \frac{1}{24}$ ounce $= \frac{1}{288}$ *as* $= 8$ *calcus,* and
1 *calcus* $= \frac{1}{8}$ scruple $= \frac{1}{2304}$ *as,* the last and smallest unit fraction.

In Roman times calculations were generally performed only as far as scruples (the word comes from the Latin *scrip-* or *scrupulum,* "a very small sharp stone" < *scrupus,* "a small sharp stone"; as in English, the Latin word also has the additional connotation of "hesitation, doubt"). The specimens of the Roman counting board or hand abacus which have been preserved have columns only for the half, third, and quarter ounce (see *hand abacus* in the Index on p. 469). It is significant that the later subdivisions of the scruple have names derived from the Greek (*obolus* for $\frac{1}{2}$, *cerates* for $\frac{1}{4}$, whence our word carat, and *chalcus* for $\frac{1}{8}$ scruple).

Such were the Roman fractions, the *minutiae* ("small quantities"). Evidently the uncial fractions or ounces were the chief fractions, the *minutiae usitates* ("useful or practical small quantities"), which were distinguished in the early Middle Ages from the abstract computational fractions, the *minutiae intellectuales* ("theoretical small quantities").

The system of fractions based on the ounce makes good practical sense. It counts the *unciae* and groups them together, 6 to every *S,* which is the abbreviation for *semis* < *semi-as* (see column 5 above); it is linguistically significant, incidentally, that the root of the word *semis* is related not to "two" but to "break." The names for the fractions are quite original (columns 3 and 4 above). There were three whole numbers of ounces (10, 7, and 5 ounces), four fractional parts of the *as* $(\frac{1}{6}, \frac{1}{4}, \frac{1}{3},$ and $\frac{1}{2}$ *as*), three subtractions from the whole $(-1, -\frac{1}{4},$ and $-\frac{1}{6}$ *as*), a very ancient sum of standard fractions $(\frac{1}{2} + \frac{1}{3})$ and the equally ancient formation *bes,* which we examined earlier (see p. 79).

In the Middle Ages one more symbol was added to the Roman *S* (6 ounces): an *S* with an additional horizontal leftward stroke at the top for 2 ounces (the *sextula*), which, with a stroke through the middle, became a symbol for a single ounce. The final result was the following sequence from 1 to 11 ounces, in which 11 ounces was written as $6 + 2 + 2 + 1$ ounces:

Fig. 33 Medieval Sequence of Ounces.

Since we shall come upon medieval uncial fractions again in discussing the counting board, we cite them here:

Fractions of Ounces

	Scruples	Ounces	Names	Symbols
1	12	$\frac{1}{2}$	*semuncia*, "half-ounce" (*semis-uncia*)	$\mathcal{L}$
2	8	$\frac{1}{3}$	*duella*, "2 sixths" (*duo-sextuale*)	
3	6	$\frac{1}{4}$	*sicilius*, "sicel" (named after the symbol)	$\backsim,2$ $)$
4	4	$\frac{1}{6}$	*sextula*, "1 sixth"	$\backsim$
5	2	$\frac{1}{12}$	*dimidia sextula*, "half a sixth"	$\curlyvee$
6	1	$\frac{1}{24}$	*scrupulum*, "very small stone"	$\mathcal{SS}$

The symbols for these fractions may be explained as follows: symbol (1) was the initial letter of the number word, but to differentiate it from *semis*, meaning $\frac{1}{2}$ *as*, it acquired a horizontal stroke, which originally cut the *S* in half; for the same reason symbol (6) was a double *S* with a slanting stroke drawn through both. Symbol (2) was computationally a double symbol (4); the remarkable form, 2, is actually two strokes, one beneath the other, which merged into one (our own numeral 2 arose in the same way; see Fig. 234, p. 413 and Fig. 241, p. 419). Symbols (3), (4), and (5) were derived from the Roman symbol for the *uncia*, which was a small circle (see Fig. 32 on p. 158): the fractions of the ounce were expressed as fractions of the circle, and half the symbol (5) indicated half the value. This list of medieval fractions offers many insights into the origin of early numerical symbols.

At this point we must forgo using these fractions for making computations in the Roman manner. One example,

$$\frac{5}{6} - \frac{2}{3} = \frac{1}{6} \quad dextans - bes = sextans \quad "-\frac{1}{6} - bes = \frac{1}{6}"$$

should suffice to show us how the very names of these fractions themselves made abstract computations according to general rules impossible, because they did not produce a conceptual image of the numbers. Thus anyone who came upon fractions in his calculations and wanted to continue using them had to rely on previously prepared computational tables.

The Roman measures of length were similarly divided into twelve subunits:

1 *pes* ("foot") = 12 *pollices* ("thumbs") = 16 *digiti* ("fingers," that is, 4 handbreadths)

so that 1 *digitus* = $\frac{3}{4}$ *pollex*. These Roman measures were adopted in England, where:

1 *foot* = 12 *inches* = 16 *digits* = 12 × 12 *lines*,

and here, too, 1 *digit* = $\frac{3}{4}$ *inch*. The word *inch* < Old English *ynce* < Latin *uncia* thus means $\frac{1}{12}$ and is explained as follows: 1 foot = 12 twelfth-foot = 12 *unciae* (−foot) = 12 inches.

The English system of weights makes use of the Roman ounce:

1 *pound* = 16 *ounces*,
  1 *ounce* = 16 *drams* (drachmas),
    1 *dram* = 3 *scruples*.

In the German system of weights, the ounce lived on into the 19th century as $\frac{1}{12}$ of a pound, with 2 *Lot* (a lead weight) per ounce; in the German system of apothecary weights the ounce has 8 *Drachmen* ("drams") and each dram 3 scruples.

As a tariff or duty, the *uncia* led to the development of the uncial letters: St. Jerome, one of the Church fathers, once inveighed against the *litterae unciales*, the "inch-high letters," in illuminated manuscripts. Today the term uncial is used for a form of medieval lettering.

The duodecimal system of fractions was a purely indigenous Roman growth, for it was unknown to the other three ancient Mediterranean cultures, i.e., the Greek, Egyptian, and Babylonian, which at various times all developed their own fractions. Only the base number 12 can be traced back to Babylonian influence (as in the zodiac and the division of the day and night into 12 hours).* Twelve also occurs very early as a mystical round number. Twelve vultures appeared as an omen to Romulus, the founder of the city of Rome, indicating that it was to last for 1200 years. The number 12 is also significant in Greek history: Homer relates that Ajax and Odysseus each commanded 12 ships; the 12 Ionian cities formed an alliance; 120 Boeotian sailors made up a ship's crew. And when we learn, further, that Menelaus was lord over 60 ships and recall the swineherd Eumeaus' 360 pigs (see p. 153), we realize that the round number 12 belongs to the same category as the other round number 60 and its multiples.

Where did these numbers come from? Here the Greek systems of weights and coinage lead us directly, and with certainty, back to the source. The ancient Babylonian system of weights and coinage was as follows: 1 talent = 60 minae, 1 mina = 60 shekels. The Greeks adopted the first of these equivalents (along with the Babylonian words themselves: Babylonian *manna* > Greek *mnâ*, "mina") and established the following table of monetary values:

1 talent = 60 minae, 1 mina = 100 drachmas, 1 drachma = 6 obols, or 1 talent = 6000 drachmas = 36,000 obols.

The talent and the mina were never issued by the Greeks as coins, but were merely numerical measures, like the pound in Germany during the Middle Ages.

THE BABYLONIAN SEXAGESIMAL SYSTEM

Was the role played by the number 60 limited to the above systems of measures? No, it was an essential peculiarity of the Babylonian number sequence.

---

* *Editor's note:* The division of the day and night into 12 hours is of Egyptian, not Babylonian, origin. Cf. O. Neugebauer, *The Exact Sciences in Antiquity* (2nd ed., Harper Torchbooks: New York, 1962), p. 81.

The Sumerians, who inhabited the southern part of Mesopotamia, soon developed, in addition to a decimal number sequence, an indeterminate positional system of numerals based on gradations of the number 60. Its origin is so unique that it warrants a brief account. This will also afford a contrasting example, from which we shall gain a deeper understanding of the essence of our own positional system of written numerals than we could derive without it.

At the beginning of their history the Sumerians, like all peoples, possessed a primitive system of written numerals — one which had a separate symbol for each and every unit, with several units arranged in order and grouped into a higher unit — a system comparable to the one we noted in ancient Egypt (see p. 42). This is shown on some ancient clay tablets, about the size of the palm of a hand, which served as the Sumerians' "paper" (see Fig. 34).

Fig. 34 Two clay tablets with Sumerian numerical symbols, from Uruk; these are among the world's earliest documents with numerical symbols. They record commercial transactions.

   (1) The number at upper left is $600 + 60 + 30 + 8 = 698$;
   (2) the large circle indicates 60^2, so that the upper left number is $1 \times 60^2 + 2 \times 60 + 3 \times 10 = 3750$.

Approximately two-thirds natural size.

These tablets show how marks were impressed into the soft clay with a round, tapered stylus which had one thick and one thinner end. The marks on the clay tablets were thus made by pressure, not by drawing or inscribing. When the stylus was held at an angle to the clay surface, the impressions were D-shaped; when it was held perpendicular to the surface the resulting impressions were circular. Our illustration gives their values as 1 and 10; 60 and 10×60 (see Fig. 35, rows A and B).

In the table on the left of Fig. 34, we read $600 + 60 + 30 + 8 = 698$, beneath that $5 \times 60 = 300$, and in the broken corner perhaps again 600, a numeral created by combining the symbols for 60 and 10. The forms of these numerals are extremely simple, since they were not inscribed, but made by pressing the stylus downward into the clay. The form of the large circle, indicating 60^2, can be seen

1		2	3	4	5
Row		Value	Original Form	Cuneiform	Sumerian number word
A	a	1			*aš*
	b	10			*u*
B	c	60			*geš*
	d	60·10			*geš-u*
C	e	60^2			*šar*
D	d	$60^2·10$			*šar-u*
	f	60^3			*šar-gal*

Fig. 35 Early Sumerian numerals and number words dating from the third millennium B.C.

in the tablet on the right in the same illustration, on which we read $36,000 + 2 \times 60 + 3 \times 10 = 3750$, and also 270, 30, and 70.

All tablets of this kind have been discovered in the vicinity of temples. Hence they are probably records of livestock and grain brought as offerings, for they consist only of numbers placed next to the names of the persons making the offerings and the nature of the offerings themselves (sheep, chickens, fish, etc.). The reverse side often shows the totals in each category itemized on the obverse. These clay tablets from the third millennium B.C. are the oldest known documents that show numerals in actual use.

These numerals have the usual properties: 10 units, represented by a small D, make a small circle O (a ten-group), just as 10 large Ds make the next group symbol, which stands for 10×60; this is made very clear by the number words themselves (see Fig. 35). But the grouping changes surprisingly from Aa to Bc, for it is not 10, but 60!

Now we come to a critical question: Why this peculiar change from a decimal to a sexagesimal grouping? At first glance, it would seem that the symbols in rows A and B originally formed two independent groupings of ten: one set of units represented by a small numerical symbol and the other by a similar but larger symbol. The former group referred to a small and the latter to a large measure. The two groups were widely disparate in scale, like the mile and the ell in Germany and the mina and the shekel in Babylonia.

The situation has been correctly explained on the basis of the significant number 60 by A. Neugebauer in 1934:* this is the connecting link between the two groups of measures. The requirements of ordinary life in Babylonia, as everywhere else, demanded the use of the common fractions of a measure, in this case $\frac{1}{2}$, $\frac{1}{3}$, and $\frac{2}{3}$. As time passed it became necessary to express the fractional parts of the larger measure as whole numbers of the smaller, for example $\frac{1}{3}$ mina

* O. Neugebauer, Vorgriechische Mathematik, Berlin 1934.

in shekels, and thus to "bond" the two separate groups of measures. The smallest common number is 6; yet this would have made the measures too close to each other (about ⅓ mile = 2 ells!), so that the distinction between the large and the small measures would have lost its purpose. Hence the number 60 was chosen because it is a multiple of ten; then 1 mina = 60 shekels and ⅓ mina = 20 shekels.

Now let us analyze the Sumerian number sequence:

1	*aš* (also *geš*)	10	*u*
2	*min*	20	*niš* (< *ni-aš*)
3	*eš*	30	*ušu* (< *eš-u* 3'10)
4	*limmu*	40	*nin* (< *ni-min* 20'2)
5	*ia*	50	*ninu* (< *ni-min-u* 20'2'10)
6	*aš* (< *ia-aš* 5'1)	60	*geš*
7	*imin* (< *ia-min* 5'2)	120	*geš-min* 60'2
8	*ussu*	180	*geš-eš* 60'3
9	*ilimmu* (< *ia-limmu* 5'4)	600	*geš-u* 60'10 (Akkadian *ner*)

This number sequence does not differ essentially from any other "normal" sequence, except perhaps that it shows an unusually large number of very "early" characteristics. Thus there is an old counting limit following the Three, for the first three number words for 1, 2, and 3 literally mean "man," "woman," and "many" respectively: *eš* is the plural ending. There are additional counting limits after 5, after 10, and after 20. The succeeding tens are made up of vigesimal groupings. Then *geš*, "the great unit," also becomes a numerical rank. But the Sumerians had another old number sequence that went up to 100, and which was tied to a specific measure for cereal grains.

Thus nothing in this rare early number sequence indicates any "deeper insight" by the Sumerians for favoring the number 60. There was some external reason, perhaps the system of weights and measures, for giving this number its special status. It was definitely not an original unit used in building up a sequence, but a number connecting two systems or scales of measurement. From this it was promoted to the status of a numerical rank: the Sumerian ranks rise by powers of 60, with decimal ranks inserted as intermediate steps within the broad sexagesimal intervals:

60^0	10	60^1	10×60^1	60^2	10×60^2	60^3
aš	*u*	*geš*	*geš-u*	*šar*	*šar-u*	*šar-gal* (= *šar-geš*)

This progression is obviously the work of the "scientific" priests. *Šar*, 60^2, is clearly a limit of counting. As so often happened in the history of number words, here too the next rank $60^3 = 60^2 \times 60$, which according to the computational principle ought to be called *šar-geš*, as in column 4 in the table of numerals above, was called *šar-gal*, the "great *šar*"; this is a word formation like our own "thousand" (see pp. 132 and 47). A noteworthy feature is the recurrence of this "great" formation at 60 and again at 60^2.

In forming larger numbers, the number following the numerical rank multiplies the latter: *šar-u* = $60^2 \times 10 = 3600 \times 10 = 36,000$. But so far nothing at all is known about the nature and origin of combinations of numbers of different ranks; *geš-u* is 60'10 = 600, but what is 70 (*u-geš*?), or 72?

Šar is clearly a limit of counting (row C in the table on p. 164) in both number words and numerals. Row D in the same table is obviously formed with *šar*: first came the 10-multiple *šar-u*, $60^2 \times 10$, then the 60-multiple *šar-geš*, $60^2 \times 60$ (circle with 10-symbol), and finally *šar-gal*, $60^3 \times 60$, the "great" *šar* (circle with 60-symbol), which did not appear in the old form but does in the new cuneiform numerals. Whereas the symbols in rows A and B are the oldest, the evolved numerals, those in row C, represent the first natural evolution of these written numerals and the symbols in row D an artificial expansion; thus we have a conceptual succession of gradations similar to the Roman numerals and to our own sequence of number words (see p. 144).

Cuneiform writing came into general use soon afterwards. The old slender cylindrical stylus now became wedge-shaped, and like an elongated right triangle in cross section. If one perpendicular of the stylus was held vertically and the other horizontally, the impression made by the stylus was a vertical "wedge." If the stylus was turned so that the two mutually perpendicular sides were inclined, the result was an oblique angle or hook, as one can readily see from the illustration of the clay tablet (see Fig. 36). From then on the

Fig. 36 A Babylonian clay tablet with cuneiform numerals in positional order, ca. 1800 B.C. It can be seen that each number is made up of two basic symbols, the wedge (for 1) and the hook (for 10), and that every "digit" is a combination (or arrangement) of these. In the middle of the second row from the bottom we can read the numbers 29 and 31, and to their right, 53 and 49. In the last column at the extreme right, the numbers from 1 to 13 (and 14, 15) are written from the top to the bottom. Size: 13 × 9 cm.
From: O. Neugebauer and A. Sachs, *Mathematical Cuneiform Texts* (American Oriental Series 29, New Haven 1945).

Babylonian numbers were written as combinations of wedges and hooks (column 4 in table on p. 164): unity was a single wedge and 10 a hook (Aa, b); the circles in column 3 were now approximated by 4 cuneiform wedges (as in column 4).

And now something happened that was unique in the history of numerals: the distinction between the large and the small symbols (which was made in rows A and B in the table on p. 164) was abandoned, and it was from this abandonment of the characteristic symbols — not from any particular intellectual deliberation — that the positional method of writing numerals arose!

If instead of the cuneiform characters Aa and Bc, we were to write X and x according to the Babylonian system, then XX xxx would clearly be $2 \times 60 + 3 = 123$. Now if we no longer distinguish between the capital X and the lower-case x, but make them all the same size, like the small wedges in Bc, then the symbol xx xxx is now governed entirely by the law of succession of magnitudes: the xx on the left is the 60-multiple of the xxx on the right. Yet the succession of ranks remains undefined, because there is no symbol for the absence of a rank, no middle or final zero; thus our number $2 \times 60 + 3$ could just as well be $2 \times 60^2 + 3$ (because there is no middle zero) or $2 \times 60^2 + 3 \times 60$ (no terminal zero). Hence we have an undefined system of positional numerals that observes the law of succession of magnitudes but not of the order of magnitudes, which then have to be deduced from the context. It was not until quite late (in the 6th century B.C.) that a sign for a gap or missing symbol first appeared within a Babylonian numeral, and still later in astronomical tables at the end of a numeral as well.

Written in positional cuneiform symbols, the number

758 is   and   3750 (see Fig. 34) are:

⟨𝕋𝕋⧏𝕎   𝕋 𝕋𝕋 ⧸

$12 \times 60 + 30 \times 10 + 8 =$ $1 \times 60^2 + 2 \times 60 + 3 \times 10 =$
$12, 38$ $1, 2, 30.$

With this new system of numerals, in which all numbers could be written with only two different symbols, the Babylonians achieved great skill in computations, and along with it a high level of mathematical development. Indeed their problems and solutions, especially in the theory of equations, are still regarded with wonder and admiration today. The Babylonians made use of multiplication tables for their computations, as the following illustration and its exact transcription show (see Fig. 37).

As the rank of the positional number decreases in this sliding succession of magnitudes — we shall see that it inclines less strongly toward its intellectual origin than the Indian — it directly begets the sexagesimal fractions: the succession glides over the ranks, from whole numbers right into fractions, without stopping. In mathematical notation (see p. 56):

$$\cdots + a_2 \times 60^2 + a_1 \times 60^1 + a_0 + a_{-1} \times 60^{-1} + a_{-2} \times 60^{-2} \cdots$$

But the most significant feature is that calculations involving fractions do not differ in the slightest degree from calculations with whole numbers (in great contrast to the duodecimal Roman fractions we described earlier), and this is what gave sexagesimal fractions

Fig. 37 Babylonian multiplication table for products from 18 × 1 to 18 × 60, from the temple library at Nippur, about 1350 B.C.

1st row: 18 (times) 1 (=) 18
2nd row: 2   36
. . . .
last row: 11   3, 18 —
that is, $3 \times 60 + 18 = 198$. The reverse side of the tablet shows 18 × 12 through 18 × 20, followed by 18 × 30, × 40 (× 50 is omitted), × 60. Intermediate products like 18 × 37 are arrived at by combining 18 × 30 + 18 × 7. For the transcription see H. V. Hilprecht, Mathematical, Meteorological and Chronological Tablets from the Temple Library of Nippur. Philadelphia 1906.

their present worldwide currency. The Greeks firmly established the succession of magnitudes in the following terms:

prôton (and) *deúteron hexēkostón* (*méros*),
"first (and) second sixtieth (part)."

The Alexandrian astronomer Ptolemy used sexagesimal fractions in his astronomical work entitled *Megálē Sýntaxis* ("The Great Compilation"), which the Arabs later passed on to the West as the *Almagest* (< *al-mégistē*, "the greatest"). Until the time of Copernicus, this was the fundamental textbook of astronomy. Latin works (translated from the Arabic) presented the sexagesimal fractions verbally as

(*pars*) *minuta prima* (*et*) *secunda*,
"first (and) second diminished (part),"

both of which still live on in our "minutes" and "seconds."

Thus not only the number 60, but also 360, came into our own culture. This was most probably occasioned by the sun's complete circuit through the heavens in roughly 360 days, from which we have the number of parts into which any circle is divided. This reflection of the celestial numbers in the terrestrial was the work of the Babylonian priests. The solar day was divided into twelve two-hour periods in analogy with the twelve-fold division of the zodiac. Moreover, the seven known heavenly bodies were used not only for classifying and naming metals, colors, and parts of the body, but also for the cyclical ordering of days (i.e., the seven days of the week are named after the planets). This arrangement rests essentially on astrological beliefs, which in every culture gave rise to astronomy and thereafter continued to live on beside it.

After the conquests of Alexander the Great, Chaldean astronomers streamed westward into the ancient Mediterranean lands and acquainted those peoples with their own culture, in which the number 12, as well as 60 with its multiples 300 and 360, were so important. Many traces remain in the Bible among other places, as in the 12 tribes of Israel, the 12 apostles and the $12^2 \times 10^3$ "sealed" servants of God of the Apocalypse (Revelations 7:4, and 14:1 & 3).

Around 2500 B.C. the Sumerian state was overthrown politically by the Semitic Akkadians who came out of the desert, but the Sumerian culture overwhelmed that of the conquerors. In addition to the Sumerian numerals, the Akkadians also took over cuneiform writing and adapted it to their own very different language. They called the number 60 *šuššu* (cf. the Hebrew word, p. 115), which the Greeks altered to *sóssos* and which shows a great resemblance to the German word *Schock* ["shock"]. From 600 came the word *ner*, Greek *néros*, and from the Sumerian 60^2 came the Greek *sáros*. Quite understandably, these borrowed words were soon eroded to indefinite round numbers, just as in the case of the Greeks and Romans, among whom *sescenti*, 600 became a round number for the same reason.

Another of the many mystical connections of the number 60 can be seen in the Platonic number:

60^4 years $= 1$ "world year" $= 360 \times 36,000$ years $= 360$ "world days" (in each of 36,000 years).
1 human life span $= 100$ years $= 36,000$ days.

Thus one day in a man's lifetime is the same as one year in a "world day." The human life span is the 360th part of a world life span or 1 degree of the circular orbit of the universe, if we equate days and years.

THE BREAK AFTER SIXTY IN THE GERMANIC NUMBER SEQUENCE

After the foregoing excursion, in which we explored the numbers 12 and 60 and the reasons for their prominence, we must go back once more to the main subject of our discussion and ask ourselves: Is this break after 60 in the Indo-European number sequence (for example, Gothic *six tigjus — sibuntehund*) due to Babylonian influence?

The reader may answer hastily: of course! If so, he fails to remember that the migration of certain aspects of Babylonian culture into the Mediterranean world, such as the Babylonian measures (time, angles, and weights) and their round numbers, took place in the last few centuries before the birth of Christ and thus in the full light of recorded history, whereas the Indo-European number sequences go all the way back to the murky realms of prehistory, long before 3000 B.C. Hence the chronological overlap of the Mesopotamian and Mediterranean cultures is subject to question. There is also serious doubt about their geographical contact: the Sumerians inhabited the southern part of Mesopotamia, while the Indo-European peoples in that period lived very far from the Tigris and the Euphrates.

Then what about the derivation of the "shock" — the relationship between the Greek *sóssos* and the Babylonian *šiššu*? If the former was really derived from the latter, then this happened in the Greek language long after the Indo-European number sequences had already been formed. But the most telling objection to this seeming bit of evidence is that even if the "shock" does have a Babylonian origin, as is widely believed, it is still unthinkable that the mere use of this one measure of quantity, which was no more frequent than the *Mandel* or the *Stiege*, could have altered the very deepest phonetic roots of an Indo-European number word. And not just the word for 60, but also the words for all the tens following 60! Since people quite frequently count beyond sixty, 70, 80, and 90 should also have acquired new names, if the hypothesis were valid. But it is definitely out of the question. The Babylonian number rank 60 was rooted in a very different conceptual soil from that of the Indo-European number words after 60.

If the break at 60 is not due to Babylonian influence, then how did it come about? Here I should like to offer a suggestion which has already been made by others (see M. de Villiers, *The Numeral Words*, London 1923). The change in Indo-European number sequences from 70 on is focused in every case on the syllable *-hund* (or *ant-*; see p. 94). And this syllable is the numerical rank "hundred," whose effect extends back as far as 70. We have established numerous instances of this sort of reverse influence of a particular numerical rank which at first stood out prominently in the advancing number sequence, and then sought to be tied down into the preceding portion. This is an event in the building of the number sequence which can be readily understood in the light of early man's conception of numbers (see p. 46). The Germanic fashion of overcounting and counting by halves, which we discussed at length, created the number word for 95, for example, by reaching back from the numerical rank 100 into the numbers preceding it (see p. 76).

But why the backward influence from 100 to exactly 70? The break in the sequence of tens mirrors a similar break in the sequence of units. The latter can be seen in Latin, for example, in which the words for 7, 8, and 9 are formed with the ending *-em* that has been taken from *decem*, 10: *novem*, *octo* (a dual form for *octem*!), *septem* (see table on p. 92). Whether this conceals an erstwhile counting back from 10, which occurs in other number sequences (such as the Finnish; see p. 75) where 9 = 1 subtracted from 10, 8 = 2 from 10, and 7 = 3 from 10, is a possibility that may require further study. Comparative philology alone can decide whether such suggested derivations for 9 as the Greek *en-ap-ken > ennefen > ennea* and Latin *(u)nu-(a)p-hem > novem* have any validity.

Now let us summarize: the reverse influence of a numerical rank upon the number words preceding it (as from 10 back to 7), as well as the reflection of a particular feature from one order of magnitude (in this case, the units) to another (the tens) — a typical example of this is the break in the Slavic number sequences after 4, which is repeated after 40, 400, and 4000 (see p. 24) — these are the results of natural forces, so to speak, which are manifested in the structure of a number sequence and thus may quite readily have affected other Indo-European number sequences as well. For these reasons the non-Babylonian explanation seems far more natural than the influence of an alien number sequence, which, though in itself quite possible, could at the very most have inserted a single foreign number word into the Indo-European sequence (just as the Aryan *sata*, 100, came into the Finnish language), but would never have had the power to alter all the subsequent number words as well.

Even if the Babylonian sexagesimal gradation is not the direct answer to our question, however, it still has deepened our understanding of the problem, quite apart from any further insight into the cultural history of numbers which we may have gained from our digression into Babylonia.

Hidden Number Words

In the course of our discussion thus far we have looked into the meanings of number words, we have tried to make out the original images that underlie them, and we have found that with the passage of time the images of the units, in particular, have suffered so much attrition that we could barely make out the few distinct traces they left.

Now we come to hidden number words! These are words to which a number word has contributed a supporting image, which sometimes appears openly, as in the German word *Zwie-licht* ("twi-light"), but may at times be fully concealed, as in *Samt*, "velvet." Words of the latter type are especially noteworthy. From the philologist's point of view it is particularly fascinating to observe how cleverly and ingeniously man has used a number to indicate the essential character of some object which he wished to name and to which no one now would attribute any numerical quality. We shall confine our examples to the most familiar languages, but the reader is urged to refer to others, especially Chinese.

ONE

The word *simple* < Latin *sim-plex*, "once-folded" (*plicare*, "to fold") is one of a large group of words whose original stem is the Indo-European root *sem*; we have already listed many of these derivatives (see p. 146). Here we can add the German word *Sint-flut* ("flood, deluge"), which in Old High German was *sinvluot*, wherein the Germanic root *sin* means "always, all-embracing"; it is a deluge which inundates everything, and has nothing whatever to do with *Sünde* ("sin"). In the same class is the German suffix *-sam* ("*als Kaiser Rotbart lobesam*"), which as an independent word once meant "of like quality"(compare the Latin *similis*, "like, similar," p. 146).

The German word for zero, *Null*, contains the other Indo-European one-form, *oinós* > Latin *unus*. The Latin diminutive ending equivalent to "-ling," or the German *-chen* is *-ulus*: *mus*, "mouse" — *musc-ulus*, "little mouse." Thus from *unulus* > *ullus*, "little one, oneling, any at all." The Indo-European negative syllable *n* combined with this to form the Latin word *n-ullus*, "none, not any." The numeral 0 obtained its name "null" because in the medieval view it was "no (numeral)," *nulla* (*figura*); its history will be discussed later. Another relative of "one" is the Latin *non*, "not" < Old Latin *ne-oinom*, "not one," as is the German *nein* < *n-ein(s)*. Other members of the same tribe are the English words *on-ly* < *one-like* and *al-one* < *all-one*. Moreover the Italian and French words for "boar" are *singolare* and *sanglier* respectively, derived from the same idea of one-ness, perhaps because the wild boar lives or ought to live alone.

The German word *Eimer*, "bucket" < Middle High German *eimbar*, Old High German *ein-bar*, *amber*, is borrowed from the Latin *amphora*, whose diminutive form *amporla* is still in current use as the word "ampoule." The Greek word *am(phi)phoreús* (< *ampho*, "both" and *pherein*, "to carry") referred to an earthenware jug with two handles. In popular colloquial speech *Ein-henkel*, "one-handled jug," was derived from *eim-bar* (from the Germanic *beran*, "to bear, to carry"; compare the words *fruchtbar*, "fruitful" and *Bahre*, "barrow, bier"). Analogously, the *Zuber*, "tub," is a two-handled vessel.

The "ace," the playing card with the value of one, comes, of course, from the word *as* in Roman fractions. There is also a large family of words derived from the Latin *pris* — *prior* — *primus*, "early" — "front" — "foremost, first," among which we shall mention only the *Primel*, "primrose," the French *prin-temps*, "spring" ("early time") and *prince*, from the Latin *prin-ceps*, "he who takes the first place" (Latin *capere*, "to take, seize").

We shall conclude, appropriately, with the "migraine," which is related not to "one" but to "half": Greek (*he*)*mi-krania*, "half-skull (pain)."

TWO

The word "two" can either join ("twin," "twain") or divide (German *Zwist*, "dispute, discord"). Man experienced both quite early in his development. Thus there is a vast family of related words in which the number word "two" is latent in various forms: German *zwi*, Norse *tvi*, Latin *bis*, *dis-*, *vi-*, and *ambo*, and Greek *dís*, *día*, and *ámphō*. The reader may recall our comments about the number Two at the beginning of the number sequence (see p. 12).

From the Old High German word for "two" come the German *zwisk*, "every two," whence *in zwisken > inzwischen, zwischen*, literally "in the middle of two" and thus "in between," and the English *betwixt* and *between*. *Twisko*, the "two-sexed," is the epithet applied in the Germanic sagas of the gods to the Earth-born ancestor who was able to produce a son without having a wife. We find another derivation of "two" in the Icelandic *tuenner*, "every two"; *tuennern* thus means something like "to make two, to double," which is the modern German word *zwirnen*, "to twist, entwine." *Zwirn*, which is *twine* in English, is hence a doubling, a string made up of two strands joined together, as in the English word *twist*; in German the word *Twist* also means a certain kind of cotton yarn. The same word *Zwist* in the Oberlausitz region refers to a cord made up of doubled strands. In the same category are the words *Zwillich*, "drill cloth," and *Zwilch*, a "two-thread cloth," for the Old High German *zwi-lich* is an imitation of the Latin *bi-lix*, "two-threaded," from *licium*, "thread." A verbal cousin of *Zwilch* is *Drilch*, "drill, denim, canvas, a cord or cloth made up of triple thread" (see p. 176).

The word *Zwist* also occurs with the meaning of "strife, discord." The explanation for this lies in the Old Norse word *tvistra*, "to divide" and thus "to break apart into two"; we shall shortly have more to say about this class of words.

The word *Zweifel*, "doubt," is clearly an offspring of *Zwei*, "two," not only in German but also in numerous other languages: for example, Old High German *zwivo*, Old Norse *tyja*; Greek *doié* (cf. "double"). The Gothic *twei-fls* expresses exactly the same idea as the Greek *di-plóos*, "twofold" — *zwie-spältig*, "split, forked, divided," expresses its meaning clearly. The concept of "dubious" is even more picturesquely expressed in the Old High German Book of Gospels of Otfried:

thiz selba uuas ihm untar zuein — "the same was to him between two."

The very same turn of speech still lives on in the Italian expression *stare intra due*, "to stand between two" meaning "to doubt." In Latin the word *du-bius* consists of the elements *du-*, "two" and *bius* < Indo-European *bhu*, "to be" (> Latin *fu-i*, "I have been"). The intensified Latin form of the same word, *dubitare*, has given rise to the French *douter* and the English *doubt*. Similarly in the Greek *en-doi-ázein*, "(to be) between two" we can distinctly make out the "two"-root, just as we can see the number word *kettö*, 2, in the Hungarian *ketelkedik*, "to doubt."

The German word *Zweig*, "twig," has the meaning of "fork," in Middle High German *zwisel*; a place known as *Zwisel* in the Bavarian Forest is named after a fork in the road. *Geweih*, "horns, antlers," is verbally a *Ge-(z)weig*, a "forked," in which the dropping of the *-z-* goes back to the two form *vi*, which is in the same class as the Gothic words *wit* and *weis* (pp. 13 & 150). Besides these we have the group of related words *zwicken* ("pinch; gripe; worry"), *zwacken* ("nip; tease"), and *zwecken* ("to pine, tack"); cf. the word *Zweck*, "goal, purpose," literally "that which is between two fingers or pincers." The *Zweck* was also a "pinched-off" piece of leather placed at the center of a target to serve as a "mark" or bull's-eye.

The numerical meaning underlying the words "twilight," "twin," and *Zwiespalt* and *Zwietracht* (both meaning "discord, dissension") is obvious. But one would probably not expect to find a "two" in the German word *Zuber*, "tub"; yet in its Old High German ancestor *zwi-bar* we see it as the "two-handled" cousin of the *Eimer*, the "jug" which has only "one handle" (see p. 171).

Both Greek and Latin have given a whole series of words to German and English as foreign loan words, whose connection with "two" is thus sometimes hard to see. Who would think of it, for example, in "diploma" and "diplomat"? The Greek *di-plóos* means "twofold, double" and is related to Latin *duplus* > French *double*, which German has borrowed as *doppel* and English as *double*. The Greek *diplóma* is "folded, doubled." The Roman emperors often granted certain rights and privileges to veteran soldiers upon their discharge from service. These were originally inscribed on the inner surfaces of *two* bronze tablets which were held together by rings through one of their long sides, like a loose-leaf notebook. The two tablets were then closed against each other and a wire was drawn through two holes drilled in the center, twisted together several times, and the two ends held with a wax seal. This was a "diploma," an imperial writ specifying certain rights and privileges. Today a "diplomat" still presents his "diploma," his written credentials, to the foreign government to which he is accredited.

The Roman writing tablets that could be similarly folded together were called a *diptychon*, and a "diptych" in art is a painting on two panels hinged at the center; a *tri-ptychon*, "triptych" is a three-paneled altar painting whose two outer wings fold over to cover the center panel (from the Greek word *ptýx*, "fold").

In the word "duel" we can recognize the Latin *du-ellum*, a "fight between two antagonists," which was an older form of the later *bellum*, "war." Here too we see the transition from *duis* to *bis*, "twice," with which the Italians still call for an encore in the theater and concert hall.

And so we come to the large and fascinating family of *bis*-words in the Romance languages. This group has two kinds of members — those in which *bis* has the purely numerical meaning of "two," and those in which it distorts or even imparts a pejorative connotation to the main root word. A good example of the first kind is the French "pushcart," *brouette* < *bis-rouette*, a "two-wheeler"; it is beautifully camouflaged in the German word *Protze*, the two-wheeled artillery caisson or guncart, from the Italian *bis-rozzo* < *ruota*, "wheel." The French *bi-cyclette* and English *bi-cycle*, on the other hand, are quite obviously two-wheeled conveyances. The *Zwieback*, Italian *bis-cotta*, French and English *biscuit*, German *Biskuit*, is "twice-baked" bread; in French a thick, strong broth is thus called a *bisque*, "twice" or "thoroughly cooked," and curiously this is also the term for a penalty point in tennis. A person who squints is known in Spanish as *bis-ojo* and in French as *bigle* (from the Latin *oculus*, "eye"), meaning "two-eyed" or "double-sighted."

Why is the intercalary day (February 29) known in the Romance languages, for instance in France, as *bissexte* and the leap year as *bissextil*, Italian *bisestile*, "two times sixth"? The last month of the old Roman year was February (see p. 25). Every four years an intercalary day was inserted in February, after the 24th, and came to be known as the "second 24th" day of February. In the Roman manner of designating the days of the calendar, this was "the sixth day before the Calends of March" (that is, before March 1, the beginning of the new year), *dies sextus ante Kalendas Martias*; hence the intercalary day was *bi-sextus*, the "twice sixth" day before the Calends. From ancient times this was popularly regarded as an unlucky day, and the French word *bissetre* still means "misfortune."

In Germany the *Bise* is a colored braid on uniforms, like the stripe along the trouser seam. The French *bis-er* is *bis* expanded to a verb, which literally means only "to do something twice" but now has the meaning of "to put on another color," especially black as in mourning; one would naturally not see this nuance in the word itself. Similar derivations from the Latin *bini*, "every two," are the Italian *binario*, "rail, track" and the French *bin-ocle* (< *oculus*, Latin "eye"), a "double glass" like binoculars or opera glasses.

Now some examples of the second *bis*-group: *bis*- can also reduce the meaning or make it pejorative. In French, it is, in this case, readily altered to *bes*, as we have already seen in the words *bissac*, "double-sack" and *besace*, "begging or beggar's sack" (see p. 14). Something which is no longer completely sour becomes *bes-aigre*, "sour-ish"; similarly in Italian *bis-lungo* is "longish," *bis-tondo* is "roundish" and if one does not really sing (*cantare*) one hums (*bis-cantare*). Whoever is "filthy as a pig" in Italian is *bis-unto* (< Latin *unctus*, "anointed"). An "oversight" in French is *be-vue* (< *voir*, "to see"); a "blinking or fluttering before the eyes" is *ber-lue*, "bad light," from the Latin *lux*, "light," whence comes the French *bluette* < *be-luette*, a "spark" which is a "side-light" and is not part of the main light. The French *ba-fouiller* is "to babble, to talk nonsense" (*fouiller*, "to dig"), and *bis-quer* is "to become vexed."

In German this change in a word's meaning is beautifully exemplified in the *Bilanz* and *Balance*, "balance, equilibrium." Both derive from the same words *bis* and the Latin *lanx*, "scale, steelyard." The first of the above German words, which became a part of commercial language by way of the Italian *bi-lancio*, means a "balance" between assets and liabilities and hence something that has achieved equilibrium and come to rest, whereas *Balance* refers to suspended objects which are not yet in equilibrium but are constantly and restlessly seeking it; for similar reasons the French call the balance-wheel in a watch the "restless."

The Latin word *bis* was derived from an old two-form *duis*, which also gave rise to the common prefix *dis-*. These two syllables impart a doubling or a pejorative sense to a great number of words that are also used in German and English: to "de-fame" is "to bring into dis-repute" (from Latin *fama*, "name, reputation, report"); "di-stance" means separation; "dif-ference" is a "carrying apart, partition" (from Latin *ferre*, "to carry"); what is "dif-ficult" is "not easy" (from Latin *facilis*, "easy"). The antithesis of *con-cordia*, "a heart (and soul)," is *dis-cordia*, "discord, dissension" (from Latin *cor*, "heart"). The corresponding French prefix is *des-*, as in *dès-espérer*, "to des-pair"; in Italian this is often shortened before a consonant to *-s* as in *s-cordanza*, "bad temper, ill humor" or in *s-fumare*, "to make dim, obscure" (from *fumare*, "to smoke"; cf. the word *sfumato*, literally "smoke-filled, smoky," used as a technical term in the fine arts to describe the soft, blurred outlines of figures as in Leonardo da Vinci's paintings).

To "set something apart" in Latin is *dis-putare*, whence our word "dispute"; *putare*, "to cut," may possibly have arrived by way of the "notched stick" at its more common meaning of "to think, consider, estimate" (see Tally Sticks, pp. 223 ff.). One who must "comprehend something through and through," or *dis-cipere* (from Latin *capere*, "to grasp"), is a *dis-cipulus*, a "disciple" or "pupil." "Dis-cipline" is then "instruction, learning" and by transference also "order." Of the many more examples that belong to this group of words, we shall cite only "di-rector," the "ruler," whose prefix echoes the old Roman maxim *divide et impera*, "rule by dividing into two parties." And *di-vid-ere* may well be the strongest word for "make two of," for it contains both the "two"-roots *dis* and *vi*(*d*).

This *vi-* < Indo-European (*d*)*ui*, "two," which we already met in the Latin *vi-ginti*, 20 (see p. 150), is formed from the Indo-European *uidh*, "to sunder," a word that recurs in the German *Witwe* and English "widow," a woman who has been "sundered" from her husband, and in the German *Waise*, the orphan who has been "sundered" from its parents, whose opposite is the "in-divid-ual" who is "undivided."

In "dialogue" and "dialect" we have the Greek form *diá* as in *diá-logos*, "speech between two" and *diá-lektos*, which is properly a conversation in which two persons take part and has by transference come to be a "dialect" in the modern sense of a specific form of collo-quial speech. The number of words formed with *diá-* in German and English is very large; we shall mention only *dia-gnosis*, "thorough knowledge" (from Greek *gignōskein*, "to know, recognize") and *dilemma*, "dilemma," a "pinch between two horns" (< Greek *lambánein*, "to take"). Greek words compounded with *dícha*, "doubled," may also have pejorative meanings, as in *dichó-nous*, "two-minded" and hence "equivocal, ambivalent, false, faithless." We may end this great family with the example of the Devil, Greek *diá-bolos*, he who "throws between" and thereby causes discord (from the Greek *bállein*, "to throw"); linguistically, the (German) *Teufel* and the (English) *Devil* came from the Greek through the Gothic *diabulus*. Moreover, the German *Daus* and English *Deuce* in dice games was originally the Two, and comes from the Latin *diavolus*, "two-throw."

There is one more "two"-form to be mentioned, which is also com-monly encountered in foreign loan-words in German and English; this is the Greek *ámphō* and Latin *ambo*, "both." An "amphi-theater" is a "double theater"; in the history of Roman architecture this was in fact made up of two semicircular banks of seats, as in the ancient theater with a stage, to make a round building without a stage but with an enclosed arena ("sand" in Latin), as in the Colos-seum in Rome, where circuses and gladiatorial contests were housed instead of dramatic performances. We have already mentioned the *ampoule* in connection with the *Eimer* or "one-handled jug" (see p. 171). As a prefix *amb-* means "from both sides": in Latin *amb-igere* is "to pull from both sides" and thus "to doubt, to dispute" (cf. English word "ambiguous"); *amb-ire* means "to go back and forth," so that *ambition* was originally an ardent or zealous moving about.

Let us look back: What a huge multitude of words man created from the number "two"! This is understandable: it is the first step past "one," from the I to the You and beyond into the world at large, where all that is one is perpetually divided into two, and the divided are joined into one.

THREE

Drillich, *Drilch*, or *Drell*, "drill," a triple-stranded material used for making army jackets, can be immediately recognized, on the analogy of *Zwillich*, as a Germanization of the Latin *tri-licium*, "three-threaded." The word *Trense*, "braid, twist," has the same origin, for the Latin *trini-care* (< *trini*, "each three") literally means "to three" and here "to braid from three strands," as with hair; in Spanish it gave rise to the verb *trenzar*, whence *trenza*, "plait, braid, cable." This word was brought to the Netherlands by the Spanish army of occupation, whence it came in the 16th century to Germany as *trensse*, a light bridle with a snaffle-bit (in contrast to the Hungarian *Kandare*, a heavy bridle with a curb-rein). The French *tresse*, Italian *treccia*, from the same root, which was thus originally a "braided" edging, was brought to Germany in the 18th century along with French fashions in clothing.

A *Dreidraht* [literally a "three-cord"] in German is a slow, dawdling, tiresome person, a bore. *Drahteln* in the Swabian dialect is "to delay, dawdle" and may very well have taken its meaning from *drehen*, "to twist, turn" and *Draht*, "wire, cord." *Dreidraht* then is probably an alliterative and cumulative doubling, formed all the more easily here because the number three serves to intensify the meaning of the word.

The foreign loan-word *trivial*, "trivial," in German goes back to the Latin noun *trivium* (< *tri-via*, "three-roads"), which in Roman times and well into the Middle Ages referred to the three basic disciplines of the curriculum: grammar, rhetoric, and dialectics. Once the student has mastered these, he could embark upon the *quadrivium*, the "four paths" of arithmetic, geometry, astronomy, and music. Together these made up the "seven liberal arts." Hence *trivial* acquired its meaning of "commonplace, ordinary, unimportant, and uninteresting." The famous Trevi Fountain in Rome, which is named after its three great streams of water, used to be called the Fontana Trivia.

The word "tribute" goes far back into linguistic history. The modern word is derived from the Latin *tri-bus* < Indo-European *tri-bhu-s*, from *-bhus*, "to be" (see p. 173) and originally meant a "third or third part" and then a "district" or "community," because Rome arose from the three Italic communities or tribes — the Romans, the Sabines, and the Albans. Thereafter *tribus* came to mean a group of related people, a clan or *tribe*, and then a tax or customs district, so that *tributum* is a "payment, tribute." A "tribune" was originally an official in charge of such a district; the raised seat from which he dispensed justice and made decisions was the "tribunal." The German word *Tribüne* (< French *tribune*) then retained only its meaning of "elevated, raised" and became the *Bühne*, or stage of a theater.

This very class of words is a significant example of the wide range of meanings a language can embrace with one simple number word. An intriguing instance of the same thing is *testament*, a word in which "three" is so well concealed that even an experienced philologist may have trouble recognizing it. "Witness" in Latin is *testis* < *ter-stis*, *tri-sto*, "to be a third party." The witness stands by as a third person, as in the illustrative anecdote about pearl-trading related in the next volume, in which such third persons are excluded by using finger-gestures under a cloth (see chapter on Finger Counting, pp. 199 ff.). "Can anyone else (a third person) hear us?" we ask when we want to discuss a private matter. One rap, a second rap, and still another rap by a third person standing by still seals a bid at a German cattle auction. In Russian the corespondent in a divorce suit on the grounds of adultery is called *tretij*, "the third party." There was an Old Latin (Oscan) word *tristaamentud* > *testamentum*, "being the third." The "last will and testament" is thus a bequest to one's heirs in the presence of a third person as witness.

It may also be worth noting that the month of May in Old Saxon was once named *trimilki*, because during that season a cow could be milked three times a day.

Why was a certain kind of small, old-fashioned pocket pistol called a *terzerol*? Allegedly because of the male hunting falcon, which was about a third smaller than the female; Italian *terzuolo* < Latin *tertius*, "third." Originally people saw in the English word *travel* only the strain and effort it entails, which is still expressed in the French word *travail*, "work," or the English "travail." *Qual*, "torment," originated in the Latin *tri-palium*, an instrument of torture which had three stakes, and still lives on in French expressions such as the *travail d'enfant*, "birth pains," and *travaille de la pierre*, "suffering from (gall-) stones."

The *Kümmelblättchen*, "caraway leaf," is actually a foreign loan, although it looks very German. It refers to a game played by card sharpers, in which three of the cards are covered. But where is the "three" hidden in this word — in the *kümmel*? Yes, because this is not a true German root, but a misunderstood assimilation of *gimel*, the third letter of the Hebrew alphabet, which we met earlier (p. 121). The Hebrews, like the Greeks, used their letters as numerals, so that *alef* was 1, *bē* 2, and *gímel* 3.

FOUR

To begin with, the square has always been expressed by the number word "four." The basic form of this is *quadratus*, which Albrecht Dürer turned into the German *Vierung*; today the word is used only in church architecture, for the square formed by the intersection of the long and short arms of a Roman basilica. Thus we come to the large clan of foreign loan-words in German which begin with *Quadr-*, above all *Quadrat*, "square," and *Quader*, "parallelepiped," from the Latin *quadrus*, "four cornered" and *quadrare*, "to square" (after *quattuor*, 4). The *Quadrant*, "quadrant," usually has the meaning of "a fourth part, a quarter"; we think of the Roman coin, the *quadrans* or quarter *as* (see Fig. 32 on p. 158), and in mathematics it refers to a quarter of a full circle; but the French "clock face," *cadran*, which is now generally circular, takes its name from the days when sundials had square faces. Then we have the fractional form with *t*: *Quartal*, a "quarter year" and *Quartier*, a "quarter (of a city)," which is a standard example of the semantic fading of number words: neither in German nor in other languages does it have any connection with an actual division into four parts. In Venice such a quarter or district is called a *sestiere*, literally a "sixth" of the city. The true meaning lives on, however, in the Dutch expression *een kwartier over sesse*, "a quarter past six." The *Quarte* and *Quarta* ("fourth, quarter"), are derived from the Latin ordinal form *quartus*, "the fourth"; the "quartet" is the same with the addition of the Italian diminutive ending *-etto*. The *r*-less word *Quatember* is derived from *quattuor tempora*, the "four seasons"; in the Roman

Catholic Church this is the name for the three holy days that intro-
duce each quarter year.

The Italian *quadro* is a painting, properly speaking a square panel,
while the same word *cadre* in French means "frame." In French
military terminology, especially in the form *carrée*, it also has the
meaning of the "core" of a troop of soldiers, which it took on from
the Italian *squadra*, "battle square" (< Latin *ex quadra*). In the 16th
century the augmentative form *squadrone* (see p. 134) also went
into the French language as *escadron*, from which the word "squadron" was formed. *Geschwader*, "squadron" (or airplanes or ships)
is a good Germanization of the same word *Schwadron*, a "squadron"
(of cavalry). The same origin might be attributed to *schwadronieren*,
"to boast, to cut open," but this word may also have come from the
Middle High German *swatern*, "to gossip." The basic word in its
unaugmented form is also present in the English *squad*, a small detachment of a dozen or so soldiers. The Italian *quadrigla* is a "four-
cornered" troop of soldiers or horsemen, whose French form
quadrille, a dance of four couples, clearly betrays the numerical
idea behind it.

The *Kartaune* < Medieval Latin *quartana* is a "quarter"-cannon
because it shot 25-pound instead of 100-pound cannonballs. The
Kaserne ("garrison"), which since the time of Louis XIV has been
the name for a very large barracks or group of barracks in which
soldiers were housed, was originally merely a shelter for "four"
soldiers on guard duty (from the Latin *quaterna*, "every four").
The same Latin word in the Middle Ages referred to a pile of paper
consisting of four large sheets, one on top of another, which were
folded in the middle and bound together, thus forming 16 pages —
the "bend" which still governs the strength of a book. *Quaternio*
became *caterna*, *quayer*, and finally French *cahier* ("notebook")
and English "quire," which in Italian is still called *quaderno*.

The "diamond" on playing cards, which is *Karo* in German, was
originally a French word: Latin *quadrum*, diminutive *quadrellum* >
Old French *quarrel*, *carrel* > *carreau* (Latin *-l* > French *-au*, as in
pellis > *peau*, "hide"); thus *kar(r)iert* means "checkered" or broken
up into small squares.

When rocks are broken up, they are also shaped into "square" or
quadratic form, so that the pit where this is done is a "quarry"
in English. "Dice" are among the most common examples of a small
"four-cornered" body. Thus we find them in Roman mosaics as
tesserae, giving rise to the English word *tesselated* (from the Greek
word *téssera*, "four"; see p. 104). Another word of Greek origin
is *trapezium*, which in German and English is a four-sided geometrical
figure, while the "trapeze" is a device used in acrobatics: the Greek
trápeza is actually *(te)trápeza*, a "four-foot" or table with four legs
(< *péza*, "foot" and *tétra*, short form of *téttares*, "four"). (In
Modern Greek *trapézi* is a "table" and *trápeza* a "bank," revealing
the latter's origin on the money changer's table — Trans.)

The Italian word for "to cut into four parts, to quarter" is *squartare*; for "to hoard, to be stingy" Italian has the amusing expression
squartare lo zero, "to divide zero into four." The Italian hangman

is called, among other things, *squartatore* from the old custom of quartering a man's corpse after hanging him.

People tend to think of the square not only as something with four corners, but also as something regular, something orderly. Thus a "square man," in Greek *tetrágonos anḗr*, is a good, a righteous man, and in Italian the expression *essere fuor di squadra* means "to be unsuitable" (literally: "outside the square"); the English "it does not square with me" is similar. The French *tête carrée*, however, is a stubborn, obstinate "squarehead," whereas the Italian *testa quadrata* also has the meaning of a capacious and hence a "clever, intelligent head."

Now for a Germanic example: the British quarter-penny was called the "farthing" < Anglo-Saxon *feorth-ing*, a "small fourth." A small "quarter" and thus a little piece of cake is a *farle* for short, a fourth of a keg (as of butter) is a "firkin" in English, and a quarter of a grain measure, the Scottish bushel, is a *firlot*. If we add to these the "fourth" of Romance origin, the "quarter" and the "quadroon" or "quarter-blood," the racially mixed person who has one white and one mulatto parent (so that his blood is thought to be "$\frac{1}{4}$ black"), we have a very interesting "quarter"-series in which the number word changes form according to the object it specifies — a phenomenon we have learned to recognize as a sign of early formation.

Let us conclude with some examples of the way in which a pure number word can become the name of an object. In the Celtic group of languages the Irish *cethir*, "four," has also come to mean "cattle" (= "four-footed"). Frenchmen use the word *quatrain* (and also *sixain*) for the goldfinch, because this bird has 4 (or 6) white spots on its tail. Similarly in Modern Greek a rose is *triantá-phyllo*, the "thirty-leaved" flower, and the nightingale in Finnish is *sata-kieli*, the "hundred-voiced" bird.

FIVE

During the 17th century, the English acquired a taste for a refreshing drink from the East Indies, which was made with five ingredients: arrak, sugar, lemon juice, seasonings, and water. To this beverage they gave the name "punch," after the Sanskrit Hindi number word *panča*, "five."

The "quintessence" is the *quinta essentia*, the "fifth being" and hence that which is "essential." According to Aristotle there were four "essences" (elements), fire, water, air, and earth; the fifth essence was the ether, the all-embracing Divinity.

The "quint" is the name for the highest string, the E-string, on the violin. Yet this is not the fifth, but the fourth string on the modern instrument; it retains the name from the old viol, which had five strings.

The French *quinte* is the jurisdictional boundary of a city with a periphery of about 5 miles; similar explanations make clear the underlying meanings of the Spanish *quinta*, "country house," *quinteria*, "tenant farm," and *quintar*, to muster all the inhabitants of a city up to the city limits for military service. There is a peculiar French expression *la quinte de toux* for a fit of coughing, from the superstitious belief that this recurs every "fifth" hour. And the French word *quinte* for "feint" and then "fancy" comes from fencing jargon, in which the "quinte" is the "fifth" thrust — three *quintes* with three different meanings. On the other hand, the French hundredweight, the *quintal*, has nothing whatever to do with "five" and is derived from Arabic (see p. 186).

The German word *Quentchen*, the "dram" or "little fifth," was originally a fifth of a ten-gram or half-ounce weight, and later became a quarter.

In the Bavarian and Austrian dialects Thursday, *Donnerstag*, is also sometimes called *Pfinztag*, whose first syllable reflects a Gothic *pinte* < Greek *pémptē*, "fifth"; thus it is the "fifth day" of the week, counting from Sunday. *Pfinztag* must have migrated with the Goths, who were Arian Christians, up the Danube River to Bavaria, for it shows the High German phonetic shift from *p* to *pf* (see table on p. 127, column II line 5); had it come from the Roman Catholic Church, it would have arrived later, and in the form *Pinztag*. This is a very impressive example of how the age and origin of a word can be revealed in its form. Thursday is *Pémptē* in Modern Greek, in which most of the days of the week have numerical names: Sunday is *Kyriaké*, ("the Day of the Lord"), Monday *Deutéra* ("Second Day"), Tuesday *Trítē* ("Third"), Wednesday *Tetártē* ("Fourth"), Thursday *Pémptē* ("Fifth"), and then Friday *Paraskeué* ("Day of Preparation" for) Saturday, *Sábbato(n)* (the "Sabbath").

SIX

Here we encounter the semester, for example, from the Latin *sexmenstris*, a "six-month" period, but also a word whose connection with "six" one would never suspect — *Samt*, "velvet." The Middle High German word *samat* was borrowed around the year 1200 from the Old French *samit*, which came by way of the Italian *sciamito* from Medieval Latin *(e)xamitum* < Greek *(he)xámiton*, from *héx*, "six" and *mítos*, "thread." In Byzantium this was a cloth woven from "six threads" and thus the finer cousin of the *Zwilch*, "canvas" and *Drilch*, "drill," which we discussed before (see pp. 172–176). In Italy the material was made with shorn upright threads and was known as velvet.

The *siesta* is the Spanish "sixth (hour)" < Latin *sexta (hora)*, which was midday in the old 12-hour day running from sunrise to sunset; in southern lands, of course, it has come to be the hour at which people take their midday naps.

An old German grain measure was the *Sechter* or *Sester*, from the Latin *sextarius*, a "sixth" (of the Roman measure of volume, the *congius*). From the same Latin word comes the *Ster*, which is still used in Southern Germany to indicate an amount of cut wood equal to 1 cubic meter. The word is also known in Italy as the bushel, which took the forms *(se)stario*, *staio*, and *satiolo*, in Italian, and *setier* in French.

In Russia waiters used to be addressed somewhat contemptuously as *Shestochka*, "little six," after the rather insignificant six-card in card games (from the Russian *šestj*, "six"). We have already mentioned the old Roman leap year, the *bisextil*, in our discussion of the word "two" (see p. 174).

SEVEN

The mysterious and superstitious significance of the number seven originated in the number of "planets" — bodies which did not have fixed places in the heavens but were "wandering" stars: the Sun, the Moon, Venus, Jupiter, Mars, Mercury, and Saturn — known at the time of the Babylonian astrologers. They were regarded as messengers of the gods, and terrestrial things like the metals, colors, precious stones, parts of the body, and later on the days of the week were ordered according to their pattern. This led to our present seven-day, planetary week whose days are named after the pagan gods; Latin designations such as *Dies Solis*, "Day of the Sun," etc., probably reflect Egyptian influence. The Jews, who brought the Babylonian seven-day week to Rome in the first century A.D., had other names for the days (see below).

The colloquial German expression *die böse Sieben*, "the evil seven," for a vicious-tempered wife is very remotely related to its object in that the seven-card in a certain medieval card game could take all the others, even face cards with the images of the Pope and the Emperor. The image on the seven-card was originally the Devil's, and later came to be that of a woman.

Septen-trion(al) is the French word for "North" ("northern"), and the Italians say *settentrione*, after the seven threshing oxen (in the constellation of the Great Bear) that the "ox-driver" Boötes constantly drives in a circle around the north pole of the heavens (from the Latin *tero, trivi*, "to grind, thresh"). Grain was threshed in the old days by driving oxen or other animals round and round in a circle over the grain spread out on the threshing-floor.

Our numerically named months Septem-ber, Octo-ber, Novem-ber, and Decem-ber also begin with seven, from the Latin *september* (*mensis*), the "seventh (month)," and so on. The discrepancy that September today is actually the ninth month, and hence should properly be named "November," stems from the old Roman way of counting the months of the year, which we discussed earlier (p. 25).

The English *sennight < seven nights* reflects the old Germanic habit of counting the nights instead of the days: *Weihnacht* ("Christmas"), *Fastnacht* ("Shrove Tuesday"), the English *fortnight* (14 days), and the German expression "8 days" for a week are all remnants of this old custom (see p. 157). The word *Woche*, "week" is common to the Germanic languages: Bishop Wulfila used the Gothic word *wiko*, "sequence to which we come" for the Christian time interval of seven days; the Old High German *wohha* is related to the word *Wechsel*, "change." The Greek word for week, in contrast, is *hebdomáda*, a "seven-ness"; correspondingly in medieval Latin it was *septimana*, whence came the Italian *settimana* and French *semaine*, but the word "weekly" is *ebdomario* in Italian and *hebdomadaire* in French, from which comes the French word for a weekly periodical.

Apart from the Sabbath, the opening day of the week, the Jews named the days of the week by ordinal numbers from the "First" to the "Sixth" day after the Sabbath. The early Christian church adopted this manner of ordering the days of the week, which was used also by the Greeks and the Goths, with the Sabbath being the holy or feast day, Sunday the first and Friday the sixth day of the week. Thus, as we have seen, Thursday, as the fifth day after the Sabbath, came to be called *Pfinztag*. Later on the Church shifted the holy day from the Sabbath to Sunday and then counted Monday as *feria prima*, continuing through Friday, *feria quinta*, followed by *sabbatum*; the Roman *feria* was a "day of celebration on which workmen and tradespeople rested" (> Italian *fiera*, French *foire*, "fair"); in the Church calendar each day is the holy or name day of a particular saint.

From the standpoint of cultural history, it is extremely interesting that the names for the days of the week were more-or-less permanently adopted by various nations; for example, the Germanic and Romance peoples generally did not keep the numerical names at all, so that in German only *Mittwoch*, "Wednesday," and *Samstag*, "Saturday," succeeded in replacing the names of the heathen gods Mercury and Saturn. Two less commonly known examples, which also reveal the borrowing of foreign number words, are the Russian and the Hungarian days of the week (see tables on pp. 98 and 113):

	Russian	Hungarian
Sunday	*Voskresenje*, "Resurrection"	*Vasarnap*, "Market-day"
Monday	*Ponedjelnik*, "not yet work-day"*	*Het-fö*, "seven-head"
Tuesday	*Vtornik*, "the second day"	*Kedd*, from *Kettö*, 2
Wednesday	*Sreda*, "middle day"	*Szerda*, "middle" (Russian borrowing)
Thursday	*Četverg*, "the fourth day"	*Csütörtök*, "the fourth" (Slavic)
Friday	*Pjatnitsa*, "the fifth day"	*Pentek*, "the fifth" (Slavic)
Saturday	*Subbota*, "the Sabbath"	*Szombat*, "Sabbath" (Greek)

* *Editor's note:* This is Menninger's translation. The Russian word means something like "the day after the end of the week."

NINE
The "nones" are the prayers offered at the ninth hour, Latin *hora nona*, or 3 o'clock in the afternoon according to the old twelve-hour day (cf. "siesta," p. 181); this is also the origin of the English words *noon, forenoon,* and *afternoon*. On a different plane, prostitutes in ancient Rome were called, among other things, *nonariae* because they did not open the doors of their houses until the ninth hour.

TEN
The tithe (*Zehnte* in German) was a tax based on an ancient Jewish form which was adopted by the Christian Church and later by the Roman authorities; it was a tenth part of the harvest and was collected in a "tithe barn." Since this tax was prevalent everywhere, we find the same formation in the English word *tithe* < Anglo-Saxon *teotha*, Old High German *zehanto*, and French *dime* from the

Latin *decima* (*pars*). *La Dîme* was the title given by the Dutchman Michael Stevin in 1585 to his textbook on decimal fractions; the same title in Flemish was *De Thiende*. In ancient Greece a *dekátē* or tenth part of any booty was set aside as an offering to the gods.

The word *decimate* goes back to the Roman custom of having only every tenth man pay a penalty instead of extracting it from the whole body of men. Thus if a Roman legion mutinied, every tenth legionnaire was put to death. The medieval saying, *Clericus clericum non decimat*, "a churchman does not decimate a churchman," is the equivalent of the German proverb, *Eine Krähe hackt der anderen kein Auge aus*, "one crow does not peck out the eye of another."

Similarly from Roman military language came the words *Dekan* in German and *dean* in English: in a Roman army camp a *decanus* was a "corporal" in charge of a tent containing ten soldiers. In the Middle Ages the term was elevated to refer to the head of a university faculty or an ecclesiastical official. The ambassador with the greatest length of service among those accredited to a government is known by the French title *doyen*, the dean of the diplomatic corps in a particular capital.

Boccaccio's *Decamerone* (Decameron) was so titled because it is a collection of tales told during a period of "10 days" (Greek *déka*, 10, and *hēméra*, "day"). The German word *Decher*, "dicker," was originally the term used in the fur trade for a bundle of 10 pelts. This designation goes back to the days when the Germanic tribes had to pay a *decuria*, a "tithe," of hides as tribute to the Romans (*c* > *ch* as a result of phonetic shift; see table on p. 127, column I, 1).

From the Latin *deni*, "every ten" came the *denarius*, the "penny" in the Carolingian coinage system (Italian *denario*, "money"; French *denier*) whose initial letter *d* became the abbreviated symbol for both the English *penny* and the German *Pfennig* (in the flowery form ₰). But the Persian *dinar*, which belongs to the same family of related words, was the name of a gold coin. "Penny" goods are called *derrata* from which are derived *denarata* in Italian and *denrée* in French.

In many countries the army drew freely on number words and even numerical ranks for the titles of various officers and commanders of units. The "corporal" or leader of 10 men in the Roman army was a *dec-urio*, just as the *cent-urio* or commander of a "century" derived his title from the Latin word for 100. In Hungarian a corporal is called *tiz-edes* (10), a captain *szaz-ados* (100), and a colonel *ezer-des* (1000), and the Turkish language and army had the exact equivalents in *on-baši*, *yüz-baši*, and *bin-baši* (*baši*, "head"); other military ranks based on the numbers of men commanded were the Greek *hekatónt-archos*, Hindi *śata-patis*, Gothic *hunda-faps*, and the Old High German *zehanzo-herostos*, the former "100-lord" or captain.

TWENTY

Who would have thought that "hussar" has anything to do with twenty? But the Hungarian word *husz* is 20, and *hus-ar* means "twenty-worth." In the 16th century the Hungarian King Mathias

Corvinus decreed that each group of twenty households in a town was to furnish one soldier, mounted and fully equipped, a "hussar," for the army.

FORTY

Among the Semitic peoples the number 40 has been a prominent and commonly used round number (as in the tale of "Ali Baba and the Forty Thieves" in the *Thousand and One Nights*). It occurs very frequently in the Bible: Moses waited 40 days for God to appear on Mt. Sinai; Jesus lived for 40 days in the wilderness; and in the Deluge "the rain was upon the earth forty days and forty nights" (Genesis 7:12). The Christian (*Dominica*) *quadragesima*, properly speaking the "40th Sunday" but actually the sixth, comes 40 days before Good Friday. From the Latin number word were formed the Italian *quaresima* > French *carême*, "Lent." In Greece Lent is still called *sarakostḗ* < (*tḕs*) *sarakostós*, the period "of 40" days.

Yet a similar derivation for the Russian *sorok*, 40, from the Greek (*tḕs*) *sarak(ostḗs*) is impossible on both phonetic and historical grounds. The word *sorok* is lacking, for example, in Church Slavonic which was strongly and directly influenced by Greek. The Russian number word was probably derived instead from the Old Norse *serk(r)*, "pelt"; this found its way into Russian popular speech through the fur trade between the Slavs and the Germanic Norsemen, who counted 40 furs to 1 *timbr* (*Zimmer*, "room") and 5 *timbr* to 1 *serkr* (cf. *Decher* and *dicker*, p. 184; for *serkr* see the German *Bär-serker* ["berserker"], "the wild warrior in the bearskin").

The quarantine period, during which arriving ships were held for observation in case there were any diseases aboard, was decreed for the first time in 14th-century Venice as *quaranta giorni*, "40 days," in reference to the Biblical number 40. Both the practice (the method of protection against the importation of diseases) and the word were brought to Germany in the 17th century by the French (*quarantaine*, "aggregate of 40 units").

FIFTY

The "fiftieth (day)" after Easter was called *pentēkostḗ* in Greek and was brought into Church Latin as a foreign loan-word: Italian *pentecosta*, French *pentecôte*, English *Pentecost*. The Greek word came into German through the Christianized Goths along the Danube. The Gothic *paintekuste* became Old Saxon *pincoston* and Dutch *pinksteren* and, after undergoing the Second or High German Phonetic Shift (see Table on p. 127), appeared in Modern German as *Pfingsten*. For its analogue in cultural history, see *Pfinztag* (p. 181). These linguistic migrations suggest that the Germanic peoples became familiar with Christianity in Southeastern Europe sometime around the 5th or 6th century, as indicated by the two Germanic phonetic shifts.

In England the same holiday has been known from the earliest days as *Whitsunday*, or "White Sunday," because Pentecost was a favored day for baptism and the newly baptized church members dressed in white. The German *Weisser Sonntag* for Roman Catholics is the

Feast of the First Communion, which comes eight days after Easter. Easter, however, is the old heathen word for the feast in honor of the goddess of spring, Ostara, whose name is preserved in both English and German (*Ostern*). The Catholic Church also took over the name *Pas-ha* with the Jewish Passover festival whence the Italian *pasqua* and French *Pâque*. The Jewish festival, of course, commemorates the Exodus of the Israelites from Egypt (Hebrew *pesach* = "deliverance").

SEVENTY

Septuaginta (Latin "seventy") is the name of the Septuagint, the Greek translation of the Old Testament, which was purportedly the work of 72 learned Jews in Egypt around 200 B.C. The Jews who lived in Egypt at that time spoke only Greek.

HUNDRED

The Latin *centum* appears in the now extinct German words *Zentgraf*, "hundred-count," and *Zentgericht*, "hundred-law": the Medieval Latin word *centa* (shortened from *centena*) was, in the legal language of the Franks, a settlement known as a "hundred"; later it came to be the fourth part of a *Gau*, a German province.

The "centner" was derived from the Middle High German *zentenaere*, from the Medieval Latin *centenarium* (*pondus*) a "100-(weight)," which is precisely translated by the English word *hundredweight*. The Dutch word *centenaar* clearly preserves the original Latin form. This weight had a remarkable history in the French language, in which it is called the *quintal*; it is *quintale* in both Italian and Spanish. The word was drived from the Latin *quintus*, "fifth," but has no discernible connection with the actual number 5. What happened was that the Latin word *centenarium* became *kintar* in Moorish Spain, and the Romance words for the hundredweight arose from this, as a result of extensive trade with the Arabs.

In ancient Greece the "hecatomb" < Greek *hekatón*, 100 + *bous*, "cattle," was a sacrificial offering of 100 cattle such as Pythagoras, among others, presented in gratitude to the gods after discovering his famous theorem.

THOUSAND

Roman distances were measured in *mille passuum*, "1000 (military) double steps or paces" of 5 feet each, and thus about 1.5 kilometers; from the Latin *milia*, treated as feminine singular, comes our word "mile." The German word *Kohlenmeiler* ("charcoal pile") is derived from the Latin *miliarius*, "containing 1000 pieces," and has been used in the Eastern Alpine region to designate the amount of wood in a layered pile of logs.

Thousand is often used as a round number in German: *Tausendschön* ("daisy"), *Tausend-fuss* ("centipede, millipede"). Thus it may seem remarkable that in other languages, other numbers have had their meanings inflated in exactly the same way: in Modern Greek, for example, the daisy and the centipede have names which literally mean "five-beautiful" and "forty-foot" respectively. Since numbers that are not numerical ranks become semantically inflated mainly for mystical reasons, their interpretation is no longer our concern. We shall merely mention the flower called the milfoil or yarrow, which is *Schafgarbe* in German, because its botanical name *mille-folium* was Germanized not to "1000-leaf" but, probably because of the very similar sound of the Greek word *meló-phyllon*, to "sheep-leaf" (*Schaf-blatt*).

We shall conclude appropriately with the artificial decimal system of measures that was devised in France at the time of the French Revolution and officially adopted in Germany in 1872. These measures use the Latin names of the numerical ranks to form the fractions of the unit measure and the Greek names for its multiples. To indicate length, for example, we have the following measures:

milli-, centi-, deci-meter ← meter → (deka-, hecto-), kilo-meter.

NUMBER

Now we have finally come to the end of our discussion and interpretation of number words. But where does the German word *Zahl* ("number") itself come from? In Germanic we find a *tal*, which appears for example in the Gothic *talzian*, "to teach," and in Old Norse as a "report, narration, story." Thus we have the English word "tale," as well as "to tell," and "to talk." In the King James version, Psalm 147:4 reads:

"He telleth the number of the stars; he calleth them all by their names."

The English expression *all told*, which means "all in all" and literally "all counted," uses the word in this sense, and a *teller* in both England and the U.S.A. is a cashier in a bank; in the British Parliament he is the voice or vote "counter." The Middle High German word *tal* likewise has both meanings, "number" and "speech, discourse." Hence it follows that among the Germanic peoples *tal* was a much-used expression related to their domestic and agricultural economy and the calculation of its yield ("accounts"). If we trace this word back through its Indo-European relatives, we come up, surprisingly, against the Latin words *dolare*, "to hew, cut, chop," *dolium*, "tub, cask, drum," *dolere*, "to grieve, feel pain," and *delere*, "to destroy," all descended from the root *del*, "to cleave, hew, cut." Thus we can understand the original Germanic word *talo* as a "notch" carved into a stick for the purpose of counting. The old German measure of length, the *Zoll*, which was originally a small notched stick, also belongs to this class of words.

But this word does not occur outside the Germanic languages. Instead, we have the Greek *a-ri-thmós*, "number," *né-ri-tos*, "numberless, innumerable," and then the Old Saxon and Old High German *rim*, "series, number," and the Modern German word *Reim* ("rhyme") — all these, through their common element *-ri-*, betray their origin in the Indo-European root *rei-* or *re-*, "to count, to arrange in order" (whence our word *ri-te* meaning "holy order"; we have already come across it in the word *Hund-ert*, see p. 130).

The Latin *numerus* (> French *nombre*) has give us the German *Nummer* and the English "number"; this goes back to a root *nem-*, "to count, order, arrange," which appears in the Greek words *némō*, "I divide" and *nómos*, "law," and the Latin *nummus*, "coin."

With this we approach the end of our study. Our hidden number words have offered us some unusually clear glimpses of the workings of early man's mind and of his verbal sureness of aim. But first and foremost, they have shown us the enormous and unsuspected importance of number words in his life and environment. He did not use his "one — two — three" just to count his sheep and cattle. To early man, number words were far more meaningful. Each had its own individual features, each was imbued with visual significance and had all the vitality of a young word still rooted in the everyday life of ordinary people.

Now that we have really come to the end of our wanderings through the realm of spoken numbers, let us pause for a while and look back over the road we have traveled. Although we may have had some notion, when we set out, that the number sequence was "made up" for the express purpose of ordering men's affairs, we have now come to recognize its most fundamental and important quality, the evolution of the number sequence.

The Evolution of the Number Sequence

The number sequence was not created or "made," it did not spring more or less fully formed from the mind of a single man of genius — it grew up and evolved slowly and randomly with man himself and his various languages. Like a frail plant, it grew and budded timidly, going from "I — you" to "one — two" and then on to three and four, which was the first of the early limits of counting. These limits became articulations in the number sequence, like the joints in a cornstalk. We found more of them at the ranks 10, 100, and 1000.

Yet the vigesimal group made up of man's fingers and toes also left frequent visible traces. Limits of counting, groupings, and changes in word form are all evidence of the slow evolutionary growth of the number sequence, as are the specifications of words meaning "many," and foreign loan-words.

We saw the surprising number and variety of "number towers" which individual peoples erected, although we had assumed at the beginning that there was only one such tower of numbers — our own — and that all the others differed only in being expressed in a different language. We saw number sequences made up of gradations of 10 and 20, continuous and interrupted sequences, cumbersome sequences which pile up layer upon layer, and fast-climbing sequences in which the pure pleasure of building up the series almost gave wings to the numbers and their meanings. This multiplicity is profoundly impressive, for it is manifested not only in the actual, practical, day-to-day existence of various peoples, but also in their intellectual and spiritual lives and in their mental creations. Although the number sequence is conceptually unique and stands alone, the many different attempts and tendencies that surround it constitute an absorbing intellectual drama which leads up to its ultimate denouement — the number sequence as we know it today. This unfolding drama casts an especially strong spell when we view it against the panorama of world history.

Number words and history! The interweaving of our number sequence with the Indo-European family of languages has enabled us to look far back into prehistory, where only shadowy outlines are to be seen, and where we learned with astonishment that many now distinct and separate peoples at one time spoke the same language. This mother tongue first built up a number sequence out of gradations of 10. We have also seen that the daughter languages became separated from their mother tongue some "thousands" of years ago, and that some of these enjoyed a rapid and rich development while others held on with difficulty to their secondary positions, and were forced to borrow from their sisters when they did not have the strength to evolve for themselves.

Then let us recall the abundance of events that have taken place in the bright light of recorded history. To mention only a very few: the Norsemen brought the Germanic vigesimal gradation to France. Decayed and hidden number words in the English language bear witness to an earlier Celtic occupation of the British Isles. Christianity with its Latin number words and habits of counting displaced the

Old Norse custom of "over-counting." The same Norse over-count
testifies to the contiguity of the Germanic peoples with the Finns
and reveals the traces of a once voluminous Scandinavian trade that
extended into Central Russia as far as the Urals; this is curiously
evidenced by the Russian number word *sorok*, 40, which the Norse
fur-traders left behind like a foundling, in the Russian number
sequence.

FOREIGN LOAN WORDS

These are among the best witnesses to past contacts and relationships
between various nations. The most magnificent example of all is
provided by the Hindu-Arabic numerals, which we too borrowed and
which are now commonly used all over the world, exemplifying the
most significant symbol of mankind's universality today.

Spoken number words offer an abundance of examples of language
borrowing one or more foreign numerical ranks because it had not
yet formed its own. For instance, Finnish (*sata*, 100), Hungarian
(*szas*, 100), and probably also Slavic (*suto*, 100) have taken over
the second decimal rank from one of the Aryan tongues (*satam*);
and the Hungarian number word *tis*, 10, very likely has the same
origin (*dasa*). There is no doubt that the Finno-Ugric peoples at one
period lived within the general sphere of influence of the Persian
language, probably in Eastern Russia (on this, see also p. 111).
The Finnish number sequence from 11 through 19 follows the
Germanic pattern.

Foreign borrowings have been even more numerous in the case of
the third decimal rank. The Germanic word for this (*þusundi*,
"thousand") was borrowed by the Lithuanians (*tukstantis*) and the
Slavs (*tysešta*), and from the latter in turn by the Finns (*tuhata*). The
Christian church brought the Greek word *chilioi* into Bulgarian and
Serbian as *hilijada*, where it stands alongside the Germanic form.
Similarly the Latin *mille* migrated into Celtic (*mile*) and Basque (*mil*).
The Hungarian language adopted a foreign word for "thousand,"
ezer < Persian *hazar*. This Persian word is very widespread: we find
it in Armenian, in the languages of the Caucasus region, in Crimean
Gothic, and even in Hindustani, into which it was introduced along
with many other Persian and Arabic words, because Persian was the
official court language of the Moslem princes of Northern India. The
Slavs in the Ukraine have not one but three words for 1000: the
Germanic *tausent*, the Hungarian-Iranian *jezero*, and the Slavic
tysesta.

The borrowings of still higher number words are even more extensive — we need only think of the almost universal word *million*;
forms of this word, in addition, have often changed their numerical
value, as with the Persian *läk*, 10^4 < Sanskrit *laksa*, 10^5 or the Persian
kurur, 5×10^5 < Hindustani *karor*, 10^7. And we should also mention the borrowing of Chinese number words by the Japanese before
we go on to examine the Gypsies' number sequence.

What sort of people are the Gypsies? Hungarians? When they first
appeared in Western Europe around the beginning of the 15th
century, they themselves gave people to understand that they were

descendants of the ancient Egyptian race; for this reason the English still call them (e)*gypsies* and the Spaniards *gitanos*. But let us take a look at their number sequence:

1 *jek*	6 *šob*	20 *biš*	70 *efta-var-deš*	7 × 10
2 *dui*	7 *efta*	30 *driganta*	80 *ochto-var-deš*	8 × 10
3 *trin*	8 *ochto*	40 *dui-var-bis* 2 × 20	90 *ena-var-deš*	9 × 10
4 *štar*	9 *ena*	50 *jek paššel* ½'100	100 *šel, jek-šel*	
5 *panč*	10 *deš*	60 *tri-var-biš* 3 × 20	1000 *deš-šel, jekzeros* or *tisikos*	

Egyptian? Hardly. This is an obvious miscellany of Indo-European number words. Wherever they may have come from initially, since time immemorial the gypsies have been wandering in Indo-European lands. The first units of their number sequence suggest that their initial home may have been somewhere in the Indo-Iranian linguistic region.

More specifically: the gypsy number words for 1 ... 6, and 10 come from the Indo-Iranian group; their 20, *biš*, also goes back to the Persian *bist* < Avestic *visaiti*, Sanskrit *vimśati*; their 100, *šel*, is closely related to the Afghanian *sil* < Iranian *satem* (wherein $t > 1$); their 7, 8, and 9 are Greek and their 30 is Roman in origin.

As the number sequence is further built up, these gypsy words continue to show a variety of patterns: their 20, 40, and 60 are old 20-groups; the intervening steps 30, 70, and 90 were formed from 10 and inserted later, drawing the 80 out of the vigesimal category into their own. 100 as an old grouping exerted a reverse influence on 50, which is ½ × 100. The word *deš-šel*, 10'100, shows that 1000 is an old limit of counting; but along with this indigenous gypsy formation there are also synonyms borrowed from Iranian (*jek-(e)zeros*) and Slavic-Germanic (*tisikos*).

Here we shall end our voyage through the cultural history of spoken numbers, with the extraordinary insights it has given us into the developing human mind and into the lives of primitive peoples and their languages!

From this fertile soil the number sequence grew naturally, much like a plant. It hesistantly tried the first steps forward, to two and then to four, then stood still, then pressed on further as far as ten, where it gathered itself and pushed boldly up to a hundred and then a thousand, and there stood still again, always keeping pace with commercial and social evolution. Then came the point at which its natural growth ceased, from which it no longer evolved unconsciously; now man deliberately drove it ahead over every hurdle and obstacle, past million and billion up to infinity, far beyond the scale or grasp of his ordinary life. Now the popular roots of the number sequence died out: it became an "abstract" sequence, an extraordinary and ingenious, but purely "technical," intellectual tool.

This artificial extension of the number sequence was the verbal analogue of the world-wide currency which the Hindu numerals, "our" numerals, have attained. These two events — the upward growth of the spoken number sequence to infinity and the world-wide spread of the Hindu-Arabic numerals — symbolically express the dominant position which numbers achieved in man's conceptual environment and have occupied ever since. It is significant that both these events occurred in the Western world and at the threshold of modern history, around the year 1500.

Although both events have the same symbolical importance, they are quite different in that the verbal expansion was inward, so to speak, and (until it reached the artificial word *million* and its successors) remained closely tied to a particular language, whereas the written numerals moved from nation to nation, from the European West to the Asiatic Far East and from the "civilized" world of white men to the lands inhabited by still primitive peoples. Number words and numerals: from the very beginning, writing has been the wandering sister of the stay-at-home spoken language. Even "our own" system of writing came from outside our own linguistic region.

This brings us to the concepts and ideas associated with written numerals. Numbers, of course, do not just trip lightly over the tongue — both economic and social necessity have forced men to write them down and to use them for keeping records and for making computations. Today we possess a complete sequence of numerals and have developed procedures for computation. But we shall find, just as with spoken numbers, that written numerals did not suddenly appear fully formed against a blank historical background; they too evolved through many different forms devised and used by individual cultures. We shall see that people, before they could conveniently make computations and carry out operations on paper with written numerals, made ingenious use of the abacus and the counting board. The ability to calculate solely with written numbers was provided for the first time by the Indian numerals.

Now, in the second section, let us turn to the cultural history of written numerals and the computations made with them.

WRITTEN NUMERALS
AND COMPUTATIONS

Preface

The next section of our cultural history of numbers deals with those written numerals and computations that have been used by ordinary people in simple domestic calculations. The main subject of this section is the development of the numerals and computations used in Western Europe. From time to time, however, some light may be shed on this development by occasional glimpses of the numerals and computational operations of other peoples, whose cultures may be either closely or distantly related to our Western civilization.

The Indian numerals which we use today in writing and computation will be discussed before we take up the Roman numerals and counting boards. The history and influence on written computations of the now forgotten counting board or abacus will be the subject of a thorough treatment, more exhaustive than any other published thus far, to my knowledge. It is not only an absorbing segment of the history of numbers and mathematics, but also and above all it forms the background for the fascinating events which accompanied the introduction of Indian place-value notation into the Western world. We come upon this seldom-trod path in the midst of the intellectual revolution from which Western man emerged to "take the world into his hands." Numbers have been one of the most important means by which he learned to master his environment.

The previous section dealt with number words — the number sequence and spoken numbers; this section concerns itself with numerals. Each section, representing spoken numbers or written numerals, can stand alone, although many themes are common to both and many developments and events will not become fully clear until light has been cast on them from both directions.

Introduction

When we speak of Roman or Egyptian or even our own numerals, we almost always mean the "official" numerals in which a state's records are kept and which all individuals are thus obliged to know. Most people are completely unaware of any other kinds of written numbers.

Yet historically every "official" system of numerals is preceded by some set of primitive number symbols, signs which individuals or associations or villages have used to "write" down numbers, in the form of notches carved in a piece of wood or of knots tied into a string cord, or in some other popular manner. A study of these pre-"official" recorded numbers will acquaint us with a multitude of now generally forgotten things that are still alive in language and custom, and above all will provide valuable insights into the "early" concepts which underlie or antedate written representations of numbers. These insights are important, moreover, because even the formal numerals frequently embody early forms which can be understood only with a knowledge of various folkways. A surprising amount of light is thrown on Roman numerals, for example, by notched tally sticks.

To "write" a number is to make it visible and record it permanently, in contrast to speaking it. Thus finger counting, the representation of numbers by various signs made with the hands and fingers, makes numbers visible but nevertheless evanescent, in a sort of intermediate stage between spoken and written numbers. These "finger numbers" are not merely child's play: they had great importance and significance in earlier times, as they still have today in many parts of the world.

These two small bits of information about finger counting and primitive number symbols already suggest that a people's formal written numerals do not weigh as heavily in the scales of culture as spoken numbers do. The written numerals are usually foreign imports; only seldom, as in the case of Egyptian and Chinese, do they grow out of the same native soil as the spoken numbers.

We should suspect from this fact that letters have also been used as number symbols. The Goths wrote alphabetical numerals in imitation of the Greeks, and so did virtually all peoples within the sphere of influence of the (Semitic) alphabetical form of writing. For many centuries they were the numerals used by Greek mathematicians.

These letter numerals and the so-called "early" numerals, whose systems were based only on the laws of ordering and grouping, were used merely to record numbers; they were not well suited to computations. For their numerical calculations people used the counting board or abacus, whose now forgotten history we shall also look into. We shall see its use in the world of the ancient Greeks and Romans, we shall find special forms of the abacus still in use in China, Japan, Russia, and even in our own culture, and then we shall turn our attention to the medieval counting boards used in Western Europe: to their different forms, to their remarkable ancillaries, the *calculi*, *apices*, and other counters, and finally to the main procedures used in making computations on the abacus.

[196]

Thus we shall have gained some insight into the way people during the Middle Ages were working with numbers at about the time the "new" Indian numerals began to penetrate into Europe by way of Italy, around 1500.

Where did the Indian numerals come from and how did they develop? What was new about them? By what route were they brought into the lands of the western Mediterranean and from there to northern Europe? The answers to these questions will unfold a vast drama of civilization extending from India through the Arab world into the West, from the quiet monastic cells of the early Middle Ages to the progressive offices of the Italian and German merchants and beyond, up to the threshold of modern times. We shall take part in this fascinating pageant of intellectual history, and thus understand how and why the Indian numerals came to be the only written numbers used by all the major nations of the world. These, and only these, enabled man to extend his capacity for calculations to undreamed-of heights. For better or for worse, they have subjected the world to the tyranny of numbers.

After this broad view we shall finally focus our attention on our narrower subject, spoken numbers and written numerals, this time in the Far East, in China and Japan. The culture of the Chinese and the Japanese is so unique and so isolated from the rest of the world that it is useful in two ways: it serves both as a comprehensive review of the main threads by which we guided ourselves through the complicated cultural history of numbers, and also as a counterpart, an analogue to our own world of numbers, from which we can learn what is extraneous or peculiar to our own numbers and what is unique and intrinsic to numbers and numerals themselves.

The common threads which we have followed through different times and civilizations will thus remind us that the common spirit of mankind first once began to evolve a single set of numbers, and then struck out along different paths represented by various separate cultures, until it finally arrived at the most highly perfected system, that of the Indian numerals. As these spread out over the whole inhabited world and came to be universally used, pushing aside all other numerals, they have become a symbol of the essential unity of mankind, which is manifested not at the outset, but only at the end of a long course of evolution.

Finger Counting

Finger Counting

Nunc mihi iam credas fieri quod posse negatur:
octo tenes manibus, si me monstrante magistro
sublatis septem reliqui tibi sex remanebunt.

"Now you shall believe what you would deny could be done:
In your hands you hold eight, as my teacher once taught;
Take away seven, and six still remain."
A Roman riddle, which was never solved during the Middle Ages
(see p. 204)

Language uses words to capture numbers. But words are ephemeral. From time immemorial, mankind has tried to find some way of making words and numbers permanent. The answer to this problem was writing.

But to fix spoken words by means of a pictorial or phonetic form of writing is a very difficult and laborious task, which only a few peoples in the world have succeeded in accomplishing. Even we, for example, though we speak a language of our own, write it with symbols that we did not invent. For "our" letters are actually Roman, and the development of alphabetical writing goes back further than the Romans, the Greeks, and the Phoenicians, all the way back to the ancient Egyptians.

With numerals, however, the story is a little different. A truly native system of writing numbers which all members of the culture had to learn and use, and which was thus an "official" set of numerals, was also developed by only a small number of peoples; the numerals used by the vast majority of cultures have been foreign imports. The digits which we commonly use today originated in India; before these, we wrote numbers in the Roman fashion. But before and quite often along with such "official" numerals, people from the earliest times have always devised their own primitive ways of recording numbers, by cutting notches, or scratching strokes, or tying knots, and, they used these means exclusively for carrying on the business of their households, or their trades, or perhaps their villages: we shall discuss these primitive numerical symbols in the following chapter.

But first we will consider still another remarkable predecessor of written numerals: finger counting, numbers formed on the fingers, which is the subject of the present chapter.

The Romans had a method of representing the numbers from 1 to 10,000 on the fingers of both hands — a form of "finger-writing." We shall call these generally overlooked arrangements of numbers "finger counting," to differentiate them from the much earlier finger-gestures of primitive cultures, which very rarely went beyond the "ten" that can be counted on the fingers of both hands.

Finger counting came to Western Europe as part of the heritage of classical antiquity, and was very commonly used during the Middle Ages; thereafter it was replaced by the Indian numerals and has now been forgotten completely. Today such "finger numbers," in somewhat altered form, are used only by Arab and Indian traders in the Middle East.

The technique of finger counting seems to have been passed on mostly by word of mouth. No textbook describing it has been preserved from Roman times. This very fact may be evidence of its widespread use among common or illiterate people, for things which do not require special teachers or schools are generally not written down.

THE VENERABLE BEDE AND HIS FINGER COUNTING
One man did write such a textbook, however. This was the English Benedictine monk who is known to history as *Baeda Venerabilis*,

[201]

the Venerable Bede, one of the greatest and most important scholars of the early Middle Ages. He died in A.D. 735 (see Fig. 247). Posterity has rightly considered him "worthy to be honored."

In those times, when the barbarian migrations over the European continent were gradually coming to an end and the Carolingian empire of Charlemagne was beginning to take form, the Church sent its emissaries, especially the Benedictine monks, to Ireland and England to lay the foundations for a Christian culture. In an age when all books were copied by hand, the monasteries were the sole refuge of learning.

This early medieval learning, apart from purely ecclesiastical and religious matters, amounted to nothing more than collecting and passing on the inheritance of Roman civilization. With all the industry and diligence which only an unworldly monk could bring to the task, this heritage was brought together from all available sources and thus handed on to the later Middle Ages: *Semper aut discere aut docere aut scribere dulce habui*, "I have spent my life pleasantly in learning and teaching and writing," said the Venerable Bede, who never once left the English monastery in which he lived. We have him to thank for the only complete record of finger counting in existence, *De computo vel loquela digitorum*, "On Calculating and Speaking with the Fingers." This forms the introduction to his work on chronology, *De temporum ratione*. To the monks of the Middle Ages this "computation of the times" meant primarily calculating the highly variable date of Easter Sunday. Spring begins on March 21. In 325 the Council of Nicaea officially decreed that Easter should be the first Sunday after the first full moon of spring, so that it would never under any circumstances fall on the same day as the Jewish feast of Passover, which is celebrated on the eve of the first full moon of spring.

Easter Sunday, in turn, established the times of still other variable Church holidays: Pentecost or Whitsuntide, the day of the Ascension of Christ, the feast of Corpus Christi, which was celebrated from 1247 on, and the feast of the Trinity, after 1334. Thus in every monastery the *computus paschalis* or "calculation of Easter" had to be performed by several monks.

Bede's flexus digitorum ("*finger-bending*"). Now we shall follow Bede's own account: first in translation with brief explanatory comments and then in the original Latin text (see p. 206). As he goes along, the reader is urged to form these numerical gestures with his own fingers. The illustrated finger-counting tables (Figs. 38 and 41) give a rough idea of how these were formed, but only a rough idea, since they are not all clear and some of them are wrong. But if the reader "spells out" these numbers on his own hands, not only will the original sources come alive for him, but he will realize their true significance. Otherwise he might well mistake these finger numbers for a form of idle amusement.

"Before we begin, with God's help, to speak of chronology and its calculation," begins the Venerable Bede, "we deem it necessary first briefly to show the very necessary and ready skill in finger counting:

Bistinctio secunda. Tractatus quartus.

left hand   right hand

Fig. 38 Finger counting from the *Summa de Arithmetica* of the Italian mathematician Luca Pacioli, the first important mathematical work to be printed, which was published in Venice in 1494. In contrast to the Venerable Bede's account, the hundreds and thousands are made on the right hand.

1: If you wish to say "one," you must bend the little finger of your left hand and place its tip on the palm;
2: for "two" lay down the ring finger next to it;
3: for "three" also the middle finger;
4: for "four" you must again raise the little finger;
5: for "five" also the ring finger;
6: for "six" you must extend the middle finger, and then the ring finger, which is called the *medicus*, alone remains bent down upon the palm.

This finger was so named in ancient times because it was believed that a vein runs directly to it from the heart. Since 6 was a "perfect" and hence a sacred number, it was deemed worthy above all the others to bear the ring, and thus was made on the "ring" finger. A "perfect" number is exactly equal to the sum of its divisors — apart from the number itself, of course. The divisors of 6 are 1, 2, and 3, the sum of which is 6. The number 12 is not "perfect," since its divisors 1, 2, 3, 4, and 6 add up to 16. The next three perfect numbers are 28, 496, and 8128.

7: for "seven" extend all the fingers and bend only the little finger over the wrist.

In contrast to 1 (as well as 2 and 3), in making 7 (as well as 8 and 9) the fingers were bent not at the middle knuckle but at the bottom joint, so that they extended far over the palm or the fleshy pad of the thumb. Pacioli failed to make this distinction (see Fig. 38; it is made correctly in Fig. 42 however).

8: for "eight" lay the ring finger down next to it (see Fig. 42).

Here we have the solution to the riddle quoted at the head of this chapter. The answer lies in finger counting: if we form the number 8 on the hand and then remove the 7, what remains is the finger-gesture for 6!

9: for "nine" place the "impudent" (middle) finger next to it (see Fig. 42).

It is important to note that only the last three fingers of the left hand are used to form the units. Now we can see how all the numbers through 9999 could be indicated just by the fingers of both hands (the names of the fingers below are represented by their initial letters):

left hand		right hand	
units	tens	hundreds	thousands
L — R — M	I — T	T — I	M — R — L

In this arrangement the person making the numbers worked backward, whereas the person opposite or facing him read the numbers in ascending order from right to left, thousands, hundreds — tens, units, as we write them today.

Actually the Venerable Bede also showed them in the same order. Pacioli's description of fingers counting, written some eight centuries after Bede's, only interchanges the hundreds and thousands; the second known description, by Jacob Leupold (see Fig. 41), adheres to Bede's:

If you wish to say 10, you must place the nail of the little finger over the middle of the thumb.

For 20, place the tip of the thumb between the index and the middle fingers.

For 30, join the nails of the index finger and thumb in a loving curve ("in a tender embrace").

For 40, place the thumb beside or on top of the index finger and extend them both.

For 50, bend the tip of the thumb inward (toward the palm) like a Greek Γ (gamma).

For 60, lay the tip of the index finger over the thumb bent as above. (Thus the illustration in Fig. 38 is incorrect: the thumb should be bent toward the palm.)

For 70, place the thumb within the bent index finger so that the thumbnail touches the middle knuckle of the index finger (Fig. 38 is correct for this).

Or: place the tip of the thumb in the knuckle of the index finger and place the latter over it.

For 80, "fill" the index finger, bent as above, with the extended thumb, so that the latter's nail thumbnail touches the index finger (with its upper side).

By "fill" Bede means that the thumb is to "close" the curve of the index finger in such a way that the thumbnail touches the nail of the index finger.

For 90, lay the index finger on the base of the thumb.

Or bend the tip of the index finger back over its own base and place the thumb over it.

Figures 38, 41, and 44 do not clearly illustrate all the prescribed gestures; some of them are wrong. Jacob Leupold is obviously interpreting Bede's description to the extent of his own understanding (Fig. 41). Bede's depiction is consistent with the well written and much later Arabic sources, so that we can be fairly sure that our translation of Bede and the explanatory comments correctly reflect the true ancient finger counting (see p. 214).

Bede continues:

Up to now you have used the left hand, but you shall make 100 on the right hand as you did 10 on the left." (Pacioli interchanges the hundreds and the thousands)
Similarly all the remaining hundreds to 900.
And 1000 on the right hand as 1 is formed on the left.
And so on up to 9000.

Now the reader himself should form the numbers 21, 75, 206, and 5327 on his own fingers. He will be surprised to find that underlying these finger gestures is a positional or place-value system: the four ranks of numbers — the units, tens, hundreds, and thousands — are made with four different groups of fingers! The gestures here play the part of digits, of which there are not 9, however, but 2×9. The reason for this lies in the nature of the system, in which first 3 and then 2 fingers are available to form each rank of numbers. "Zero" was indicated by the normal relaxed position of the fingers.

Thus the place-value concept took form even in ancient times, when the Indian system of numerals — the one we use today — was not yet developed, and it was used in the early Middle Ages, when the Indian numerals had not yet been brought to the West. We shall meet the positional concept once more on the abacus, again quite independently of our modern place-value notation. The Hindu-Arabic numerals were therefore not the first in history to embody the concept of a place-value notation.

Digiti and articuli ("fingers" and "joints"). In medieval books on numbers and mathematics, all numbers were divided, after the manner of Boëthius (who lived in the 6th century A.D.) into three classes: the nine *digiti* or "fingers" from 1 to 9, the *articuli* or "joints" meaning all the numbers divisible by 10, such as 20, 700, 850, etc., and the *numeri compositi* or "combined numbers" consisting of both previous classes, like 23 and 857 (see Figs. 162 and 227).

Where did these remarkable names come from? There is no doubt that they were derived from finger counting, in which the nine units were formed with the three "whole" fingers and the tens (to 90) with the index finger and thumb together (except for 50) and in such a manner that these two always touched at the "joints." These names, to be sure, did not appear in finger counting itself, but were technical

Fig. 39 The old classification of the natural numbers into *digiti*, *articuli*, and *numeri compositi*, from a medieval manuscript of the 14th century. Bayrische Staatsbibliothek, Munich.

Fig. 40 2000 shown on the fingers; an illustration from a copy of Bede's work on finger counting, dated 1140. From D. E. Smith, *History of Mathematics*, Boston, 1928.

terms connected with the abacus, on which they served to designate the number of places in multiplication.

Example: we wish to multiply 40 × 70, which means placing 4 tens × 7 tens on the abacus; since 4 × 7 = 28, the medieval rule reads:

decenum per X si multiplicaveris dabis digitis C et articulis mille,

"give to the digitus 8 the value 100 and to the articulus 2 the value 1000" (result: 2800).

In the Salem Algorism, a 12th century manuscript on arithmetic which we shall deal with extensively (see Fig. 232), 35 and 67 are to be added. When learning to add the units, school children say for 7 + 5 = 12, "put down 2 and carry 1"; pupils in medieval monastery schools learned:

scribe digitum 2, transfer articulum 1,

"write down the digitus 2, carry over the articulus 1."

Today the units are still called *digits* in England and *doigts* in France.

The words of Bede's original text will help the reader who knows Latin to visualize these curious finger gestures, and at the same time give him a sample of the style of medieval learned writing (from 2 on, *quum dicis* is abbreviated to *q. d.*):

1: *Quum ergo dicis unum, minimum in laeva digitum inflectens, in medium palmae artum infiges.*

2: *q. d. duo, secundum a minimo flexum ibidem impones.*

3: *q. d. tria, tertium similiter afflectes.*

4: *q. d. quatuor, ibidem minimum levabis.*

5: *q. d. quinque, secundum a minimo similiter eriges.*

6: *q. d. sex, tertium nihilominus elevabis, medio duntaxat solo, qui Medicus appellatur, in medium palmae fixo.*

7: *q. d. septem, minimum solum, caeteris interim levabis, super palmae radicem pones.*

8: *Juxta quem, q. d. octo, medicum,*

9: *q. d. novem, impudicum e regione compones.*

10: *q. d. decem, unguem indicis in medio figes artu pollicis.*

20: *q. d. viginti, summitatem pollicis inter medios indicis et impudici artus inmittes.*

30: *q. d. triginta, ungues indicis et pollicis blando conjuges amplexu.*

40: *q. d. quadraginta, interiora pollicis lateri vel dorso indicis superduces, ambobus duntaxat erectis.*

50: *q. d. quinquaginta, pollicem exteriore artu instar graecae literae Γ curvatum, and palmam inclinabis.*

60: *q. d. sexaginta, pollicem, ut supra curvatum, indice circumflexo diligenter a fronte praecinges.*

70: *q. d. septuaginta, indicem, ut supra, circumflexum pollice immisso superimplebis, ungue duntaxat illius erecto trans medium indicis artum.*

80: *q. d. octoginta, indicem, ut supra, circumflexum, pollice in longum tenso implebis, ungue vindelicet illius in medium indicis artum infixo.*

90: *q. d. nonaginta, indicis inflexi unguem radici pollicis infiges. Hactenus in laeva.*

100: *Centum vero in dextera, quemadmodum in laeva facies.*

200: *Ducenta in dextera, quemadmodum Viginti in laeva.*
300: *Trecenta in dextera, quemadmodum Triginta in laeva.*
   Eodem modo et cetera usque ad DCCCC.
1000: *Item mille in dextera quemadmodum unum in laeva.*
2000: *Duo millia* — — — *Duo* — —
3000: *Tria millia* — — — *Tria* — —
   Et cetera usque ad novem millia.

Bede goes on to show how the numbers from 10,000 to 1 million
are formed using finger gestures which sound artificial and were
probably rarely if ever used. Perhaps Bede himself extended the limit
of counting here. Leupold's 18th-century mathematical compilation
depicts these higher numbers (see Fig. 41):

Fig. 41 Bede's finger counting illustrated in Jacob Leupold's *Theatrum Arith-
metico-Geometricum*, published in 1727 — a thousand years after Bede.

10,000: *Porro quum dicis decem millia, laevam medio pectori supinam appones, digitis tantum ad collum erectis.*

20,000: *Viginti millia quum dicis, eandem pectori expansam late superpones.*

30,000: *Triginta millia quum dicis, eadem prona, sed erecta pollicem cartilagini medii pectoris inmittes.*

40,000: *Quadraginta millia quum dicis, eandem in umbilico erectam supinabis.*

50,000: *Quinquaginta millia quum dicis, eiusdem pronae, sed erectae, pollicem umbilico impones.*

60,000: *Sexaginta millia quum dicis, eadem prona femur laevum desuper comprehendes.*

70,000: *Septuaginta millia quum dicis, eandem supinam femori superpones.*

80,000: *Octoginta millia quum dicis, eandem pronam femori superpones.*

90,000: *Nonaginta millia quum dicis, eadem lumbos apprehendes, pollice ad inguina verso.*

100,000: *At vero centum millia, et ducenta millia, et cetera usque ad DCCCC millia, eodem quo diximus ordine in dextera corporis parte complebis. Decies autem et centena millia, quum dicis, ambas sibi manus insertis invicem digitis implicabis.*

10,000: If you wish to say ten thousand, place the back of your left hand against your chest with the fingers extended, and point to the neck.

20,000: place the left hand with the fingers spread out over the chest (shown differently in Fig. 41).

30,000: point with the thumb of the extended hand to the cartilage in the middle of the chest.

40,000: you must place the back of the hand over the navel.

50,000: point with the thumb of the extended hand to the navel.

60,000: grasp the left thigh from above.

70,000: place the back of the left hand over the left thigh.

80,000: place the palm of the hand over the thigh.

90,000: place the left hand over the small of the back, the thumb pointing toward the groin (and thus forward).

Then for 100,000 through 900,000 do the same on the right side of the body in the same order. For 10 times 100,000 (which is 1 million), place both hands together with the fingers interlaced.

So much for the Venerable Bede. Now that he has introduced us to finger counting, we shall have something to say later on about its significance (see p. 212).

FINGER COUNTING IN ANTIQUITY

As we follow our present path, we come to a very important milestone of cultural history, St. Jerome's translation of the Bible into Latin. Jerome died in A.D. 420. This Church father retains a place in popular affection as a result of the legendary story about the lion and the thorn which he removed from its paw. Albrecht Dürer's famous picture shows the Saint in his little house, bent over his writing desk and completely absorbed in his translation of the Bible into Latin.

Jerome's *Vulgata*, the "Vulgate" or "popular" Bible, is still used by the Roman Catholic Church as its official Holy Writ.

St. Jerome also wrote commentaries on the Bible, including one on Matthew 13:8 which tells the Parable of the Sower:

But others fell into good earth, and brought forth fruit, some an hundredfold, some sixtyfold, some thirtyfold.

Jerome's comment on this passage is so curious that it is worth quoting here in the original Latin:

Centesimus, et sexagesimus, et tricesimus fructus, quanquam de una terra et de uno semente nascitur, tamen multum differt in numero. Triginta referuntur ad nuptias: nam et ipsa digitorum conjunctio, quasi molli osculo se complectens et foederans, maritum pingit et conjugem.

Sexaginta ad viduas: eo quod in angustia et tribulatione sint positae, unde et in superiore digito deprimuntur, quantoque major est difficultas expertae quondam voluptatis illecebris abstinere tanto majus et praemium.

Porro centesimus numerus, quaeso diligenter Lector attende, a sinistra transfertur ad dextram, et iisdem quidem digitis, sed non eadem manu, quibus in laeva manu nuptiae significantur et viduae, circulum faciens exprimit virginitatis coronam.

The hundredfold and the sixtyfold and the thirtyfold fruit, though they may have sprung from the same soil and the same seed, nevertheless differ greatly in their number. Thirty is a symbol of marriage: for this manner of placing the fingers, which are joined and interwoven as in a sweet embrace, represents both man and wife.

Sixty stands for widowhood, because the widow endures trouble and tribulation that weigh on her, just as (the thumb) is pressed down by the index finger which lies above it (in forming the number sixty). But the more difficult it is to abstain from the pleasures that were once enjoyed, the greater will be the reward.

But now attend carefully, dear Reader: the number hundred is transferred from the left hand to the right, where it is formed with the same fingers but not on the same hand as the hand which symbolized marriage and widowhood. The circle now formed on the right hand stands for the crown of virgin purity.

These interpretations of St. Jerome's have very graphically preserved the finger gestures for the numbers in question (see Fig. 44). They also bear witness to the popularity of finger counting in the fourth century, since the Church father has used their images to make explanations and draw analogies which would be completely incomprehensible to anyone without a knowledge of finger counting.

Moreover, today in Naples (and perhaps not only there) the finger position for the number 30 still has the same symbolic meaning as in Jerome's commentary: *molli osculo se complectens,* a "tender embrace." And a Roman writer tells us that the same gesture, but made on the right hand, served as an appeal to Venus!

Universal familiarity with finger counting, at least in late Roman times, was also demonstrated by another Church father, St. Augustine, who died in A.D. 430 when he was Bishop of the town of Hippo in North Africa, and who also provided us with much evidence about finger counting. One passage from his writings shows that not only numbers, but also calculations, were made on the fingers. He is explaining Verse 21:11 of the Gospel according to St. John, in which Simon Peter cast his net at Christ's command and

brought up 153 fishes. St. Augustine tries to see a mystical significance in this number: it is made up, says he, of all the numbers from
1 through 17; but 17 in turn is composed of 10, the number of the
Ten Commandments, and 7, the number of the Holy Spirit, and is
thus particularly significant. Then Augustine continues:

> *Apud vos numerate: sic computate. Decem et septem faciunt centum
> quinquaginta tres: si vero computes ab uno usque ad decem et septem et
> addas numeros omnes — unum, duo, tria: sicut unum et duo et tres faciunt
> sex; sex, quattuor et quinque faciunt quindecim: sic pervenes usque ad decem
> et septem, portans in digitis centum quinquaginta tres.*

> Count for yourselves and figure as follows: 10 and 7 make 153, for
> if you count from one to seventeen and add together all the numbers —
> 1, 2, 3: 1 and 2 and 3 are 6; 6 and 4 and 5 are 15; you finally arrive at 17
> and carry 153 on your fingers.

Thus as one counts up to 17, the fingers constantly join together
the previous numbers until they finally "carry" 153. This example
is very important, for it reveals the main purpose of finger counting:
to record temporarily the intermediate sums in mental arithmetic:
$1 + 2 = 3$, $+ 3 = 6$, $+ 4 = 10$, $+ 5 = 15$ and so on.

Macrobius, a Roman contemporary of Augustine in his own writings, drew heavily on older works, relates that

> Many people regard Janus as the god of the Sun; thus he is often shown
> in statues as forming the number 300 with his right hand and the number
> 65 with his left (*manu dextera trecentorum, et sinistra sexaginta quinque
> numerum retinens*). This symbolizes the days of the year, which is the
> Sun's chief creation.

This passage is taken almost word for word from Pliny the Elder.
Pliny is referring to the statue of Janus in Rome, which was shown
forming the number 365 on its fingers as an attribute identifying him
as the god of time and of the year (*digitis figuratis CCCLXV dierum
nota*). With Pliny, who died in A.D. 79, we arrive in ancient Rome
where, as he informs us, there was a statue in the forum of Janus, the
two-headed god of all beginnings and endings ("January"), which
showed him representing the 365 days of the year on his fingers.
There could scarcely be better evidence of the universal familiarity
with finger counting in Rome than this passage, which describes it
casually as nothing at all out of the ordinary.

The Roman satirist Juvenal (died A.D. 130) also betrays his knowledge of finger counting when he says of Nestor, the oldest and wisest
of the Greek princes at the Siege of Troy:

> *Felix nimirum qui tot per saecula mortem
> distulit atque suos iam dextera computat annos,*

> Happy is he who so many times over the years has cheated death
> And now reckons his age on the right hand.

For the fingers of the right hand were used to form the hundreds!
Another confirmation is provided by Quintilian (1st century A.D.),
the famous teacher who was celebrated even in the Middle Ages, and
who said: "The uneducated man betrays himself less by his fear of
the answers than by mistakes in computation because of the un-
certain and unclear numbers he forms on his fingers."

And with Firmicus Maternus, who wrote a textbook of astrology in A.D. 340, we look into a schoolroom:

Vides ut primos discentes computos digitos tarda agitatione deflectant,
Do you see how awkwardly beginners in computation bend their fingers?

The finger numbers from 1 through 15 have been preserved on some very rare Roman *tesserae* (counters); of these, two beautiful ivory specimens showing the numbers VIIII and VIII formed with the fingers, are illustrated here (Fig. 42). They come from the collection of Lord Hamilton, the English ambassador to Naples during the Napoleonic Wars. Since such *tesserae* have been found only with numbers up to 15, they were probably counters in some kind of game.

Fig. 42 Two Roman counters showing the numbers VIIII and VIII formed by finger gestures. Ivory, 2.9 cm in diameter, 2 and 4 mm thick. Probably 1st century A.D. British Museum, London.

More evidence of the similarity between individual numbers in Roman finger counting and those described by the Venerable Bede could easily be cited, but let us turn our attention for a while to finger counting in ancient Greece. Here the evidence is extremely scanty, but the 5th-century B.C. comic dramatist Aristophanes does say once in *The Wasps*: first the income of the city of Athens is to be counted, and then the expenditures of the judicial bench, "not indirectly by using counters, but right on the hand":

μὲν λόγισαι φαύλως μὴ ψήφοις ἀλλ'ἀπὸ χειρός.

This passage is as vague as the expression *pempázein*, which refers to nothing more specific than a numerical grouping of five stemming directly from the hand (see p. 41).

But why is Greek testimony so scarce? The Greek evidence comes from early writings; the Roman from writers who lived during and after the time of Christ. The Roman period is historically connected with our own, and its character has been preserved to the smallest details. We are probably not far wrong if we attribute the advancing use of finger counting to the expansion of the Roman Empire.

Many barbarian tribes, speaking a multitude of foreign languages, lived outside the Empire but within the sphere of Roman culture and trade. As the commerce of Rome expanded, the merchant's needs also grew: the numbers with which he worked became larger and so did the computations he had to master. This is where the counting board or abacus was useful. Intermediate computations could be recorded in some way by a slave, as St. Augustine shows in a very informative example. In the absence of paper, the numbers were "written down" on the fingers at once and thus made visible. There was always a division of labor between the computer and the recorder in a counting house, as we see very clearly from medieval illustrations: one man calls out the sums and amounts, another makes calculations, and a third writes (see Fig. 193 and compare with the picture of the Roman gravestone, Fig. 138).

The use of finger counting was also stimulated from the outside, by real problems that arose in trading with peoples who spoke other languages. For numbers formed on the fingers could be universally understood by merchants, even without speaking the language of their barbarian partners.

One even senses that finger counting became a kind of technical jargon of commerce. Were the numbers formed by the fingers the source of the number words that grew up out of the earliest stages of mankind's development? Of course not. Did they spring from a single creative mind, as a sort of offspring of learning? No, although there was something artificial about them. They had elements of both: a custom that grew up anonymously to meet the needs of ordinary life, and a deliberately thought-out answer to a particular problem; they were a sort of commercial language in themselves, like the pidgin (= "business") English used in the Far East and the South Pacific.

This interpretation finds support in the custom of

FINGER COUNTING IN ARABIC AND EAST AFRICAN COMMERCE TRADE

In the seaports and market places of the Red Sea, Arabia, and East Africa, merchants have evolved a finger language that is understood in every market of every country in the region. Buyers and sellers come to terms underneath a cloth, a fold of garment or a strip of muslin from a turban, by touching the fingers of each other's hand and thus bargaining in complete privacy. Our illustration shows two Indian pearl traders settling the price of a pearl under a handkerchief (see Fig. 43).

The privacy thus achieved is one of the main purposes of this commercial finger language, for in these countries all transactions are concluded in the open market place. If idle bystanders and others are not to be let in on the price, seller and buyer must come to an understanding both silently and secretly. It also has the advantage

Fig. 43 A pearl merchant of Southern India bargaining with a buyer by means of finger gestures made under a cloth. This rare picture (taken in 1956) by the Indian photographer Nimal Abeyawardene was very kindly lent to the author by Th. Martens of Munich.

of allowing traders to deal with each other directly, without letting a possibly untrustworthy interpreter come between them. Not only the European, Indian, Arab, and Persian merchants, but also those from the interior of the continents, the Abyssinians, Somalis, and Bedouins, all understand the finger language of the shores of the Indian Ocean.

One of the rules of Arab and East African commerce is the following: if the buyer's hand touches the seller's extended index finger, this means either 1, 10, or 100. Both parties to the transaction are agreed about the numerical rank of the price; if not, they first specify the coin in which they are bargaining. This resembles our own habit of saying, for example, 5, 6, or 12 when we actually mean 5000, 6000, or 12,000 dollars.

Correspondingly, if the buyer touches the seller's 2 (or 3 or 4) first fingers, this means, 2, 20, or 200 (3, 30, or 300; 4, 40, or 400); touching the whole hand then means 5, 50, or 500. The little finger alone means 6 (60 or 600), the ring finger alone 7, the middle finger alone 8, the bent index finger 9 and the thumb 10 (or 100 or 1000). If the buyer strokes the index finger from the middle joint knuckle toward the tip, he is saying $-\frac{1}{2}$ or "one half off"; if he strokes the index finger toward the knuckle, this means $+\frac{1}{2}$. The fractions $(-)$ $+\frac{1}{4}$ and $(-)+\frac{1}{8}$ are similarly indicated; these exhaust the "vocabulary" of this Oriental language of trade. In simple terms:

2500 = 2 × the thumb and 1 × the whole hand grasped;
   $4\frac{1}{2}$ = 4 fingers touched and the index finger stroked toward the
      middle knuckle;
   76 = ring finger and then middle finger touched (a succession of
      numerical ranks!).

This finger language is "spoken" very fluently and rapidly, and mistakes are virtually never made.

An agreed and mutually understood finger language is used today, not between people who speak different languages, but in the world's largest cattle market in Chicago. Like that of the Middle East, this custom reveals something about the essence and purpose of finger counting, even though the Roman method is not used. But we also have some late evidence that our Roman finger numbers continued to be used for a very long time in the Arabic world of the Mediterranean.

In the year 1340 the Byzantine scholar Nicholas Rhabdas of Smyrna wrote his Greek "Treatise on the Finger Measures" (*Ekphrásis toû Daktylikoû Métrou*), which is almost the same as the Venerable Bede's, even to the smallest details. The former could not have been paraphrased or copied from the latter — the differences in place and time, some six centuries later, weigh against such an idea, as do other considerations. Even the Arabs knew the Roman finger numbers, as Arabic and Persian documents from the 14th century show. All their gestures are the same as Bede's, except that the right and left hands are reversed, with the units beginning on the right hand in accordance with Arabic script which is read from right to left. When praying, say these sources, the believer should place his right hand on his leg as if he were forming the number 53 (thus the index finger is the only one extended).

How is the bowstring grasped when shooting with a bow and arrow? With the tips of the index finger and the thumb, "as the number 30 is formed."

One Arab poet very ingeniously mocked another, Khalid by name, who had become rich: "Khalid went with 90 and came back with 30" — he came back poorer, apparently. But if the reader forms these numbers on his fingers, he will see that for 90 the thumb and index finger are pressed tightly together to form a "thin," "lean" figure, whereas for 30 they form a full, "rich" circle.

We may conclude with the Arabic counterpart to Juvenal's reference to Nestor, two lines by a poet with the charming name of Sheref-ed-din Ali al Jezdi:

Were I to recount all the wonders of the world,
I should count as many as the numbers which the left hand forms.

ROMAN FINGER COUNTING IN THE WEST

In the summer of 822 I undertook, with Tatto's guidance, to study arithmetic. He began by explaining to us the books of the Consul Manlius Boëthius concerning the various kinds and classes and meanings of numbers. Then we learned to count on the fingers, from the books which Bede wrote on this subject.

Thus wrote Walafried Strabo, who in the year 842 was Abbot of the Monastery of Reichenau on the Bodensee. We infer from his account that finger counting continued to be in use in the West as well. But here it had moved from the market place into the scholar's study. Finger numbers were no longer merely the universal language of commerce; now they began to serve the higher art of mathematical computation. This, indeed, was the role we found them playing in the Venerable Bede's writings (see p. 202).

Now we are in a position to understand the words of Brother Berthold of Regensburg (1220–1272), one of the most celebrated and popular preachers of the Middle Ages:

"man zalte in der alten ê an den vingern. des kunnet ir ungelerten liute niht, wan es ist der gelerten vil, di es niht kunnent. man zelt also, so sin sehzic wirt, so leit man den dumen in die linken hant."

In former times people counted on their fingers. But unlettered people could not do this, and even among the educated there were many who could not (count on their fingers). Counting was done thus: if the number was sixty, for example, one began with the thumb in the left hand.

Here Berthold himself erred, for the number he describes is not 60, but 50.

The first secular book, or at any rate the first printed one, in which we again come across finger counting, is another scholarly mathematical compendium, *Summa de Arithmetica, Geometria Proportioni et Proportionalita* by the Italian Luca Pacioli, which was published in 1494; the illustration of Fig. 38 is taken from this book. That finger numbers were adopted by a mathematician, who actually chose them because of their suitability for his purposes (unlike the monks, who merely described them because they were handed down to them), shows the important position that finger counting attained in medieval mathematics. *Abacus atque vetustissima veterum Latinorum per*

Fig. 44 Finger numbers from a 13th-century Spanish codex. The numbers from 100 through 900 are shown on the left, and those from 1000 through 9000 on the right; below are the gestures for 10,000 and 20,000. These hands with their slender, elongated fingers are drawn in the Byzantine style.

digitos manusque numerandi consuetudo — "The Abacus (= counting board) and the Age-old Custom of the Ancient Latins of Counting on the Hands and Fingers": such was the title of the book published by the German writer Aventin in Nuremberg in the year 1522. Here we see the connection between finger counting and the abacus. We shall learn the techniques of making computations on the abacus later; but finger counting here clearly serves the purpose of recording temporarily the intermediate results of these computations. The Italian mathematician Leonardo of Pisa (1180–1250),* whose important contribution to mathematics are still highly regarded, testifies to this. At one point he refers to finger counting by the expression *servare* — that is, "to keep, store" — meaning what our schoolchildren mean when they say "carry." And he goes on to say, very clearly,

ponens semper in manibus numeros ex divisione euntes,

constantly keeping in his hands the numbers which result from the division.

But the most remarkable thing of all is that Leonardo here wants to use finger counting not in conjunction with the abacus, but with new Indian numerals, whose chief proponent he had become in the West. The relevant passage, which is of the greatest importance for the history of mathematical computations, reads as follows:

Predictis figuris Yndorum earumque gradibus secundum materiam superius descriptam cum frequenti usu hene cognitis, opportet eos qui arte abbaci uti voluerint, ut subtiliores et ingeniores appareant, scire computum per figuram manuum, secundum magistrorum abbaci usum antiquitus sapientissime inuentum.

If these numerals of the Indians and their place-value notation are to be thoroughly mastered through constant practice, it behooves those who would become adept and expert in the art of computation to learn to count on the fingers, which the masters of computation according to the old manner once found to be invaluable.

Leonardo's words *ars abbaci,* moreover, refer to the art of computation in general and not to the counting board alone (see p. 425).

Our final witness for the use of finger counting in computations is the "royal" passage concerning the number 73 formed on the fingers, for it specifically mentions the *abacistae,* those "learned men" who knew how to make computations on the *abacus* or counting board. The author of this passage was the Hohenstaufen Emperor Frederick II (died 1250), who constantly studied this absorbing subject while struggling with the Pope for mastery over the civilized world. He never ceased to be a friend and patron of learning, of the arts — and of hunting. The last of these was so dear to him that he even wrote a famous book about falconry: *De arte venandi cum avibus* — "On the Art of Hunting with Birds." In this book Frederick explains how the falcon is held by a skilled huntsman:

Manum vero non plicet interius neque exterius, sed in rectitudine brachii teneat, conjugendo pollicem extensum indici, et replicet indicem ad extremitatem pollicis, et erit modus, secundum quem abacistae tenent septuaginta

---

* *Editor's note:* also known as Leonardo Pisano or Fibonacci.

cum manu, et alii digiti eiusdem manus replicentur in palmam sub illis duobus digitis, ut firmius sustenentur, ad similitudinem tenentis numerum ternarium, et sic ex replicatione indicis super pollicem, et trium digitorum in palma sub illis, teteat manum ad formam abacistae tenentis septuaginta tria.

The hand is held facing neither inward nor outward, but extended in the same direction that the arm is extended. The index finger is then laid over the extended thumb and bent forward over the end segment of the thumb: this is just the way in which the masters of computation (*abacistae*) form the number seventy with their fingers. The other fingers of the same hand are bent over the palm beneath these two fingers (the index finger and thumb), to support the latter, just as the number three is formed. Thus the index finger is bent over the thumb and the three other fingers beneath it, in the manner of a master of computation forming the number seventy-three."

The reader may recall the intriguing Arabic counterpart of this passage, which states that the bowstring is to be drawn with the fingers held so as to form the number thirty.

In the 16th century, however, as the Indian numerals came to be increasingly deeply rooted in the West as a result of printing and because of their concise and convenient forms which at last made purely written computations possible thus assuring victory over the counting board and the Roman numerals, the days of finger counting were coming to an end. Although Fig. 41 is taken from an 18th-century book, this was one of the very last works to describe finger counting, and it was even then mentioned merely as a curiosity and no longer treated as a useful skill. Soon thereafter no one knew anything about finger numbers: they finally disappeared forever from the history of civilization.

There are still a few places where finger counting lives on today or where it has only lately died out; it was used by peasants in the Auvergne, in Walachia, in Bessarabia, and by gypsies in Serbia. But even these people used it not in the old Roman manner, but as a means of computing.

COMPUTATIONS ON THE FINGERS

Up to now we have regarded finger numbers as a form of numerals, which in a sense served just to record numbers temporarily. But the fingers can also be used as a calculating machine, so to speak, with the numbers to be operated on "put in" and the final result read off after a simple intermediate calculation. Finger counting was used in this manner for the small multiplication table from 5×5 to 9×9 and for the large multiplication table from 10×10 to 15×15. This was the method:

Let us multiply 6×8, for example. Hold the hands extended upright and on each of them form the excess over 5 in both numbers ($6 = 5 + 1$; $8 = 5 + 3$), so that one finger is bent on the left hand and three on the right. Count the bent fingers: $3 + 1 = 4$; these will be the tens of the final result: 40. Multiply the standing fingers: $4 \times 2 = 8$; these will be the units of the final result. Answer: 48. Thus there is no need to memorize the multiplication table beyond 5×5.

To multiply 13 × 14 as an example from the large multiplication table, the procedure is as follows: bend the fingers on each hand to represent the excess over 10 — thus three fingers are bent on the left and four on the right hand. Their sum, 3 + 4 = 7, gives the tens: 70; their product, 3 × 4 = 12, gives the units; 70 + 12 = 82; this is the excess over 100, and the final answer is thus 182. Here too one need know the multiplication table no further than 5 × 5.

Actual computation on the fingers is surprisingly effective, especially for simple people with little or no schooling. Remember that we ourselves would be utterly unable to make calculations had we not learned the multiplication table. Today pupils memorize it at the very beginning of their schooling, but in the Middle Ages this was such a difficult matter that people employed special tables called *mensae Pythagoricae*, from which such products as 6 × 8 could be read off, just as we might consult a table to find the product of 16 × 18 (Fig. 45).

One gets the impression that each new such "labor-saving device" in mathematical computation was greeted with joy. Perhaps this was also how the method of finger-multiplication was received; at any rate, Leonardo of Pisa writes:

... multiplicationes in manibus addiscendo semper utantur colligere, ut animus pariter cum manibus in additionibus et multiplicationibus quorumlibet numerorum expeditior fiat,

... multiplication with the fingers must be practiced constantly, so that the mind like the hands becomes more adept at adding and multiplying various numbers. (Compare the passage quoted from St. Augustine, p. 210).

Other medieval and early modern writers, however, although they wrote about finger counting, say nothing about computations using the fingers; perhaps they were not too frequently used.

"Complementary operations" were, of course, quite readily employed in the Middle Ages, as we shall see when we come to the abacus. Instead of actual numbers, such as 6 × 8 for example, medieval arithmeticians worked with their "complements" within some rank, in this case 4 and 2 up to 10. In many textbooks on arithmetic we find not multiplication tables as we know them, but a description of the following kind of operation:

6 4 6 × 8; subtract the 10-complement of one number from that of
8 2 the other: 6 − 2 = 8 − 4 = 4; after the remainder write the
4 8 product of the complements 2 × 4; answer: 48.

Was this peculiar method of multiplication inherited from the Romans, or was it an independent popular invention of the Middle Ages? That is hard to tell. As evidence in favor of the former hypothesis we may cite its use among common people in the Auvergne and Wallachia: both these regions were ruled and influenced for many centuries by Rome. One could also point to the excess over 5 that is inherent in Roman numerals: 6 = VI, 7 = VII, 8 = VIII, 9 = VIIII. But that is all. Only more thorough and accurate anthropological research in areas where "early" computations are still or were recently used can resolve this question one way or the other.

Fig. 45 Multiplication table from an arithmetic book of the 16th century. Compare the early medieval multiplication tables (Figs. 113 and 162).

Moreover it must be kept in mind that for a primitive computer the
multiplication table itself represents a significant advance, and that it
can always be adjusted by addition. We need only think of the
Egyptians, who in general could perform such multiplications as
$25 \times 43 = 1075$ merely by multiplying by ten and by doubling the
factor 43 and then adding:

/ 1	43	
10	430	
/ 20	860	The numbers in the lines marked by a slash are added
2	86	together.
/ 4	172	
25	1075	

SOME FORMS OF FINGER COUNTING USED BY PEOPLES OF OTHER CULTURES

In the Egyptian ell, which contained 28 "fingers," we have an
example of the individual units of a measure being counted in the
middle: I, II, III, (symbols for) 4, 5, 6, 7 (Fig. 46). But the Egyptian

Fig. 46 Egyptian system of measures with the numbers 4 through 7 "counted
on the fingers." Drawing of the first half of the scale (with the beginning, I,
omitted). Original about 25 cm long.

4, 5, 6, 7 are clearly finger numbers: 4 is a pictograph of a hand with
the thumb closed over the palm, 5 a hand with thumb extended away
from the hand, 6 a hand with thumb extended and fingers bent inward (to augment the 5 formed on the other hand), and 7 symbolized
in some manner that is not quite clear. Quite remarkably, right next
to these we find the "true" Egyptian number symbols, such as ⋮⋮ ∩
= 16, etc. (see Fig. 8); the symbol ⊂⊃ relates to fractions of the ell.
The "early" finger numbers were still so tightly interwoven with the
ell, a measure that arose from the ordinary lives of common people,
that the inconsistency in this system of numerals was not felt at all.
In addition, the ancient Egyptian symbol for 10,000, a pointing
finger, also very likely goes back to some ancient form of finger
counting (see Figs. 8, 9, and 10).

Ancient China offers another instructive example. The Chinese
represented the numbers 1, 2, and 3 by those numbers of strokes
respectively; the strokes can readily be interpreted as numbers once
formed on the fingers. There is nothing special about that. But the
Chinese symbol for 4 is a square with two small vertical lines drawn
within it (see Figs. 47 and 265). Yet this was originally the picture of
a hand (without the thumb). Ancient Chinese coins show, in addition

Fig. 47 Chinese numerals for 4, 8, and 9; these were originally representations of numbers formed on the fingers. These were found on ancient coins (cf. Figs. 48 and 280–281); for their modern forms, see Fig. 264.

Fig. 48 Chinese coin bearing the inscription: "Series 19." The number 10′9 is on the left side of the coin. This is a rare piece dating from the beginning of the Han Dynasty, around 200 B.C.
From the collection of R. Schlösser, Hannover.

Fig. 49 Statue of Buddha with sacred finger gesture. From the highest terrace of the Borobodur temple in Java. Ninth century A.D.

to the four simple vertical strokes, two perpendicular horizontal strokes joining them; these were clearly suggested by the form of the human hand (see Fig. 279). From this the symbol gradually was eroded to its modern form.

The Chinese word *pa* means "to split, cleave, cut," and the Chinese number word for 8 is also *pa*. The old symbol for this number consists of two "separated" signs which may represent a single gesture for "to split." This would mean that the similarity between the two words was extended to the written characters, which may well have happened, as shown by other examples. But could this not also be a picture of "two hands," 4 + 4 (see Fig. 47)?

At any rate, it is quite certain that the modern character for 9 is derived from the picture of a shaking hand, which is a number gesture. This appears on an ancient coin, on the left side of which, next to the square hole in the center, is the number 19, referring to the series in which this coin was minted (see Fig. 48). In Southern China a sudden movement of the hand to the right ear still stands for the number 9.

But what was still clearly recognizable in ancient Egypt was transformed in China by the masters of calligraphy into a matter of good style. Here, however, we have one of the rare examples which reveal the way in which "early" finger numbers were incorporated into the "mature" system of numerals.

Finger counting? At first the reader may well have looked on it with great suspicion: What could it be but a form of trivial amusement, once it did more than represent the numbers 1 through 10? Now we have seen that numbers counted variously on the fingers have always existed and are still in use today. The hands "speak," the fingers form the individual "letters." For the deaf and dumb, this is the only means of communication. But over and above these practical methods are the "sacred" finger gestures with which we shall conclude this chapter.

In Italy, Byzantine mosaics predominated in ecclesiastical art well into the 12th century. The image figure of Christ looms up in the vault of the apse (as for example in the church at Cefalù in Sicily), bending the ring finger and the thumb of his right hand in a solemn gesture to form a circle. A deep symbolism undoubtedly underlies this "Christian" gesture, as it also does the arm and finger movements of the Siamese temple dancers and the *mudras* of the priests on the island of Bali.

The Indian Buddha indicates a level of spiritual achievement (see Fig. 49) by the gesture he forms with his fingers. These finger gestures are as numberless as the various stages on the road to perfection; just as in Romanesque and Gothic churches every column capital is different from every other, no two gestures made by the often hundreds of Buddha images in an Indian temple are the same. These gestures of Buddha's do not represent numbers, of course, but they do show that "finger writing" has played a leading role in other cultures as well. Moreover, Indian Buddhist writings divide the art of computation into three stages or levels: counting on the fingers, which is called *mudra*, mental arithmetic, and higher computations.

Folk Symbols for Numbers

Tally Sticks

"The notched tally stick itself
testifies to the intelligence of our
ancestors. No invention is simpler
and yet more significant than this."
J. Möser, *Patriotische Phantasien*, 1776

Numbers formed with the fingers vanish in an instant. How was a merchant to keep his accounts, or even to make his calculations, with these alone? To represent the next number he must first destroy the previous one. Finger counting shares the visual quality of writing, but it differs in its impermanence which it has in common with number words. This is what gives numbers counted on the fingers their intermediate position between spoken numbers and written numerals.

EARLY WRITING AND READING

Today we are so accustomed to reading and writing that we can scarcely imagine what it was like when most if not all people could do neither. The letters of the alphabet are the very foundation stones of our education. What is the first thing we learn in school? "Dot your i's and cross your t's" — in other words, how to write properly! It is of profound significance that one of the greatest and most difficult achievements of a culture has now become the most elementary prerequisite of an education.

Suppose someone were to do away with all numerals, even the Roman? Then we would have some idea of the state of written numbers in the early Middle Ages. To be sure, Roman numerals found their way into the monastic cloisters along with other aspects of the culture inherited from ancient Rome; of course, Roman numerals were later brought by the students from these monastic schools into the world of literate people, of merchants, of scribes in chanceries. But the farmer, who never went to a monastery school, also had to keep track of his crops and livestock and harvests. The peasant could also be a creditor or a debtor. Where did he get the numerals to reckon with?

From no one but himself! He developed his own number signs, which were adequate for his own ordinary dealings, and which no one but himself could read and understand. Who, for instance, without being told how, could read the symbols on the tally sticks of Fig. 58 or on the magnificent Swiss notched stick of Fig. 52, with the carving of a cow at the top? Who knows what numbers are signified by the notches on the war club from the Fiji Islands (Fig. 6)? Perhaps a member of the tribe which carved the club could still read them, just as many Swiss can identify a particular hamlet or valley from the carvings on a notched stick. But their significance is always personal and valid only within a limited area: these are folk-signs, number symbols devised by simple people themselves to meet their everyday needs. They are not learned and taught in school as part of the general culture. In their level of development, of course, these are primitive numerals, since they naturally draw on the simplest laws of written numerals, ordering and grouping (pp. 39 ff.).

The primitive nature of these folk symbols is also revealed in the way by which they are made: they are carved or cut or scratched. In the history of language, this is the earliest forerunner of the art of writing.

The word *schreiben* came into the German language from the Romans: Old High German *scriban* < Latin *scribere*. But the Latin word goes together with the Greek *skariphdomai*, which, like the Greek word for "write," *gráphein*, originally meant "to scrape, scratch." Yet the Germanic peoples did not originally learn to inscribe symbols from the Romans, for they had their own word for this: the modern German *ritzen* ("to scratch"), Anglo-Saxon *writan*. This word replaced the Latin loan word in the English *write*, and is preserved in Modern German in the word *um-reissen* ("to sketch, outline"); a *Grund-riss* in German is a line drawing. In his *Germania* (Book X) Tacitus, the Roman historian, describes the customs of the earliest Teutons.

Virgam frugiferae arbori in surculos amputant eosque notis quibusdam discretos super candidam vestem temere ac fortuito spargunt. Mox si publice consultetur, sacerdos civitatis, si privatim, ipse pater familiae precatus deos caelumque suspiciens ter singulos tollit, sublatos secundum impressam ante notam interpretatur.

From a sapling the Teutons cut off a twig, cut this up into small pieces, put certain marks on the sticks and then toss them at random upon a white cloth. If anyone wishes to know what the fates hold in store, the priest of the community, praying to the gods like the father of a household, his face raised toward the heavens, three times picks up a stick and interprets it according to the marks which have been carved upon it beforehand.

From this old Germanic custom of casting lots derives the modern German expression *Buchstaben lesen* ("to read letters") (= *auflesen*, "to pick up the lots"), which in English is *read* (from *raten* ["to guess, solve"]) the "secrets" hidden in the symbols carved on the sticks (Old Norse *run* > "rune"). These, of course, were not the ancestors of the letters we use today.

TALLY STICKS WERE UNIVERSAL

Paper, which was a Chinese invention, first came to Germany in the 14th century. In the beginning it was too expensive for common use; like parchment in the monasteries, it was used only for the most important documents. Moreover great skill was required for writing on it. The common people's "paper" was wood, its writing instrument was a stylus or sharp knife and its letters were notches or grooves. For this reason, we find some form of tally used everywhere.

Even the various European names for these tallies provide some very informative glimpses into the cultural history of early writing and computation. In Middle and Low German there were *Kerbholz* ("notched stick"), *Dagstock* and *Knüppel* ("cudgel"); in Bavaria and the Tyrol people used the *Span* ("chip"), the *Kärm* and the *Raitholz* (= *Rechenholz*, "reckoning" or "tally stick"). The German Swiss had *Tessele* (< Latin and Italian *tessera*, "dice," then "four-cornered tablets" and finally "mark, stamp"; cf. p. 179) as well as the *Alpscheit* (literally, "Alp log") and the *Beile*, the latter word probably derived from the Medieval Latin *pagella*, "scale," from which the German words *Pegel* ("water gauge") and *peilen* ("to take soundings, measure the depth of water") and the English word *pail* have also most likely descended.

Austria and Vienna have their *Robitsch* or *Robasch*, a borrowing of the Slavic word *rovaš*, "notched stick." "*Es ist mir in die Rabuse gegangen*" ("I've had to go into notches"), a Bohemian would say regretfully when he had to borrow money. This peculiar word comes from the Slavic *rubatj*, Russian *rubitj*, "to cut, notch," to which the Russian ruble (*rublj*) is also related; the ruble was originally a piece of silver, about the thickness of a finger, cut off from a long bar of silver. These Slavic words, along with the Old High German cognate words *ruaba*, "number, count" and *ruabōn*, "to count, reckon," betray the connection between "cutting, notching" and "counting, reckoning," which crops up time and time again. It is most obvious in the English word *score* (< Old Saxon *sceran*, "to shear, cut"), which in addition to "cut" and "keep count" has also taken on the meaning of the number word for "twenty" (see p. 50); it is no less evident in the original Germanic word *talo*, the "notch in a piece of wood" which also means "number" (see p. 187), in the Serbian *broj*, "number" < *britj*, "to cut" and, to take an example from a primitive culture, in the Bantu word *vala*, which means "to notch" but also "to reckon, to count."

In Sweden there was the *karvstock* and in Holland the *kerf*. The Romans used the word *talea*, a "cut twig," for a "staff" or "stave." From this were formed the Medieval Latin *talare*, "to cut," which became Italian *tagliare*, Spanish *entallar* and French *tailler* (French *tailleur*, English "tailor"). Thus the notched stick was called a *taglia* or *tessera* in Italy, a *tarja* in Spain, a *taille* in France and a *tally* in England.

This category, finally, also includes the words for "book," or *codex*, as the medieval manuscript volume was called (for instance, the *Codex argenteus*, see p. 259). The book takes its name from the *Buche*, or "beech wood," from which the wooden writing tablets were originally fashioned, just as the Latin *liber* is named after the "bark," and the Greek *biblos* after the papyrus, on which it was written. The material on which the writing was done thus gave its name to the written product. The Latin *caudex*, too, was the "log of wood" that was cut up into *tabulae* or *tabellae*. Such wooden tablets were then tied together at one side into a bundle, and the resulting bundle was called the "beeches" < Gothic *bokos*; the Middle High German expression *lesen an den buochen* meant "to read the wood tablets" — the singular form of the word *Buch* not appearing until later. Shortly thereafter the herdsmen of Graubün-den tallied up the amount of milk they got from their cows in wooden "books" of this kind: each "book" consisted of a wooden writing tablet enclosed between two hinged wooden covers (see Fig. 50). Such wooden "books" were the earliest ancestors of the modern book; but even earlier they were bundles of notched sticks linked together at one end, of which we can find examples from Switzerland, Russia, and China (see Figs. 61, 64, and 75). The Romans later covered the wooden writing tablet with a layer of wax upon which they

Fig. 50 Wooden book of the Graubünden herdsmen.

inscribed letters with a stylus; this became the Roman writing tablet (cf. also the "diploma," p. 173). We find the same curious connection between wood and computation again in the Roman habit of using the word "codex" to refer to the "book" in which they kept their dispatches in successive order — *nomina in codicem digerit*.

But the most striking bridge between "cutting" and "computing" is made up of the Latin words *putare*, *imputare*, *deputare*, and *computare*. *Putare* means literally "to cut" (cf. *amputare*, which Tacitus uses in the same sense as the modern English word in the passage quoted on p. 224); *imputare* is "to cut in, indent," so that "to incise a notch for someone" takes on the figurative meaning of "to assign a debt to someone"; *deputare* is then the exact opposite of "indent" with its figurative meaning of "impute," since whatever had been tallied up to someone's account was later "cut away" from the notched stick after the debt was paid or the obligation discharged. *Computare* then clearly became "to compute, to reckon," as shown by the word *computus*, which in the early Middle Ages meant a "record of time." The word *putare* thence acquired its standard meaning of "to reckon, think, consider, believe."

Now after all this, what does a tally stick actually have to do with computation? We shall find the answer to this question in the following section.

VARIETIES OF TALLY STICKS

The Simple Number Stick

A contemporary of Pieter Brueghel's (died 1569) has an anecdote about the famous Flemish painter:

While he was in Antwerp, he lived with a young girl whom he would have married if she had not had the unfortunate habit of constantly telling lies. He made a pact with her, that for each lie she told he would cut a notch in a piece of wood; he took a good long stick, and when the stick was filled with notches, there could be no longer any question of marriage; this happened before very long.

One could scarcely find a better example of the meaning of the German phrase, *etwas auf dem Kerbholz haben* (literally "to have something on the notched stick" = "to have a skeleton in the closet").

From the standpoint of writing numbers, what exactly did the astute Pieter Brueghel do? Every time this daughter of Eve bent the truth beyond the breaking point, he took his knife and tallied one up to her account: for each lie he cut one notch, in one-to-one succession, so that one lie = one notch, again one lie = one notch and once more, one lie = one notch, ..., just as the Fiji Islander did with his club (see Fig. 6, p. 39). The notches were thus a supplementary quantity whose number value, because of its abstract, "colorless" nature, was used to represent the lies, with all the emotional burden they carried. A notched piece of wood of this kind is a simple *number stick*. We can identify it from the one-to-one succession of the notches — their ordering, in other words — which we now recognize as one of the earliest and most basic laws of the number sequence.

Number sticks have been used in every period and by all peoples.

Just as the Fiji Islander notched the handle of his club, the woodsman counts up the bundles of brushwood he has produced and, in a very different sort of occupation, the vineyard worker uses his vine-cutting knife to mark on his stave the number of basketfuls of grapes he has delivered. Many primitive peoples keep a record of the passage of time on their so-called calendar sticks in just the same manner.

The inhabitants of the Nicobar Islands in the Indian Ocean often have to count the numbers of coconuts they have picked. Since a plain number stick would never suffice for this purpose, they cut themselves a rod half a meter long from a bamboo stalk and split one of its ends into a number of lengths, which remain connected like the brush of an old-fashioned broom. Then the number of coconuts is marked by notches carved in the split lengths at the end (see Fig. 51).

There is no more impressive evidence of the universal use of simple number sticks in all periods of history than the illustration of the two carved bones of Fig. 51, with their notched edges. There seems to be very little difference between them, but they are separated by thousands of years: the bone at left was notched by a prehistoric man, but the one on the right by a Swiss farmer in our own time.

Number sticks have the force of legal documents when the number of days worked or the amount of goods tallied up on them serve as a record for compensation; they then become a sort of check, or "commercial paper." In ordinary life they once played a role whose importance we can scarcely imagine today. Tradesmen, innkeepers, bakers, and smiths kept their accounts on tally sticks, as they do today in ledgers, and they often bound their flat notched sticks together to form something like a book (see Figs. 61 and 64).

When in Schiller's play, Piccolomini's health was being toasted in Wallenstein's camp, during the Thirty Years' War, the barmaid presented a bottle to the company with the words:

Das kommt nicht·aufs Kerbholz. Ich geb es gern.
Gute Verrichtung, meine Herrn!

'This doesn't go on the tally; I give it free.
Success in battle may you see!"

An Italian saying goes:

Come e bello vivere in questo regno:
si mangia, si beve e si segna
tutto su un pezzo di legna!

"How pleasant it is to live in this land:
We eat, we drink and we just tally
It all up on a piece of wood!

But the innkeeper just laughs and says: "In this inn, unfortunately, there is no chalk and no tally stick." In other words, no one is allowed to eat and drink on credit.

When a person chatters on endlessly without having anything to say, the Germans say that he *redet aufs Kerbholz,* "talks on the notched stick." *De kerfstock loopt to hoog,* complains the Dutchman, "The bill is getting too high." And if he wants to send away someone who doesn't pay his bills, then *is de kerfstock ijzeren* ("turned into iron"): no more notches can be carved in the tally stick.

Fig. 51 Number sticks. *Left:* an English brushwood cutter's tally stick (Germanic Ethnological Collection, Berlin); *top:* a number stick of split bamboo used to count coconuts; *bottom left:* a prehistoric; and *bottom right:* a more recent notched bone, both from the Museum of Ethnology in Basel.

Thus the notched stick appears in figures of speech as a "debt" or a "bill." From an 18th-century writer comes this fine passage: "When I look up and think back, I see a great tally stick, which reminds me that I have sinned against you a thousand times more."

How common bookkeeping on tally sticks was, even in such a progressive commercial town as Frankfurt-am-Main, is shown by this quotation from the Weavers' statutes of the beginning of the 14th century:

Item so will auch die gemeyn geselschaft, dass alle rechenmeyster nit sollenn die kerben einem andern leihen, das eynem wirt zusteet.

It shall likewise be the rule for the entire guild that no master calculator shall lend to one person the tallies which are owed to another.

The use of tally sticks as bills or records of obligations can be traced all the way back to the time of the early Germanic tribes. A feature of the common law of the Franks and the Alemanni was the *festuca*, a rod that was exchanged between the parties to a transaction in matters of law, especially those involving promises of payment. According to Salic law, the debtor had to turn over to his creditor his *festuca*, on which his own identification symbol and the sum involved in numbers were marked (as in Fig. 52). In this *fides facta per festucam*, this "promise made by the *festuca*," the rod in question was nothing more than a notched stick on which the debt was tallied. The custom was continued into the Middle Ages, when a purchase was sealed "*mit hant und mit Halmes*, "with hand and stick," as was then the law and the custom.

Once the debt was paid or the obligation otherwise discharged, either the tally stick was burned or else the sum was cut off — that is, the creditor cut the notches off the stick, "he made the tally (actually the wood) smooth"; this is shown beautifully by the old Russian tax rods, which as a result of repeated addition and removal of notches finally became very thin (see Figs. 61 and 62). The same was true of the Romans' *tabula rasa*, the "erased (smoothed-out) tablet"; this expression has by transference come to mean "an old debt cancelled" and hence "a new leaf." On the wax-covered writing tablets which the later Romans used, the writing was erased merely by rubbing it over with a finger or the flat end of the stylus.

Now we can understand the lines in which Fischart, the most satirical portrayer of his own times, ridicules a mendicant nun:

Sie will kurzum Gott nichts schuldig pleiben,
sondern will das Kerbholz rein und glatt abkerben.

In short, she wants to owe nothing to God;
but wants the tally stick clean and smooth.

Fig. 52 *Stiala de latg*, a tally stick carved by a cowherd to record the amount of milk produced by the whole village. The one on the left is "cut off," i.e., the account is settled. From Bündner Oberland, Switzerland. 12 and 15 cm long.

Martin Luther, addressing a friend to whom he had not written for a very long time, began his letter thus: "I must cut off the tally stick, for I have not answered your letter for a long time."

Each installment paid on a debt was also recorded in this manner, as a municipal document dating from the year 1588 indicates: "A woman who is a debtor is to pay 12 gulden per year, until the whole debt has been paid; and each time this payment is to be cut off from her tally stick."

A sale of goods in 1453 is documented by a court judgment:

... *und haben beide gebeden das gericht, das is soliche gut dem Becker abesnide von siner kerben und Clase Riess zusniden an sine kerbe.*

... and both /parties/ have requested this judgment, /to the effect that/ Becker shall remove these goods from his tally and Clase Riess add them to his own.

It is not clear from this citation whether the amount of the debt was being transferred to a lease of the goods or whether this was a transfer from one household to another. But the words do show very clearly that tally sticks played an important role in the lives of peasants.

The whole significance of tally sticks and people's familiarity with them from the very oldest times is summed up in a nutshell by Skirnir's Song in the Norse Edda, in which the "scratched" magic spell is broken by erasing the incised mark (*thurs*, "scratch," was the name of the evil rune *þ*, which always brought bad luck; see Fig. 95, column 12, line 9):

I scratch a *thurs*   and then three runes:
Lust, sorrow   and pangs of love;
I scratch them out   as I scratched them in,
When they were needed.

The other kinds of tally sticks we shall now examine show how significant and practical these wooden numerical records were to our immediate ancestors.

Stialas de Latg

In these fascinating popular "milk sticks" of the Tavetsch Valley in the Bündner Oberland district of Switzerland, we shall find a kind of tally stick used to keep the "books" of a coöperative enterprise. Thus their numerical value and legal validity is extended from the individual to a whole village community.

From a piece of alder or ash wood the Alpine herdsman each day would carve a 5- to 8-sided rod, some 15 to 20 cm long, which he colors with red chalk; thus the light-colored uncolored carving stands out quite vividly (see Fig. 52). Then for every peasant who owns some cows in the common herd, an individual symbol is carved on the rod (in this case, two symbols on each face). Below each such symbol runs an incised line, the individual peasant's "account," on which the cowherd tallies off the amount of milk yielded by his cows. And when the herdsman adds a special ornamental carving, like the reclining cow in our illustration, these *stialas* are no longer primitive records, but a gay and colorful form of folk art.

Swiss cattle spend the whole summer at pasture in the high Alpine meadows. A particular herd may consist of cattle belonging to a number of different owners, let us say seventeen. The milk which the cows give every day must be processed and delivered at once. Since the mountain peaks in the Tavetsch area are close to the villages, each day a farmer, A or B or C, etc., according to an accepted and fixed order, goes to the mountain pasture and makes cheese, using the milk not only from his own but from all the cows in the herd. Obviously records of the amounts of milk used have to be kept; this is the chief herdsman's duty. Below the individual symbols of each of the 16 farmers who did not make cheese on a particular day, or all the farmers except A, notches are cut for the amount of milk from his cows that was used that day:

one notch crossing the long groove means 10 *Crennen* ($= 10 \times \frac{3}{4}$ pounds),
one cut (without removing the wood) means 5 *Crennen*,
one notch in the bottom right edge means 1 *Cr.*,
one cut in the same edge without removing the wood means $\frac{1}{2}$ *Cr.*,

smaller fractions are not counted. Thus farmer R, for example, has "lent" $4 \times 10 + 4$ *Crennen* of milk to farmer A on a given day. Finally farmer A, who made the cheese that day, carves his own personal sign on the surface of the butt end, which is the bottom surface as the *stialas* are shown in our illustration, and takes them home with him as a record for the milk he has borrowed from the other farmers in his village.

After the cycle has been run through and each of the seventeen farmers, A, B, C, etc. has made cheese on one day, they meet on a Sunday after church service to "combine the milk" — that is, to settle accounts. Each brings along his *stiala*. The amounts which farmer A has recorded on his own *stiala* are what he owes; the amounts tallied up to A's account on the other *stialas* are what the others owe him.

With this simple form of bookkeeping there is no possibility of cheating: no one would add notches to his *stiala* because that would increase the amount he owes; and no notches can be removed, because this would immediately show up on the red wood. If A on his day in the Alpine pasture used 60 *Crennen* of B's milk, and B on his day used 90 *Crennen* of milk from A's cows, B owes A the difference of 30 *Crennen*. A cuts away or "strikes out" his 60 *Crennen*, since he owes nothing to B, and writes down on a piece of paper that he still has 30 *Crennen* more coming from B. The differences are either paid then and there or else carried over into the next round.

We may well admire the clear and simple method of recording these very complicated debit and credit transactions. On these *stialas*, moreover, for the first time we find number symbols with different values — for $\frac{1}{2}$, 1, 5, and 10 units. They are not really numerals like 1, 2, and 3, for they do not follow in order; they are actually signs for groupings: one notch stands for a "group" of two cuts.

The *stialas*, which continued in use well into the 20th century, were the records of transactions among the many "shareholders" of an Alpine community. Generally, however, there are only two parties to a transaction: buyer and seller, or creditor and debtor. Both must keep records of the obligation undertaken and discharged. But receipts and promissory notes call for paper and writing, which simple people neither understand nor like. Then how can the transaction be recorded on a tally stick so that no one can cheat? This was accomplished by the split or

Double Tally Stick

A long piece of wood is cut lengthwise almost to the end; the part with the large end is the "stock" (the main stick) and the split-off portion is the "inset" (the piece laid on the main stock). In Vienna these were called *Manderl* and *Weiberl* ("little man" and "little woman") respectively.

In the case of the beautifully made Finnish double tally sticks, stock and inset are identical and are held together by wooden pegs (see Fig. 53).

Fig. 53 Finnish tally stick to record units of work done. Size 25 × 2 × 1.5 cm.
Kansallis Museum, Helsinki.

When a payment or delivery is either made or received, the debtor inserts his inset in the stock, which the creditor generally keeps, and notches are cut into or removed from both pieces at once. Then both parties take back their own pieces and keep them until the final settlement. In this marvelously simple fashion the "double bookkeeping" makes any cheating impossible. These "countersigned" records are called *contretaille* and *tacca* (notch) *di contrasegno* in French and Italian, respectively.

Until quite recently a three-part tally stick was used by the snow-removal men in Vienna (see Fig. 54). The middle portion with the head was kept by the driver, one of the side pieces by the foreman at the loading platform and the other by the foreman at the unloading platform; thus there was a record in triplicate. Since both foreman had as many different side pieces as there were drivers, one for each, they would hang them together on a string around their necks. All three parts of a tally stick were marked with the same number, here 174.

Here, too, no cheating was possible. For this reason the double or triple tally sticks always had the force of legal documents.

And if anyone is not good at settling his accounts by writing and reading, he shall be satisfied with crudely made tallies or tickets: Then if one party brings to court a tally or ticket as evidence of his debt, and the other party produces the corresponding or matching ticket or wooden tally, and they are found to be the same, credence shall be given to them and the amounts stated on them shall be acknowledged,

says the Basel statute book as late as 1719. Even Napoleon's book of statutes, the *Code civile* of 1804, which was in force in a number of German provinces as well as in France, still had the provision that

es tailles correlatives à leurs échantillons font foi entre les personnes qui ont dans l'usage de constater ainsi les fournitures qu'elles font ou coivent en détail.

Fig. 54 Tally stick in three parts, used by snow-removal men in Vienna.
30 × 4 cm.
State Museum of German Ethnology, Berlin.

The tally sticks (= insets) which match their stocks have the force of contracts between persons who are accustomed to declare in this manner the individual deliveries they have made or received.

The absolute security against deception which is the special advantage of stock and inset was transferred, as early as the 10th century, from tally sticks to written documents:

Am Tage St. Andreä anno 1594 sind zwey gleichlautende, einer Hand-
schrift aufeinander ausgeschnittene Briefe, denen jede Parthey einen zu
sich genommen, ausgefertiget worden, [weshalb man] nicht nötig gehabt,
dergleichen Kontrakte zu unterschreiben, sondern es haben solche, wenn sie
ordentlich aufeinander gepasset, eben die Kraft des Beweises gehabt, wie
denen ausgeschnittenen Kerbhölzern beigeleget wird.

On St. Andrew's Day in 1594 two identical letters were prepared, one written above the other and the two cut apart, of which each party took one copy; [wherefore] it was not necessary to sign the contracts, but if they matched one another properly, they had the force of legal documents, just like the matching portions of a carved wooden tally stick when they are fitted together.

These were the notched tickets mentioned in the Basel statute just quoted; they were also known as notched letters, split papers, or split tickets. In Medieval Latin they were called *cartae partitae* or *dentatae*, "divided" or "toothed papers" (see Fig. 55).

Fig. 55 Notched tickets: these documents written out twice and then, after the manner of the double tally sticks, cut apart along a zigzag or wavy line. Only an exactly matching half was accepted as true and correct. Kunsthistorisches Museum, Vienna.

The contract was written out in full two or more times on a sheet of paper or parchment, letters were drawn in between the two or more copies of the text, and the parts were then cut apart through the letters along a jagged zigzag or wavy line The wavy cut has been preserved here and there in the coupons of stock and bond certificates or other commercial paper, at the end of the booklet from which the coupons or cheques are detached. A trace of the old usage also persists in the English word *charter*, meaning to "rent or hire a ship or boat," for the *charter party*, the contract between the merchant and the ship operator, is literally the old Latin *carta partita*, a "paper cut apart";

in the Hanseatic League of medieval European trading towns the German term *Zertepapier* and the French term *chartepapier* were used. The English *indenture* preserves the serrated or denticulated edge of the notched paper in its very name (< Medieval Latin *indentare*, "to cut teeth into"; *dens*, "tooth").

The check as a certificate of demand for payment goes back to the notched tally sticks. The English Royal Treasury, as we shall see in the Exchequer tallies, kept its records of income, such as taxes, and expenditures on notched tally sticks (see p. 236). But the Exchequer also issued the stocks of double notched sticks or *tallies* (perhaps, for instance, with a £20 notch) to its citizens as certificates of payment. The holder could then go to the Royal Treasury, which held the inset or matching piece, and if the stock and inset agreed, the stock would be redeemed in money. In English *to check* still means to compare an original document or piece of writing against a copy to see if the copy is correct. Hence the written certificate or money order, which was to be presented for remittance and checked against its security, later came to be called a "check" or "cheque"; we shall see later (p. 347) how this word is related to "chess" and "checkerboard."

A *check* is thus generally an "identification mark," just as the Greek word *sýmbolon* originally meant a "distinctive mark" (< Greek *sym-bállein*, "to throw together") — that is, a broken shard, for the most part with writing on it, which fitted the piece from which it had been broken. Another such mark of identification or recommendation was the Roman *tessera hospitalis*, the "guest-countersign."

In addition to this verbal evidence for the long survival of the notched tally stick, there is also the very interesting and rare instance of a written testimony: the word for "contract" in Chinese is symbolized by two characters at the top, one for a tally stick (stick with notches) and one for a knife, and another at the bottom which means "large" (see Fig. 56). A "contract" or "agreement" in Chinese is thus literally a "large tally stick"!

Popular speech has naturally also taken note of the peculiar quality of the double tally stick. "To keep a tally with someone" was the German commercial jargon for a business relationship; in a figurative sense it then came to mean to be close to someone in a personal sense, just as German country people say, "*Sie hat's mit ihm*" ("She's in love with him" — literally: "She has it (a tally) with him").

The English language is very rich in expressions of this kind, because in England the notched stick (the *tally*) itself played a dominant role in the state's finances until well into the 19th century (see the *Exchequer tallies*, p. 236). *To keep tally with somebody* is the equivalent of the German expression just mentioned. *They were tallies for each other*, i.e. "they were like double tally sticks to each other," by transference means "one was the spitting image of the other." From this came the use of *tally* in the sense of "counterpart" and of the verb *to tally* with the meaning "to match or fit together": *the account does not tally* means "the calculation is not correct." The reader can figure out for himself why *to live tally* means "to live in sin" and just what a *tally-wife* is.

Fig. 56 "Contract": Chinese character made up of a tally stick and a knife; the character at bottom means "large."

The *tallyman* or "junk-dealer" got his name from his habit of carving notches in a stick. He is the proprietor of a *tallyshop* where second-hand goods are bought on credit and paid for in installments; the Frenchman says *acheter à la taille* when he means "to buy on credit." Thus in the *tally* the original wooden notched stick has become paper, so to speak.

Special Kinds of Tally Sticks

The notched number stick and the double tally stick, which we have just discussed, may well appear to have exhausted the forms of notched sticks used to record numbers. Certainly these were by far the most common. But especially in some remote valleys of Switzerland tally records of such peculiar kinds have been preserved almost to the present day that we must look at at least three examples.

An *Alpscheit* (literally "Alpine wood-billet") from the Lötschental district (Fig. 57) was a villager's certificate showing that he was

Fig. 57 An *Alpscheit* or "Alpine billet" with insets, dating from 1752, on which the pasture rights of the peasants in a village are recorded. It is triangular in section, 1.30 m long and 9 cm wide, weighs more than 3 kg and has more than 70 cut-outs into which corresponding inset pieces were fitted.

entitled to let his cattle graze on the communal pasture. Small pieces were cut out at regular intervals from each of the three surfaces. The peasants kept these inset pieces, or *Beitesseln*, as evidence of their individual pasture rights, for the notches and cuts carved into them, when they matched those on the cut-outs, were the proof of their particular "cow" rights:

one long groove is 1 cow right,
one short groove is $\frac{1}{2}$ cow right,
one long cut is $\frac{1}{4}$ cow right,
one short cut is $\frac{1}{8}$ cow right.

One cow was considered the equivalent of 10 sheep, in terms of pasture rights, but one had to have $\frac{1}{8}$ cow right to be allowed to pasture one sheep. The peasant would keep his small inset pieces in a small, ornately carved box, while the cowherd would keep all his "Alpine billets" strung together on a cord.

Here we see a special use of the notched tally stick: the inset piece which was fitted into one of its grooves was not a promissory note but a certificate of rights that gave its owner a share or a fraction of a share in the community pasture. Yet it was not the exact equivalent of a stock certificate, because it did not entitle the peasant to fractional ownership, but merely to proportional use of the pasture. One cow right in a good mountain pasture was considered to be worth 1000 Swiss francs. Thus these tiny carved bits of wood had the same effect as securities or bonds.

In contrast to the cow rights, however, the *Kapitaltesseln* ("capital tallies") of Visperterminen, Switzerland, were the exact equivalents of bonds. Quite apart from their economic importance, they are all the more interesting from our standpoint because of their remarkable numerals. The community had funds which it would lend out to individual farmers. As a promissory note for his debt, the borrower gave the municipal government a tally stick with his personal mark carved on it and the amount of his debt marked on the opposite side. These tallies were strung on a cord through the holes in their ends (see Fig. 58) and kept until the debt was paid off.

Fig. 58 Capital tally sticks: on the obverse side are carved notches showing the debt owed by a peasant; on the reverse is his personal symbol. From Visperterminen, Switzerland.

The Swiss water tallies are also of interest because of their numerical symbols (see Fig. 59). In the area of Wallis, Switzerland, the aqueducts leading down from the mountains play an important role in the local economy, because the arable land becomes very dry in summer. Thus the streams of glacial water are carefully collected and led through gullies and ravines, often over long distances, to the cultivated fields where the crops are grown. These glacial streams are the property of the community which may, for example,

$= 4$ hours
$= 2$ hours
$= 1$ hour
$= \frac{1}{2}$ hour
$= \frac{1}{8}$ hour
$= \frac{1}{16}$ hour

$= \frac{1}{4}$ hour
$= \frac{1}{8}$ hour

Fig. 59 Water-tallies from Wallis, Switzerland.

have 34 such aqueducts. The farmers in the area must buy the rights to share this summer water. Like the pasture rights, the water rights are recorded and certified by carved tally sticks.

Ethnologists also include among tally sticks the so-called "messenger staffs" or "turn tallies" on which are carved the personal symbols, for example, of the men who take turns in ringing the church bells. Whoever has the stick has the duty that time around. We shall not describe or illustrate these, however, since the notches carved on them have no numerical significance.

Now that we have briefly seen a few examples of number symbols carved in wood, let us take a look at the most outstanding example of the tally stick.

THE BRITISH EXCHEQUER TALLIES

These are unique in cultural history. In the Exchequer tallies the ancient number stick of obscure and common origin achieved its highest development: it became an official government record.

Some years ago, when repairs were being made in Westminster Abbey, several hundred "exchequer tallies," or notched sticks used by the British Royal Treasury, were found together with documents and the remains of the leather sacks in which they had probably been kept (see Fig. 60). These dated from the 13th century.

1	2×10000 £
2	100 £
3	2 *score* £
4	10×1 £
5	17 *s*
6	11 *d*
7	$4\frac{1}{2}$ *score* £
8	$16\frac{1}{2}$ £
9	100 £ + 16 £ + 9 *s* + 8 *d*
10	$(20 + 6\frac{1}{2})$ £ 3 *s* 4 *d*

Fig. 60 Exchequer tallies from the 13th century. On these sticks the treasury officials carved the amounts of tax owed and paid. These continued in use until 1826; the method of carving the notches remained the same until that year. In the case of Nos. 2 and 3 the notches are in the bottom surface, and the head of the tally points to the rear.
Society of Antiquaries, London.

Ever since the 12th century the English Royal Treasury had kept its records in books and on tally sticks, by methods that remained virtually unchanged until the 1820's. Thus from *tally* was formed the old English word for "tax," *tailage* or *tallage*, which resembles the related French word *taille*.

The Court of Exchequer, which was the state tribunal of audit, consisted of several departments. Here we are concerned with the central office, where the sheriffs, who administered the shires or provinces of England, settled their accounts with the Crown. The table in this office was covered by a counting cloth with a square checkerboard pattern, from which the whole court of audit took its name (see p. 347). Over this table the sheriff reported to the Treasurer, item by item, and item by item the Calculator laid out the amounts in *calculi* or counters on the chequered cloth and thus arrived at the final sum. Then the Cutter carved a *tally* to record the amount paid or owed to the Crown. The whole procedure was observed and audited by members of the court and higher officials. The use of tally sticks and counters enabled everyone to understand without having to know how to read and write.

This is how such an account was typically settled: if the official in charge of a district owed the Crown £100 annually in taxes and duties, he would make the first payment at Easter, let us say £40. In testimony of this a *tally* for £40 would be carved, of which the official retained the *stock* (or *stipe*) as a receipt for his payment and the court of audit kept the inset piece (the *foil*) for its own records. The name of the payer and the nature and amount of the payment were incised on both pieces. Then on Michaelmas-day (September 29) the official had to make good the entire annual levy of £100. To do this, he would submit his *stock* for the amount of £40 and pay the rest. The *stock* was very carefully compared (*checked*; see p. 233) with the *foil* in the Treasury's possession; if they agreed, the amount was written up to the payer's account. If any deception or cheating was detected, the official would be arrested on the spot. The payments made by the county officials were entered in the receipt book, and when their obligations were discharged their accounts for the year were closed with the phrase: *et quietus est* or *and he is quit* (of his debt).

As early as the year 1300 the *tallies* were issued by the Royal Bank to be used as a medium of exchange in England. For example, King Edward I's steward once received a *tally* as a money order in place of a payment owed by one of the inhabitants of London. From this man, who himself was a debtor of the Royal Treasury, the steward could demand and receive the sum of money marked on the tally. In this manner the state avoided the burdensome collection of debts owed to it and at the same time satisfied its own creditors: a cashless system of exchange, using notched tally sticks as cheques. From the middle of the 14th century on this system was in constant use, and did not begin to decline until the rise of banking institutions in the 18th century. But this use of tally sticks is no longer to be seen anywhere.

The banking term "dividend" probably also goes back to the English tally, which from sometime in the 13th century was also frequently called *tallia dividenda*, or simply *dividenda*, "a stick to be divided," especially in using the insets (foils) in the King's possession to make purchases for the Royal Court. In return for his goods, the tradesman was given a *dividenda* which he would later redeem at the Treasury, just as today a bond-holder, in his "coupons," has a certain share in the earnings of the enterprise which issued the bond. After the 18th century in England the "dividend" referred not to the tally or coupon but to the actual share of the earnings, and from there it acquired its present general commercial usage.

Because of their official nature, the carving of the notches in the tallies had to be uniformly regulated. Fortunately we are adequately informed about this and about the procedures of the court of audit by the *Dialogus de Scaccario* ("Dialogue Concerning the Chessboard"), written in 1186 by the then Royal Treasurer Richard, Bishop of London. The rules prescribed here, and above all the *tallies* themselves, remained in force for many centuries. In 1782, to be sure, it was commanded that tallies be no longer issued, but they remained valid until 1826. Then in 1834, when the state's official collection of cancelled *tallies* was being destroyed, a vast number of them were burned with such excessive zeal in the furnaces below the Houses of Parliament that the Parliament buildings themselves went up in flames.

Now, what does the *Dialogus* have to say about the "cutting of the tallies" — *de incisione talearum*?

In summo ponunt m. li. (librae) sic ut incisio eius spissitudinis palme capax sit . . .

The notch for £1000 is placed at the end and is as large as the hand is wide (see No. 1 in Fig. 23);
for £100 the notch is as large as the thickness of a thumb, and to distinguish it from that for £1000 it is not straight but curved (see No. 2);
for £20 it is as large as the thickness of the little finger (No. 3);
for £1 it has the breadth of a ripe barleycorn (No. 4);
for 1 shilling it is smaller but still large enough to be seen as a notch (No. 5); whereas
for 1 penny only a cut is made, with no wood being removed (No. 6);
for a half of any of these units a notch or cut half the length is carved: one cut slantwise, one perpendicular to the edge (Nos. 7, 8).

This prescribed succession of magnitudes was to be strictly observed: the largest number is always on the outside and the others follow in order. Examples 9 and 10 in Fig. 60 show how simply and easily the tally cutter avoided any possibility of confusing the larger notches with each other: The highest units were carved into the bottom surface (like the notch for £100 in No. 9), but the smaller ones at the top from left to right, so that the lowest ranks stand above the highest, from which they are very clearly differentiated!

Now we can make an observation both revealing and fruitful for the cultural history of numbers. On the tallies we see that the intermediate rank between £100 and £1 was not £10, as we might have expected, but £20: — we meet the old vigesimal grouping again in the *scores* see p. 49)! *A score of pounds* is £20. But what is a *score*? The old Saxon word *sceran* is similar to the English word "shear," to cut, cf. German *Schere*, scissors. Thus a *score* is something which has been cut or carved — a notch! This was incised on a tally stick whenever a group of 20 was to be counted (such as 20 sheep), and from this notch the vigesimal grouping acquired its English name *score*. There is a Finno-Ugric counterpart: The Lapp word *tseke*, "notch," likewise came to be the number word for 10 (see p. 113).

Now we clearly understand the significance of the English expressions *to run into scores*, meaning "to go into debt," and *on score*, meaning "on credit" (on the tally stick, without paying). Previously we saw many English expressions in which *score* was used with the meaning of 20 (pp. 49 ff.); but we can also find many passages in Shakespeare in which the word *score* still has its original meaning of "notch" and "tally stick":

Our forefathers had no other books than the score and the tally.
(King Henry VI, Part Two, Act 4, Scene 7.)

Then from Scene 2 in the same act we have:

I thank you, good people. There shall be no money, all shall eat and drink on my score.

Of Macbeth it is said at the end of the tragedy (Act 5, Scene 7) that

He parted well and paid his score.

The English language shows another interesting development. If someone lent a sum of money to the Bank of England, a *tally* was carved to record the transaction, of which the Bank retained the inset piece or *foil* (from the Latin word *folium*, "leaf") and the creditor received the *stock*. Thus he became a *stock-holder* and possessed a *bank-stock*, which had the very same worth as paper money issued by a government. From this custom comes the modern expression "stocks" for "shares in a business enterprise" or "paper money."

Notched tally sticks used to keep official records of state finances! After this rare case in a country with a high level of culture, we may conclude with a simple "early" example such as could once be found virtually everywhere. This is a Russian "tax book" for a Cheremysian village. The Cheremysians lived along the banks of the Middle Volga, and spoke a language belonging to the Finno-Ugric group.

For each of the 13 households in the village the tax collector had a tally stick. On the stick he carved an identifying symbol at the head and the number of inhabitants (see Fig. 61; half a notch for a child), and the amount of tax due annually at the end. Then the amount paid was "cut off." From Fig. 62 one can see clearly that this was done repeatedly and that the tax collector did not make a new tally stick every year.

Now let us glance back over the ground we have covered: what have we learned from the tally sticks? For one thing, they represent the oldest form of bookkeeping. They do not simply show the amounts involved but also record the transaction between buyer and seller or creditor and debtor, without giving either party any opportunity for cheating, even for such highly complicated exchanges as those

Fig. 61 A Russian "tax book" from the Middle Volga area. For each household there is a tally stick on which are carved an identifying symbol, the number of inhabitants and the amount of tax to be paid. One tally stick is about $17 \times 2 \times 1.5$ cm in size.

Kansallis Museum, Helsinki.

involved in the milk-and-cheese coöperatives of the high Alpine meadows. They were account books in which the finances of a household or a village community were recorded in a generally very sensible fashion, and here and there they even rose to the status of "legal tender" or "share certificates" or medium of exchange. Their close connection with the economic life of mankind, which remained unchanged from the very earliest times until just recently, makes the notched tally sticks awesomely impressive as documents.

We finally have the secret of the "earliest" writing which was born of the common people. As uninitiates, we cannot read it. But we do know one thing: It was a method of writing numbers, and it consisted of primitive numerals.

Now let us examine the "numerals" or number signs themselves.

Fig. 62 A tally stick from the Russian "tax roll" of Fig. 61. The paid debts were "cut off" and new sums incised in the same spot.

THE NUMBERS ON THE TALLY STICKS

The simplest tally sticks merely bear rows of notch after notch; they form the amount to be counted by means of the supplementary quantity of notches carved. The rows of notches are organized by groups. Systems of written numbers formed according to this rule, like the Egyptian or the Roman numerals, we shall call "early" numerals or row numerals (see Fig. 8).

Number Notches

On the tally sticks we see groupings of notches requiring some sort of symbol different from the simple unit notches. At first we may be inclined to think that a notch is a notch. Yet the tally sticks we have seen thus far show a great variety of different kinds of notches (see Fig. 63): notches carved transversely and perpendicularly into the

Fig. 63 Various forms of notches.

wood, incisions, half-notches and half-incisions, slanting notches and incisions, notches and incisions cut into the edges of the stick, round notches, beveled notches, straight and round notches carved in the middle of the wood surface, etc.

But all these various special forms of notches were avoided in the English Exchequer tallies. They used only simple notches, differentiated from each other by size and position, and thus surprisingly arrived at a consistent and regular system of measures.

Quite often we have seen notches in the form of an X or a V (see Figs. 53, 58, 59, and 64). Anyone not familiar with tally sticks might think that these were Roman numerals transferred to the wooden sticks. Although this did happen here and there, did the peasants in some remote mountain village really have to wait until Roman numerals were introduced before they could carve X and V. Aren't these shapes actually the easiest ones to whittle with wood and knife? To ask such a question is to answer it. One hardly needs to see or refer to prehistoric notched bones with such notches (see Fig. 51) to realize that the other commonly used Roman numerals L, C, M, or ∞ rarely if ever occur on tally sticks, and that one likewise does not see back-counted formations like IV or IX. Carpenters also form the numerals I, V, and X with their axes when they number the beams and timbers they have fashioned.

Now let us ask the same question the other way around: Are the Roman numerals, at least I, V, and X, actually forms of notches cut into tally sticks? I believe that further facts will confirm that this is so, apart from mere similarity of form and the fact that such notches are very easy to carve.

At the beginning, the Roman numerals observed the basic and simple laws of ordering and grouping, like the notches in tally sticks. X is a group-form. How did this come about? By a crossing of the usual symbol for unity. We shall see this simple device over and over: crossing two signs makes a group-form. In the Indian Kharosthi numerals X = 4 is clearly the symbol for a grouping (see Fig. 12). Excellent examples are again provided by the Swiss tally sticks, such as the number 42 incised on the second stick from the left in Fig. 58.

Fig. 64 A bundle of Alpine number billets, small flat sticks some 20 cm long on which are carved the cow-rights to which their owner is entitled; the owner's name or symbol is on the reverse side. The most ornate of these sticks, the one at the extreme right showing the number 122, gives the total. This bundle, from Saanen in the Canton of Bern, is dated 1778.

For the most part a crossing of straight-line symbols means 10, so that if we wish to give a name to it, we can call it a "ten-stroke." A striking example is to be seen on the chief tally stick of Fig. 64, which shows the total, where even the V notch is crossed (in the number 122; see *b* in Fig. 65). Here we see that the unit notch which stands by itself is also crossed through into an X notch; the number 122 could also have been formed as in *c* in the same illustration.

a   b   c

40 2   50 50 10 10 2   50 50 10 10 2
  42   122   122

Fig. 65 The ten-stroke in notches on Swiss tally sticks.

A double crossing then indicates the group 10 × 10, or 100 (see *a* in Fig. 66). Thus we find even such numbers as 190 and 256 on tally sticks (see Fig. 58).

a   b   c   d

100   100 100 40   10 10   4   100 100 50 5 1
           ─────   ──
            2   2
          190   19   256

Fig. 66 Double crosses and half-signs formed by notches carved in Swiss tally sticks.

These notched numbers carved in tally sticks from a remote Alpine valley suggest a possible explanation for the extraordinary and hard-to-trace Roman numeral L for 50. The rules governing the formation of numbers on tally sticks are of a very early date, for the amounts recorded by them are ordered in a series and grouped. But if the quantity makes up only half a grouping, only half the symbol for the particular group is carved! For example, the V symbol on the ten-stroke is half an X and hence means half of 100, or 50. This is very readily seen on the right-hand tally stick in Fig. 58 in the numbers 19 and 256 (see also Fig. 66, *c* and *d*); in the case of the last number the half unit-notch on the ten-stroke stands for 5.

A study of tally sticks reveals two very important things about early written numerals:

The crossing of a carved symbol most frequently means a grouping, usually 10. (From the "crossed" Roman numeral X comes the Latin word *decussare*, "to cross through" from *decussis*, "the number X," from *decu-* for *decem* 10 used in compounds).

Half the symbol stands for half its normal value (see the Roman symbols for fractions, p. 161).

Now since they have played an important role in our culture for hundreds of years, let us again examine the

ROMAN NUMERALS

primarily the symbols I, V, X, L, C, D, and M. If the unit I is crossed it produces the symbol X for 10, half of which, V, has the value 5. In Fig. 67 the Etruscan coin at the left shows the bottom half of the ten-symbol, ∧, whereas in Rome the upper half, V, was commonly used (see Fig. 66).

But what about L for 50? This numeral has nothing to do with the letter "L," which it coincidentally resembles. Our illustration shows

Fig. 67
ROMAN COINS (top)
Left: with the value of 60 (sesterces), 215 B.C.;
right: with the series number 70 of the minting (90 B.C.). This is one of the so-called "serrated coins" (*nummus serratus*) which Tacitus says the Germanic tribes prized especially highly (see p. 358).
ETRUSCAN COINS (bottom)
With the symbols ∧ and X; the coin on the right bears the Gorgon's head (5th century B.C.).
Landesmuseum, D_rmstadt.

:wo Roman coins, on which an arrow-shaped symbol and an up-
side-down T can be recognized. Moreover, an inscription carved on a
Roman milestone contains the numbers 51 and 74 (see Fig. 71, lines
4, 5, and 6). Now the development is quite clear: The two oblique
strokes on the arrow-shaped symbol were incised as a curve or
a horizontal straight line, until finally this numeral came to be
written just like the letter L.

What was the origin of the initial arrow-shaped form? It may be a
half-symbol, like the V of X, in this case the upper half of a double
or twice-crossed numeral I, such as we have seen on wooden tally
sticks (Figs. 58, 59, and 66). Thus it may have developed from a
deeply rooted early form (see Fig. 68).

Another possible explanation is that it is an old form of the Greek
etter Ψ. The Romans took their alphabet from the Etruscans, and
they in turn from the Greeks (see columns 7 and 9 in Fig. 95 on
p. 265). Since they did not make use of the letter Ψ, this was free
to be used as a numeral. Thus the Roman numeral for 50 is thought
by some to have developed from an arrow-shaped variant of the
Greek Ψ (Fig. 95, p. 265, line 26).

Correspondingly, this hypothesis (which is Mommsen's) sees the
Roman numerals for 100 and 1000 as originating in similarly "free"
Greek letters of the Greek alphabet. Since the Etruscans had replaced the aspirated Θ and Φ by the unaspirated T and P, the
former two were also available for use as numerals Actually, there
was an old Φ-form, a circle with two diameters at right angles to each
ther, which resembles the Roman numeral for 1000 (compare Figs.
9 and 70 with Fig. 10). The Θ supposedly developed into the Roman
umeral C (as shown in Fig. 70).

This transformation does not seem too convincing, however,
ecause the use of foreign letters looks too "learned," especially
these particular letters, because their respective numerical values of
and 500 which they had had in Greece now became 100 and 1000
in Roman civilization (see Fig. 95, p. 265, lines 9 and 24). Couldn't
the Romans themselves or their ancestors have invented symbols of
their own?

The symbols on the small tablet which the computer holds in his
left hand in the Etruscan cameo illustrated in Fig. 134 are certainly
numerals. We see here a few familiar forms, and also a circle with
an oblique cross within it. This may have been the old sign for 1000.
The cross may well first have right itself to assume a vertical-and-
horizontal position (b in Fig. 70) and then dropped the horizontal
bar (c); then it would have strongly resembled the Roman numeral
for 1000. This derivation is quite surprisingly supported by an
analysis of the symbol (d): D for 500 is doubtless the right half of the
symbol; this leaves the left half, C, as the numeral for 100! This last
step cannot be documented, yet the "breakdown" does find some
confirmation in the 500-symbol on a Roman milestone inscription
in which it still bears the horizontal bar of the original full form (see
Fig. 71, fourth line from bottom; cf. d in Fig. 70).

Was the Etruscan symbol, the circle with an oblique cross, an independently invented form or was it in turn derived from the Greek

Fig. 68 The Roman 50-numeral L as
a half-symbol for 100.

Fig. 69 Coin of the ancient city of
Pherai in Thessaly, with the initial
letter *phi* in the old form (instead of
Φ), which the Romans may have
adopted as their numeral for 1000 (cf.
Fig. 10). 4th century B.C.;
1.8 cm in diameter.
Staatliche Münzsammlung, Munich.

Fig. 70 Possible origins of the Roman
numeral C for 100:
top: from the Greek letter Θ;
bottom: from a supposed Etruscan
numeral for 100 or 1000 (cf. Fig. 97).
If the second hypothesis is accepted,
then *c* could be the Roman symbol for
1000 (as in Fig. 10, p. 44). Its left
"half," *d,* may then have become the
Roman 100, while the right "half" D
became the Roman numeral for 500.

Fig. 71 Inscription on a Roman milestone on the Via Popilia in Lucania. This shows the old form of the numeral for 50 (lines 4, 5, and 6 from the top) and a peculiar form for 500 (line 4 from the bottom). The inscription is 74 cm wide, with letters 3 cm high. Around 130 B.C.
Museo dell Civiltà Romana, Rome.

The inscription reads: "I built the road from Regium to Capua and erected all the bridges, milestones, and postal stations on it. From here to Novceria the distance is 51 miles, to Capua 84, to Muranum 74, to Consentia 123, to Valentia 180, to the statue on the seashore 237 and from Capua to Regium in all 321 miles. Likewise as Praetor in Sicily I tracked down Italic fugitive slaves and returned 917 persons (to their masters). Moreover I was the first to establish that on public lands hersdmen must yield to farmers. At this place I built the forum and other public buildings."

theta Θ? At the present time there is no way of answering this question.

At any rate, neither of these two hypothetical derivations is inconsistent with the idea that the arrow-shaped 50 form was originally half of a horizontally crossed $\times$ standing for 100. This "early" numeral could well have disappeared later on. Or perhaps, instead of being crossed with a horizontal stroke, which would have required a considerable degree of accuracy in carving, a curve or parenthesis was placed next to the $\times$, so that ($\times$ stood for 100 and ($\times$) for 1000. Then the Etruscan symbol would have had a completely indigenous, "early" derivation, as we have seen in the case of the tally sticks. Even the unique form ∞ for 1000, which appears in the inscription of Fig. 73, could then easily have been derived directly from ($\times$).

Now let us get on with our analysis of the Roman numerals themselves.

The numeral for 100 was later written as C, which (quite by coincidence!) was also the initial letter of the Latin number word *centum*, 100, just as the numeral for 1000 developed into the form M, with which (again, purely by coincidence) the number word *mille* began, while the symbol for 50 came to be standardized as L, which does *not* correspond to any Latin number word. By contrast, in the early Greek number symbols from time immemorial the initial letter of the name for a numerical rank served as the numeral for that number: for example, *Δ* was the numeral for 10 because it was the first letter of the Greek word *ΔΕΚΑ*, "ten" (see Fig. 97). This method of writing numbers was very common in the Middle Ages (as, for instance, the Indian numeral 2 became identified with Z probably because of the word *z-wei*, "two"; see Fig. 118) but never appeared at all in Roman times. For 1000 the Romans could also use the curious form ∞, which ever since the English mathematician Wallis proposed it in 1655 has been accepted as the mathematical symbol for infinity (see Fig. 73).

All the other Roman numerals are quite clear: D for 500 is half the symbol for 1000. The higher numerals, those for 10,000 and 100,000, are artificial extensions of the already available symbol for 1000; this is more evidence that 1000 was an old limit of counting. The numerals for 5000 and 50,000 were at one time the halves of the symbols for 10,000 and 100,000 (see Fig. 72, also Fig. 31).

Fig. 72 The Roman half-symbols for
500   5000   50,000
These are the "halves" of
1000   10,000   100,000
shown in the top row. At the bottom
are the transformations that appear
on Roman denarii (see Fig. 31).

"Caesar sesterces 100 millions"

Fig. 73 The Roman numeral for 100 million here consists of the symbol ∞ placed within a frame, which itself originated from an extension of the numeral for 100,000 (see Fig. 10). From an inscription dating from the year A.D. 36. Letters 12 mm high.
Scavi di Ostia (Excavations), Rome.

And what about the million, the ⨯ or X standing beneath a gate? The gate is nothing more than a rectilinear version of the curved sign for 100,000, which we have seen already on the *Columna rostrata* (Fig. 10). It is an abbreviation of *centena milia*, the last of the verbal numerical ranks for which there was a full name. By exact analogy to the verbal expression *decies c. m.*, "ten times a hundred thousand," the numeral was written as an X beneath a curve (see p. 44, where we spoke of this numeral in a different connection).

In medieval manuscripts a numeral often had its value multiplied by a thousand by the addition of a superior bar, for example $\overline{II}$ for MM, 2000 (see Fig. 44); this was not common in Roman texts, however, where for the most part it served only to identify the symbol as a numeral, i.e., to distinguish it from the letter of the same form (see p. 281).

Now let us summarize. The Roman number symbols did not evolve continuously and smoothly any more than did the Roman number words. They began with very ancient forms on tally sticks, certainly up to X and perhaps up to 100. However the symbol for 1000 may have originated, whether from a foreign letter or an indigenous carved form, it was at any rate treated as an early numeral: it was halved to produce a numeral of half its value and by the same token multiplied 10 and 100 times by *decussatio*. In this manner the limit of counting was pushed forward as far as 100,000. Then when the breakthrough came in calculations involving millions, the limit remained at the numeral for a million, which was no longer ordered in series but counted and thus built up in gradations.

Thus the Roman numerals are the "sisters" of the Latin spoken numbers. Like the latter, the former were not invented all at once by one person, but were evolved gradually by the people who wrote them.

The Chinese Han Sticks

If we closely examine inscribed Chinese wooden sticks (Fig. 74) from the time of the Han Dynasty (200 B.C. to A.D. 200), we come across the "ten-stroke" not only at 10, but also at 20, 30, and 40, just like the insets on the tally stick from one Alpine village (see Figs. 58 and 64).

Fig. 74 The ten-stroke on Chinese Han sticks:
10   20   30   40   the same, fused together.

This ancient method of writing 20 and 30 has been preserved even to our own time as a special form in China and even more in Japan. The interesting feature is this: a single Chinese character could also have only a single one-syllable pronunciation. The usual Chinese way of writing 20 (2 × 10) has two characters, and is thus pronounced "*erh-shih*" "two-ten." But the old special form of 20 has the special monosyllabic name, *nien*; the Japanese, in contrast, follow the rule in saying *ni-ju*. The same applies to 30 (Chinese *sa* instead of *san-shih*; see p. 450). Hence, there are two Chinese number words for 20, each corresponding to a numeral.

These Han sticks are among the very oldest documents that record events of ordinary or commercial life. They were found in a guard post of a part of the Great Wall of China which projected far into the wasteland, where the Gobi Desert meets the eastern spur of the Tarimbek Mountains. The characters written on them deal with the duties and the life of the border troops; the ones illustrated here (see Fig. 75) record the payments made for baked bricks used in building the Great Wall (1, 2) and the distances traveled in the course of these operations (4, 5, 6).

The Han sticks are important not only because of the historical events they record but especially because they represent a significant intermediate stage between the tally stick and written numerals. Are these slender pieces of wood tally sticks? Naturally; and like all tally sticks they are marked and read from top to bottom. The later numerals for the tens, however, have not yet replaced the older

Fig. 75 Wooden number sticks from the time of the Han Dynasty, with early Chinese characters. These were found on the Great Wall and are about 2000 years old. Each stick is about 40 × 2 cm in size; translations (except for 3) are at right; *Li* is "mile."

method of writing with notches crossed through, except that here the brush replaces the knife.

The memory of the tally sticks remained alive for thousands of years. Even today, the Chinese write their characters in vertical columns and always only on one side of even the thickest paper. What is the Chinese word for "book"? A bundle of sticks. Thus the Chinese written character for "book" is a cluster of tally sticks strung together (Fig. 76), and appears to be an Asiatic relative of the Russian "tax book" and the Swiss "Alp book" (see Figs. 61 and 64). The word "law" in Chinese is then very interestingly conveyed by a Chinese character representing a "book" of tally sticks solemnly elevated on a table (see bottom character of the figure).

Now, back to the notches themselves.

Notched Numbers Tied to Specific Measures

These are definitely numerals — that much is certain. But they are not abstract numerals, for the numbers they represent are tied to the objects numbered. The unit notch which means "5 *Crennen* of milk" on one tally stick has the value of "4 hours of water" on another and of "1 franc" on a third. Thus the specific measure is present as a silent partner in these early number symbols.

Fig. 76 "Book" and "law" in Chinese characters.
Top: "book" (new form, with old form below), a bundle of tally sticks as in Fig. 61;
bottom: "law," a bundle of tally sticks lying on a table.

The notched tally sticks of a Wallachian shepherd are beautiful examples of the way in which a numeral is still tied to a particular object or objects. From the first, second, and third peasants, let us say, the chief herdsman receives a certain number of sheep for which he is responsible and records these numbers on a long tally stick. The peasants in turn record the number of sheep they have handed over on their own tally sticks. Now the herdsman is obligated to remit certain amounts of cheese to the sheep owners. The amounts of cheese are recorded by the shepherd on the other side of his tally stick, using quite different number symbols for this purpose. In the case of written numerals, this is the exact counterpart of the classes of spoken numbers: The numerals depend on the nature of the objects whose quantities they represent (see p. 31).

Fig. 77 Shepherd's tally sticks on which the kinds of notches used to record amounts of cheese differ from those that represent peasants and sheep.

Thus we have found the same formations and the same early laws in both spoken number words and written or carved numerals. Just as number words were liberated by stages from the objects they counted, the early numerals were first freed from their association with the wooden tally stick, then the forms of the notches were written instead of being carved, and here and there the resulting numerals also were removed from their content and became "abstract" numerals that could be applied to any or all objects to be counted. But some did not extend their domain beyond the village or the rural district; they remained valid only for peasants and their farm economy. These were the peasant numerals.

Peasant Numerals

Gottfried Keller, in his well-known novel, *Der Grüne Heinrich*, tells the following story:

In the house opposite there was a dark, open hall completely filled with various kinds of junk.... Far in the background there sat a heavy-set, aged woman dressed in old-fashioned clothes.... She could read printed letters only with great difficulty, and could neither read nor write Arabic numerals; instead, her whole mathematical arsenal consisted of a Roman One, a Five, a Ten, and a Hundred. As she had absorbed these four numerals in her early youth, in some faraway and forgotten land, handed down to her after being used for a thousand years, she manipulated them with remarkable dexterity. She kept no books and had no written accounts, yet she was quite able at all times to keep track of her whole trade, which was often carried on in several thousand very small transactions, rapidly using a piece of chalk to cover a table top with large columns of these four numerals. When she had posted all these small amounts in this manner, from memory she would simply arrive at their sum by erasing one row after another with her moistened finger as quickly as she had written it down, at the same time posting the results at the side. Thus new and smaller groups of numbers appeared whose meaning and purpose no one but herself knew, since they invariably consisted of only the same four bare figures which to outsiders looked like some kind of ancient magic writing. She was never able to carry out the same operation with pencil or pen, nor with a slate-pencil on a slate, because she not only needed all the space on a table-top, but also because she could only draw her large symbols with a piece of soft chalk.

We could scarcely find a better representative to demonstrate peasant numbers than this magnificent old woman junk-dealer of Keller's. What does she need any newfangled numerals for? They are just as mysterious and magical to her as her own are to others; but to her own clear and simple mind her peasant numerals are also perfectly clear and simple. Whether or not they were actually Roman numerals is less important than the fact that she chose these particular forms and clung to them steadfastly. L and D and M did not appear in her calculations any more than they do in the following examples:

Fig. 78 An Alpine tally stick on which the herdsman has recorded in peasant numerals the numbers of different animals (cows, oxen, goats) he has pastured in the Alpine meadows during a summer. At the top are the initials of the owners' names and the year 1813; below the herdsman has written "Jöri Bregetzer made this *Scheita* ("billet")". Length 35 cm. Museum für Schweizer Volkskunde, Basel.

The Alpine tally stick carved by the herdsman Jöri Bregetzer in the year 1813 shows the numbers of various animals in his charge during that summer (see Fig. 78). According to his account, these were 80 cows, 35 *Galtiere* (oxen) and 200 goats (cf. Fig. 66).

[249]

Fig 79 "Bar-figures," developed from the tally sticks used in Switzerland.

Fig. 81 Peasant numerals from Carniola. The numerals, which are different for gulden and groschen, were written "on the line." The ten-stroke appears in 10 and 5 groschen; half-symbols in the middle; at the bottom the sum of 256 gulden and 7 groschen, is written on the line, which to prevent cheating is terminated at the extreme left by a small hook.

Fig. 82 Peasant numerals used in scoring the game called *Jassen*. Place-value indicated by position on curved line, groups of 2 and 5.

These numbers are still carved into the wood, but this Alpine billet is more than just a tally stick. The numerals on it have now been separated from their objects and are used to count anything. This development of the numerals on tally sticks can still be seen, even in our own times, in Switzerland. Here each shareholder in an Alpine meadow in a particular district had a long, flat wooden billet on which was recorded the amount he contributed to the communal cheese-making. From this *Beilen-Rechnung*, or hatchet-figures (see p. 224), developed the written *Stangenrechnung* or "bar figures," in which the *Stange*, the "bar" is our ten-stroke (see Fig. 79).

The diagrams can be understood without further explanation. At the right are recorded the amounts of milk contributed during the day by the community members A, B, C, and D; the daily amounts are then combined on the bars at left. Here, too, we see an instance of back-counting in the writing of these numerals.

Time and again the ten-stroke crops up in combination with the old tally numbers (see Fig. 80):

a ┤─Y 115   ⌐V┴┴ 252   *b* XVHHH| 191

Fig. 80 Peasant numerals with ten-stroke: (*a*) from the Prätigau; (*b*) from the Wallis district.

The symbols for money in use in Carniola well into the 19th century (see Fig. 81) are of interest because of their great antiquity. Here we see the writing "on the line" (at the top in the Figure at left), which to make cheating impossible has a small hook at the beginning (bottom left), the half-symbols for half-values (middle) and the ten-stroke which combines the individual single-unit gulden into groups of ten. (In the figure, *Gul* = gulden and *Gro* = groschen).

When even the denominations of money and the groups are to be differentiated by special numerals, as is done in *Jassen*, a card game played by peasants in the south of Baden, this is done very cleverly by keeping them apart by a line (see Fig. 82). This line is drawn in a large S-shape and thus provides three separate segments, for the 100-, 50-, and 20-groups, the only ones used in scoring this game; on the first two segments the numerals are grouped by twos and on the last by fives.

Until the middle of the last century people in the Canton Uri were taught to add up such sums as 457 + 60, for example, in peasant numerals, as shown in Fig. 83. In this operation numbers were written down, erased and others written down again, just as the old woman did in Gottfried Keller's book (see p. 249).

The "line" or "stroke" was originally an image of the old tally stick itself. In several places in Switzerland, as grain and potatoes were delivered in sacks a tally was kept by making marks on a vertical line, as illustrated in Fig. 84; every group of five strokes was made alternately on the right and left of the vertical line, so that the total (which in our example is 13) could be read off more easily. Hence the German expressions, "to have someone on the line," meaning to have a debt someone owes marked up on a line (= tally stick), as well as "to put someone up on the board," meaning that from then on all his sins will be marked down or chalked up to him exactly.

Our series of examples ends with a Styrian peasant calendar of the year 1398, from which the month of September is reproduced here (see Fig. 85). The forms of the written numerals obviously resemble the notches carved on tally sticks, so that they look rather like Norse runes. Again we see the numbers written "on the line" along with the clear groups of 5 and 10. Even if calendar sticks were no longer preserved today, ethnologists and anthropologists would have no reason to doubt that these early peasant numerals were derived from the tally and only later became independent of it.

Fig. 83 The addition 457 + 60 = 517, as performed in peasant numerals.

13

Fig. 84 Counting "on the line."

Fig. 85 Farmer's calendar from Styria, with peasant numerals which clearly recall the notches carved on tally sticks. The letters stand for the days of the week. The numbers 1 through 19 refer to the nineteen-year lunar cycle, in which every year has one number, the so-called "golden number." The bottom row reads:

2 10 18 5 15 4 12 1 9 ... 16 5 13.

With this we come to the end of our discussion of tally sticks, which have given us a fascinating glimpse into the simple business dealings of unlettered people in earlier days. The tally stick was the peasant's and the tradesman's and the merchant's account book; on it, with knife on wood, they learned to carve the primitive numerals which were sufficient to manage their own affairs, and which they took with them and kept on using even as they were learning to "write" properly. But for the cultural history of numbers it is important to understand that the early rules which govern the arrangement of spoken numbers apply to written numerals as well. It could even be argued, in fact, that the simplest forms of written numerals preceded spoken number words and that these laws of ordering and grouping were transferred from the numerals to the number words. We see, moreover, that these rules are universal: we find them used all over the world, even in China, in the symbols that make up an "abstract" system of numerals. We have been able to arrive at this understanding only by traveling over the long path we have followed thus far in the present volume, since today tally sticks have disappeared completely in our civilization and have been replaced by the sophisticated form of writing now in use. But we must admit that even primitive people can indeed "read and write."

Knots Used as Numerals

Before concluding our consideration of folk symbols, we must look at just one more curious form of primitive numerals, or recorded numbers, one that has no trace of writing in it: the knots. Knots have of course, never evolved into "written" numerals.

The custom of counting both days and objects by tying knots is found all over the world. The Tibetan prayer-strings and the rosary are both forms of number-strings on which the prescribed numbers of pious exercises is recorded in knots. King Darius of Persia gave his subjects a cord with 60 knots tied into it when he set off on his attempt to conquer ancient Greece; each day they were to untie one more knot, and if he had not returned by the time the last knot was untied, they were no longer to wait for him (see p. 153). Since such number-strings are the exact equivalent of the number or tally stick, there is no need to say much more about them. But there have been a few curious forms of number-strings with knots with which the reader is probably not familiar and which go beyond the mere lining up of one knot after another (see Fig. 86).

In the Ryukyu Islands in the Pacific Ocean, between Japan and Taiwan, workmen braid strands of straw or reeds with fringes in various different forms to indicate the wages they have earned. Each form of fringe signifies a particular unit of value, so that together they make up a kind of place-value notation. A free end stands for one unit and a knot for five units.

The knotted cords of Peru, called *quipus*, the only form of "writing" known, played a very important role in the Inca Empire because they recorded all the official Inca transactions concerning the land and subjects of the empire.

From a main strand (which was some 50 cm long) hang the often variously colored strings (each about 40 cm long) into which the knots have been tied (see Fig. 87). One of these strings would be used to show the number of sheep, for example, another for goats, and a third for lambs or kids, in the same manner as Bolivian herdsmen today still keep track of their flocks. The knots themselves indicate the numbers.

Three different kinds of knots were tied on the *quipu*: single (overhand) knots (No. 1 in Fig. 88), double or figure-of-eight knots (No. 2), and slip knots with 2 to 9 loops (No. 3). As the photograph (Fig. 87) also shows, these knots are not distributed at random over the strings but are arranged in a decimal gradation so that the hundreds are closest to the main strand at the top, the tens follow below in a second horizontal row, and the units are tied into the ends of the strings at the bottom; the units alone may be represented by a double or figure-of-eight knot (for 1) or a slip-knot (for 2 to 9). Thus 235 is "written" with 2 knots in the top row, then 3 single knots in the middle and finally a slip-knot with 5 loops at the bottom end.

If the reader already suspected that numbers underlie these knots, his suspicion will have been strengthened by the slip knots with various loops and by the obvious order in which the knots are arranged, and it will have been changed to certainty by the "head

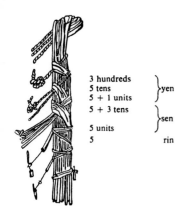

3 hundreds
5 tens
5 + 1 units } yen

5 + 3 tens
5 units } sen

5 rin

Fig. 86 A string of plaited reeds on whose fringes workmen of the Ryukyu Islands have recorded the wages owed to them. Here they amount to 356 yen, 85 sen, and 5 rin.

Fig. 87 Peruvian *quipu*. The knots are tied in order of numerical rank, with the units at the end; it can be seen clearly that the knots representing each rank are not just spaced at random, but that they run horizontally across the *quipu*, each row at a particular height. In this example the ends have been destroyed. About 40 cm long.
Linden Museum, Stuttgart.

strings" marked by the letter K in Fig. 88. These are threaded through the loops at the top of a number of strings, and the knots tied in them give the sums of the numbers shown by the knots in the strings.

The most striking feature of the *quipus* is their close similarity to the Exchequer tallies: in Peru, as in England, a very early and primitive form of numerals achieved official recognition and was used to keep the state's financial records. The only difference between them was that without exception all accounts of the Inca Empire were kept in these *quipus*, because apart from these knots no other system of written numerals was known. In every Inca settlement there were four official *quipu* keepers, known as *camayocs*, who tied the knots in

Fig. 88 The various knots in a Peruvian *quipu*: 1: single or overhand knot; 2: double or figure-of-eight knot; 3: a kind of slip-knot with three loops; 4: the head string, marked K in this illustration, which is run through the loops at the head of the other three strings showing the numbers 150, 42, and 231, bears their sum, which is 423.

these strings and submitted them to the central government in Cuzco. There is no doubt that this perhaps intentionally obscure manner of recording numbers, which only initiates could read, was a strong support for the monarchical absolutism of the Inca ruler. An example of the opposite situation is provided by ancient Athens, where the government of the city-state was obliged to reveal its records openly to the citizens so that every official was subject to "democratic" criticism (see p. 269) — dictatorship and democracy in bookkeeping!

At first glance, it seems puzzling that the Incas could also "write down" their history, their laws, and their agreements contracts on *quipus*. Garcilaso de la Vega, son of a Spaniard and an Inca princess, gives a very revealing account of the reception of the Spanish ambassador by the Inca:

Among the common people and the nobility who attended the Inca in the audience hall there were two official historians who recorded Hernando de Soto's message and the Inca's reply in knots.

How this was done is revealed to some extent in a second passage in which Garcilaso complains of a poor translator:

His translation was not good and not accurate; this was unintentional, of course, since he did not understand the meaning of what he had to translate. Instead of the three-in-one and single God he read three gods plus one make four (!), adding the numbers so as to make the expression intelligible to himself.

This quotation shows how things and events which are not numerical could be translated into numbers as an aid to memory in handing them down by word of mouth: a primitive form of quantification of non-numerical data, quite familiar in the modern world, where it is carried out to the most minute details by data-processing machines.

These people, the "scribes" or ayutas, formed a high-ranking stratum of Peruvian society at the time of the Spanish conquest. We really do the same thing, if we want to consider the knots we tie in a string or a handkerchief as a reminder.

Is the *chimpu* of the Bolivian and Peruvian Indians descended from the *quipu*? Almost certainly, for it too records numbers (such as 4456, for example in Fig. 89) on strings in a form of place-value notation, but in a different and surprisingly sensible manner, using fruit seeds strung like beads. The strings are first knotted together at the top. Then each numerical rank, thousands, hundreds, tens, and units, is indicated by the appropriate number of seeds. The four seeds representing the number of thousands are threaded on four strands, four more for the hundreds on three strands, five seeds for the tens on two strands, and finally six seeds for the units on just one strand. Then the four strands are knotted together at the bottom (u − − − − − u′ in the Figure) and the number 4456 is now "written down" and can be stored away.

The Chinese too at one time used knots to record numbers. Lao-tse, the Chinese philosopher who lived in the 5th century B.C., urged his compatriots of Tao-te-king to go back to the simple ways of doing things: "Let people again tie knots in cords, and these will serve as writing."

After seeing the *quipus*, we can understand to some extent what Lao-tse meant by this admonition and what kind of attitude he was recommending. A paper war in knots is inconceivable. But wherever writing has been made too easy, it will rage, as our civilization has learned since the invention of the typewriter.

Some scholars suspect the peculiar fashion, characteristic of the Indian Devanagari script, of arranging the letters symmetrically on each side of a vertical line (see Fig. 243), may well be a vestige of an old *quipu*-like form of "knot-writing."

Even German folkways offer an interesting example of "knot-writing" in the *Miller's Knots*, used by millers in their transactions with bakers up to the beginning of the twentieth century (see Fig. 90).

Fig. 89 The *chimpu* of the Peruvian and Bolivian Indians, a descendant of the *quipu*. This one shows the number 4456.

Fig. 90 Miller's knots used to indicate amounts and kinds of flour.

The particular knots illustrated here were used in the province of Baden.

The miller had to "write down" somehow the amounts and the kinds of flour and meal contained in the sacks he delivered. For this purpose he would use the draw strings that tied the mouths of the sacks. The quantities and measures were indicated by knots (1–7 in the Figure) and the kinds of meal or flour by loops or tufts (8–12). The miller's measure of flour was the *Sester*, an earlier measure of volume containing 10 *Mässel* (for the origin of the word, see p. 181).

1 *Mässel* = one simple overhand knot	(1)
2 *Mässel* = the same knot with a strand drawn through it	(2)
or tied in the bight of the draw string	(3)
5 *Mässel* = $\frac{1}{2}$ *Sester*, and	
10 *Mässel* = 1 *Sester*, both represented by special knots	(4, 5)
2 *Sester* = the same 1-*Sester* knot tied in the bight or with a	
strand drawn through it	(6)
6 *Sester* likewise represented by a special knot	(7)

Here for the first time we see that different numbers can be represented not only by series of the same knot, but also by special individual knot-symbols, which thus become knot-numerals: 1, 2, 5, 10, 20, and 60 measures. The intermediate numbers are then generally formed by combinations: $8\frac{1}{2}$ *Sester* = $6 + 2 + \frac{1}{2}$ *Sester*.

The manner of indicating the kinds of meal and flour delivered by the miller, for instance "hog-mash, rye, barley, 1st and 2nd grade wheat-flour," can be seen from the illustration (8–12). Barley rye, for example, was represented by a loop tied into a *Sester*-knot.

Like the tally sticks, these ingenious miller's knots are striking evidence of the inventiveness of simple, unlettered people.

Alphabetical Numerals

The Gothic Numerals

"Let him that hath understanding count the number of the beast: for it is the number of a man; and his number is Six hundred threescore and six."
Revelation 13:18

Notches, knots, and peasant numerals have only a limited, or even purely personal, validity. For this reason their elevation to the status of government records in the case of the British Exchequer tallies and the Peruvian *quipus* was exceptional — a historical curiosity, so to say.

During the Middle Ages the "official" method of writing numbers in Western Europe was the Roman numerals. These eventually made their way north via the monasteries. Here they found no opponent of equal strength, for the popular numerals could make no claim to universal or widespread dominance, and they were, moreover, quite compatible with the Roman. Nowhere north of the countries bordering the Mediterranean was there a generally used system of written numerals, nor any universal form of writing, for that matter. Thus the Roman numerals continued to predominate in northern Europe even after the fall of the civilization which invented them; thanks to their "early," primitive nature they met with such ready reception that they were never felt to be alien, and in the 16th century they were stubbornly defended as "*de düdesche tall*" (the German numbers), as against the Hindu place-value notation that was then invading Europe.

The Greek numerals, on the other hand, never came to be deeply rooted among the Germanic Goths.

In their southeastward migration to the shores of the Black Sea and the lower reaches of the Danube, the Goths entered the sphere of Greek culture. The Visigoth Bishop Ulfilas, who died in A.D. 381, translated the Bible for the benefit of his Christian congregations. This magnificent monument of Germanic linguistic history is best preserved in the famous *Codex argenteus* (of which 187 out of 330 folios still exist). This manuscript, with its letters of silver on purple-colored parchment, which after long wanderings has finally found a permanent home at Upsala in Sweden, was written around A.D. 500 in Italy, most likely by an East Goth copyist.

Ulfilas, a highly educated man who knew both Greek and Latin, invented this Gothic alphabet expressly for his translation of the Bible (see Fig. 91). For this purpose he chose 17 Greek and 3 Latin (h, r, s) letters, as well as 7 Germanic runes (those for j, u, f, o, and perhaps also q, hw, and *þ*); the last of these may have been derived from the Greek *thēta*, θ, by removing the horizontal stroke in the center and placing it vertically at the side. Two letters of Greek origin, ϙ and ↑, did not correspond to any Gothic phonetic value, but served only as the numerals for 90 and 900, respectively. With this we come upon the remarkable fact with which we are concerned here:

The 27 letters of the Gothic alphabet served simultaneously as numerals.

Ulfilas arranged the Gothic number words according to his Greek model (see p. 270). The 27 = 3 × 9 different Gothic letters are used to represent the group of nine units, the group of nine tens, and

[259]

Numerical value	Gothic	Derived from	Phonetic value	Numerical value	Gothic	Derived from	Phonetic value	Numerical value	Gothic	Derived from	Phonetic value
1	A	⋏	a	10	I ï	I	i	100	ᚱ	ᚱ.	r
2	B	B	b	20	K	K	k	200	S	S.	s
3	Γ	Γ	g	30	Λ	λ	l	300	T	T	t
4	d	Δ	d	40	M	M	m	400	Y	Y	w
5	E	E	e	50	N	N	n	500	Ϝ	Ϝ:	f
6	u	q:	q	60	G	ς:	j	600	X	X	x
7	z	Z	z	70	Π	Π:	u	700	Θ	Θ.	hw
8	h	h.	h·	80	π	π	p	800	Ω	Ω:	o
9	φ	Ψ:	þ	90	Ч	ς	–	900	↑	↑	–

Fig. 91 The Gothic alphabet consists of 17 Greek and 3 Latin (.) letters and 7 Germanic runes (:). Bishop Ulfilas' Gothic Bible, the *Codex argenteus* (see Figs. 92 and 93) was written in this alphabet.

the group of nine hundreds. Thus, for example, we read the parable of the sower (Mark 4:8)

"*... jah wahs jando. jah bar ain .l.*
jah ain .j. jah ain .r. jah qaþ: sae..."

[And other fell on good ground, and did yield fruit that sprang up and increased;] and brought forth, some thirty, and some sixty, and some an hundred. [And he said unto them]...

The *Codex argenteus* is written in uncial letters with the words run together and without punctuation, except that periods are used to mark the ends of sentences. When letters serve as numerals, they are covered by a horizontal stroke and occasionally set off by periods at

Fig. 92 The numerals x̄ ⳝ ℞ for 30, 60, and 100, from the Gothic Bible (folio 294). In this photograph the silver letters appear white against the dark background of the purple parchment; thus the negative, with black letters against a light background (as shown in Fig. 93), is easier to read.

each side. The Gothic alphabetical numerals are used especially to indicate the verse numbers in the margin (see Fig. 93). References are also made to the corresponding Bible passages within the decorative arches at the bottom of the page. In addition, some number words, especially the thousands, are also written out in full (see Figs. 27 and 28). But the Goths made no computations with these alphabetical numerals, nor did people use them to keep their ordinary, daily accounts. Of the Goths' own early written numbers we know absolutely nothing.

With these Gothic alphabetical numerals we have touched upon a subject which very seldom appears in the cultural history of numbers, but which is nevertheless of the very greatest significance. Let us now turn our attention to this.

Letters and Numbers

Fig. 93 Left margin of Folio 125 of the Gothic Bible, showing verses No. 156, 157, 158, and 159 and references to the same verses (Gospel According to St. John) in the arch at the bottom. Folio size 25 × 30 cm, letters 0.5 cm high, text itself measures 17 × 14 cm.

Numbers seek form as words and as symbols. In essence, however, letters are the embodiment of words, not of numbers. Yet letters also come into contact with numbers, both indirectly through the medium of number words and directly in themselves, as we have just shown in the Gothic alphabetical numerals.

Letters and numbers meet in number words in two different ways. The first and most frequently encountered is the writing of numbers as number words. This is easily understandable when no computations are to be made and there are not many numbers. We have seen from our study of spoken numbers that every culture began by expressing its numbers as words. But even in their books on arithmetic the Arabs, for example, although they both knew and used the Indian numerals, also wrote out their numbers in words. Here we may recall the Indian symbolic numbers (see p. 120). The Chinese ideographic numerals are the only known instance in history in which the written number word has been the same as the numeral (see pp. 458; also p. 53).

The second way in which letters and numbers come together indirectly is in the use of abbreviations for the written number words as numerals. In the early Greek system of writing numbers, for example, the numerals were the initial letters, like the Δ of $\Delta EKA = 10$. Another and less familiar example of the same thing is the Arabic *siyaq* script, which is still used in Persia. This script uses the abbreviated and stereotyped written number words as numerals in a peculiar form of place-value notation (see p. 276).

Now for the direct contact between letters and numbers. One of our most priceless possessions is the established sequence of the alphabet with its twenty-odd letters in their fixed order. One need only recall the great convenience of our dictionaries and directories, which the Chinese with their ideographic characters do not enjoy. The fixed order of the letters in the alphabet may have originated in astrological concepts. Perhaps it goes back to the 30 or so constellations through which the moon passes on its travels. At any rate, numbers are indissolubly linked to the fixed sequence of the alphabet: the first letter stands for the first number, the next letter for the next number, and so forth.

The first to recognize and take advantage of this numbering quality of the letters were the Greeks, who did not invent the alphabet themselves but took it over from the Phoenicians. It was also the Greeks who realized the three possible ways in which letters can be associated with numbers:

1. The continuous one-to-one sequence in which the numbers 1 through 24 were represented by the letters A through Ω respectively. In this manner the Alexandrian scholars numbered the books and lines of Homer's epic poems and the ancient Greek masons numbered their blocks of marble. If numbers higher than 24 were called for, then $AA = 25$, $AB = 26$, $A\Gamma = 27$ and so on were used. Even today we still occasionally use this system of numbering, although we rarely if ever go beyond the first few letters of the alphabet. A

similar method of numbering was used by the Indian mathematician Aryabhata.

2. The graduated succession is that which we saw in the Gothic alphabetical numerals. The alphabet, since it had fewer letters, was expanded to $27 = 3 \times 9$, so that each group of 9 letters would represent the units, the tens, and the hundreds. How the thousands were formed in this system we shall discuss later, when we shall also see that the Greek mathematicians (even Archimedes and Diophantus) used these alphabetical numerals for their computations. This system of symbolizing numbers was retained in Greek culture (Byzantium) until it was gradually replaced during the 14th century by the "Indian" place-value method.

3. The place-value method. Here the first nine Greek letters, α through ϑ, served as "numerals" in the Indian manner. Along with a zero sign, these letters were the exact equivalents of the Indian numerals 1 through 9 plus 0 (see p. 273). This method is still in use today in a Javanese and a few Southern Indian number systems.

The Greeks both thought of and used all three of these methods of representing numbers by letters. Thus we have a remarkable record of the numerical use of letters all the way from the beginning of Classical Greek to the end of Byzantine Greek culture.

But what was the origin of the alphabet, which played so important a role in the development of numerals? We shall briefly sketch the history of the alphabet as a background to discussing its use by the Greeks to represent numbers. Then we shall add to this a survey of the other forms in which the connection between letters and numbers has been manifested.

HISTORY OF THE ALPHABET

One of the greatest wonders of the history of human culture has been the world-wide spread of the 22 symbols which the Phoenicians, a minor trading people, devised to represent the consonants of their Semitic language. For not only did the Hebrew and Arabic alphabets, which are used to write these other Semitic languages, derive from the Phoenician, but the entire groups of Turkish-Mongolian and Persian-Indian (by way of Aramaic) and above all the European alphabets (by way of Greek) as well (see Fig. 94).

Trade and Writing: The Phoenicians were merchants who sailed the Mediterranean Sea and traveled through the lands bordering on it; and it was their later successors, the Arab traders, who carried the Indian numerals from the Orient to the West!

The alphabetical form of writing is thought to have developed approximately as follows: at the outset every form of writing (such as the Egyptian, the Babylonian, or the Chinese) was made up of picture-symbols for words (like the picture of a head standing for "head"). Then when the image of the head was replaced by the initial sound h, this had become a letter.

An alphabet, in which each sound is, generally speaking, represented by one symbol, is the last and highest stage in the development of writing, which usually runs from the ideographic through the syllabic to the phonetic form. The alphabetical form of writing follows

Fig. 94 The origin and spread of the Phoenician alphabetical writing. The dashed line (running down from the Greek alphabet) shows the direction in which the Greek alphabetical numerals migrated.

a major intellectual principle: All the innumerable words that make up the language are resolved into a limited number of some 20 to 30 individual phonetic symbols. This principle originated in ancient Egypt, whence it was passed on by the so-called Sinaitic inscriptions, and probably also through the influence of Crete and Cyprus, to be finally perfected by the Phoenicians. It was in Phoenicia that the Egyptian hieroglyphic image of "house" was streamlined down to a "letter" which merely sounded the initial sound *b* of the Semitic word *beth*, "house."

In column 1 of the table of Fig. 95 are some of the documentable Egyptian hieroglyphs; column 2 gives their Phoenician letter-forms and column 6 their Hebrew names (derived from the Phoenician). The reader is urged to examine the table so that this segment of the history of writing may come to life for him; he will also better understand the numerals that have been derived from letters of the alphabet.

The Phoenician form of writing consisted of 22 symbols which — and this is a peculiarity of all Semitic alphabets — stood for the consonants only (see p. 116). Around the 11th century B.C. the Greeks became acquainted with the Phoenician alphabet and adopted it for themselves. In so doing, they took over the already established sequence of the names, forms, and phonetic values of the signs (compare columns 7, 8, and 10 with columns 4 and 6 in Fig. 95).

Thereupon the Greeks made two important changes: to adapt the alphabet to their own language, they altered some of the letters to vowels (*alpha, epsilon, iota, omicron, upsilon,* and *omega*), and they added the three letters *phi* φ, *chi* χ and *psi* ψ. The Ionian-Milesian alphabet — for there were many early Greek scripts — was officially adopted in Athens (with the forms shown in column 7 of the table). All western alphabets are derived from this.

Thus we come to the third important achievement of the Greeks: the letters have numerical values according to their fixed order in the alphabet (column 9). This tying of the letters to numbers which

Line	EGYPTIAN 1	PHOENICIAN 2	HEBREW 3 Letter	4 Phonetic value	5 Number-word	6 Name	GREEK 7 Letter	8 Phonetic value	9 Number-word	10 Name	LATIN 11	RUNES 12 Letter	13 Number value
1	ℬ	≮	א	'	1	alef 'Rind'	Aα	a	1	álpha	A	ᚠ	4
2	▭	9	ב	b	2	beth 'Haus'	Bβ	b	2	bēta	B	ᛒ	18
3		⌐	ג	g	3	gimel 'Kamel'	Γγ	g	3	gámma	C	-	-
4	⊲	Δ	ד	d	4	daleth 'Tür'	Δδ	d	4	délta	D	ᛗ	24
5		∃	ה	h	5	he	Eε	e	5	e-psilón	E	ᛗ,ᛁ	19,13
6	Y	ᚤ	ו	w	6	waw 'Nagel'	Fς	-	6	vaû	F	ᚢ	1
7		Iᚻ	ז	z	7	zajin 'Waffe'	Zζ	z	7	zēta	(G)	ᚷ	·7
8		ዘ	ח	h	8	heth	Hη	ä	8	ēta	H	ᚺ	9
9		⊕	ט	t	9	teth	Θϑ	th	9	thēta	100?	ᚦ	3
10		1	י	j	10	jod 'Hand'	Iι	i	10	jōta	I	ᛁ,ᛋ	11,12
11		≯	כ	k	20	kaf 'offene Hand'	Kκ	k	20	káppa	K	ᚲ	6
12		∠	ל	l	30	lamed	Λλ	l	30	lámbda	L	ᚱ	21
13		ᙢ	מ	m	40	mem 'Wasser'	Μμ	m	40	mŷ	M	ᛗ	20
14		५	נ	n	50	nun 'Fisch Schlange'	Νν	n	50	nŷ	N	↑	10
15			ס	s	60	samek	Ξξ	x	60	xî	-	-	-
16		O	ע	'	70	ayin 'Auge'	Oo	o	70	o-mikrón	O	ጸ	23
17		?	פ	p	80	pe 'Mund'	Ππ	p	80	pî	P	ᚳ	14
18		ſ	צ	s	90	sade	-	-	-	-	-	-	-
19		φ	ק	q	100	qof	Ϙϙ	-	00	kóppa	Q	-	-
20		٩	ר	r	200	resch 'Kopf'	Pϱ	r	100	rhô	R	ᚱ	5
21		W	ש	š	300	šin 'Zahn'	Σσ	s	200	sigma	S	ᛋ	16
22		X+	ת	t	400	tau 'Zeichen'	Tτ	t	300	taû	T	↑	17
23			ך	-k	(500)	(kaf)	Yυ	ü	400	ŷ-psilón	V	u: ᚢ	2
24			ם	-m	(600)	(mem)	Φφ	ph	500	phî	1000	w: ᚹ	8
25			ן	-n	(700)	(nun)	Χχ	ch	600	chî	X	ng ᛜ	22
26			ף	-f	(800)	(fe)	ΨΨ	ps	700	psî	50?	-z Y	15
27			ץ	-s	(900)	(sade)	Ωω	ō	800	ō-méga	-	-	-
28							Ϡ	-	900	sampî	-	-	-

Fig. 95 The development of alphabetical writing, from the Egyptian (column 1) through the Phoenician (2) to the Hebrew (3), Greek (7), Latin (11), and Germanic (12) alphabets. The Greeks gave their letters numerical values (9), and the Hebrews (5) adopted a similar pattern. Column 13 shows the numerical values assigned to the Runic characters.

the Greeks invented then was handed back to the Semitic donors of the letters themselves: Hebrew adopted the alphabetical numerals (column 5), and they went from there to Syriac and thence to Arabic. Two coins (*shekels*) (Fig. 96) dating from the time of the First Uprising of the Jews (A.D. 66–70) show, above the cup in the center, the old Hebrew/Phoenician signs for 2 and 4, meaning "in the second (and fourth) years of the Uprising." With the aid of Fig. 95 we can

Fig. 96 Hebrew coins. Above the cup they show the alphabetical numerals for 2 and 4 (1st century A.D.). State Numismatic Collection, Munich.

very easily decipher the inscription, which runs from right to left, as *šql yšr'l,* "shekel of Israel." The Hebrew letters shown in column 3 of the table are the so-called "square writing"; with the addition of the terminal forms *-k, -m, -n, -p* (or *-f*) and *-s* the sequence of numerals, which once ran only as far as 400, was extended to 900.

An elderly Jew once told the present writer that in his youth he used to hear the Jewish cattle dealers trading as follows: "*mem shuk* (40 marks), *gimel shuk* (3 marks), *lamed gimel* (33)" and so forth.

The connection between letters and numbers gave rise among Judaic and early Christian scholars, as well as the Greeks, to *gematria.* This word is very likely a corruption of the Greek *geometria,* which the Jews wrote as *gmtr,* and referred to the study of numerology in general. Since each of the letters in a word had a numerical value, the word itself also had its own individual number. Thus "Amen," Greek $\alpha\mu\eta\nu,$ = $1 + 40 + 8 + 50 = 99$; hence this word at the end of a Greek prayer would often be written as $\kappa\vartheta$, 99.

The Jews, so as not to take the name of the Lord in vain, would write the number 15 not as *yh,* but as *tw* (= $9 + 6$), for *yh* are the initial letters of the name "Jehovah" (*yhwh* = *Yahweh*). For similarly "sacred" reasons the Irish would avoid the Roman numeral X = 10 on account of the name $XPI\Sigma TO\Sigma$, and the Chinese Boxers during their rebellion against the foreigners (= Christians) removed the symbol + for 10 from their coins because of its fancied resemblance to the "Christian" letter X (see Fig. 264). Because the Greek word for death, $\Theta ANATO\Sigma$, begins with a *theta* $\Theta = 9$, the Greeks instead of the number 9 would write the circumlocution $8 + 1$ or $4 + 5$; there is a counterpart to this superstitious custom in Japan (see p. 453).

Persons or things the letters of whose names had the same total numerical values were also thought to be mystically related. In the Middle Ages, for instance, people would "calculate" the outcome of a dual or individual combat from the arbitrary sums of the letters or digits in the combatants' names; the one with the larger sum would be the "predicted" winner. If the numerical value of Siegfried's name was 238 and the conventional sum of this was 4, and if the numerical value of Hagen's name was 65 and the conventional sum of this was 2, then in a duel between Siegfried and Hagen, gematria predicted that Hagen was sure to be vanquished. By the reverse process, names were often thought to underlie numbers: the most famous example of this is in the Book of Revelations (13:18):

Let him that hath understanding count the number of the beast: for it is the number of a man; and his number is Six hundred threescore and six — *kai ho arithmòs autoû hexakósioi hexêkonta héx.*

There can be no doubt that a name is latent here. In early Christian times this number from the Apocalypse was deciphered, among many others, as *Neron kaisar*, "Caesar Nero" — an interpretation that is quite possible if the Hebrew writing of this name, *nrun qsr*, is transcribed into Greek and then "encoded":

nun	*reš*	*waw*	*nun*	*qof*	*samek*	*reš*
50	200	6	50	100	60	200 = 666.

The "numbering of letters" and the "lettering of numbers" was the art of *isopsephy*, as this sort of toying with the numerical values of various words was called (from the Greek words *ísos*, "equal," and *pséphos*, "pebble, counter, number") in all seriousness. As late as the 16th century the German mathematician Michael Stifel (died 1567), who is highly regarded by posterity, valued his "word calculations," as he called them, above his more important work in mathematics.

Now we shall turn to the numerals used by the Greeks: how did ordinary people, not to mention the great Greek mathematicians, write numbers, and how did they make their computations?

The Two Greek Sets of Numerals

The Greeks had two different systems of numerals, an early one that arranged the numbers in order and grouped them like the Roman numerals (these will thus be called here the Greek "row numerals"), and a later erudite system of alphabetical numerals which first appeared in the 5th century B.C. but was not adopted as the official system of numerals in Athens until the 1st century B.C.

The Greek row numerals had individual symbols for the numerical ranks 1, 10, 100, 1000, and 10,000; these (except for 1) were the initial letters of the corresponding number words. They could be grouped by fives by using the initial letter Π of the word *pénte*, 5, as shown in Fig. 97.

$$\Delta \text{EKA} \quad \text{HEKATON} \quad \text{XIΛIOI} \quad \text{MYPIOI}$$
$$\Delta_{10} \quad\quad H_{100} \quad\quad X_{1000} \quad\quad M_{10\,000}$$
$$\mathsf{P}^{\Delta}{}_{50} \quad\quad \mathsf{P}^{H}{}_{500} \quad\quad \mathsf{P}^{X}{}_{5000} \quad\quad \mathsf{P}^{M}{}_{50000}$$

$$\text{HH}\Delta\Delta\Delta\text{IIII}_{234} \quad \text{X}\mathsf{P}^{H}\text{HHHH}\mathsf{P}^{\Delta}\Gamma\text{II}_{1957} \quad \mathsf{P}^{M}\text{MXHH}_{61\,200}$$

Fig. 97 Early Greek numerals (row numerals). Like the Roman numerals, these are patterned on a decimal 10-grouping interrupted by a quinary 5-grouping. The units are represented by vertical strokes. The 5-grouping appears in the initial letter of the word *ΠÉNTE*, "five." This illustration shows the higher ranks, whose numerals are the initial letters of the number words (see also Fig. 128).

This system of written numbers thus made use only of the symbols for the old decimal grouping, which is evidently combined with a quinary 5-grouping.

These early Greek numerals have been given the very unfortunate name of "Herodian numerals" — unfortunate, because the grammarian Herodian lived around A.D. 200 in Byzantium, and thus more than five centuries after the appearance of the numerals named for him. Moreover Herodian mentioned them only once in passing. He was in no sense their inventor. Thus calling the early Greek "row numerals" Herodian has as little justification as naming the Hindu numerals we now commonly use after Adam Riese.

These Greek numerals were used in Attic inscriptions from the middle of the 5th century until well into the 1st century B.C., chiefly in the public tribute lists and the accounts which the financial officials of the city state presented to the citizens. The inscriptions of the first category listed the amounts paid annually by subjects who owed taxes to the Athenian treasury. Figure 98 shows a fragment of a square marble block, $1 \times 0.4 \times 3.6$ m in size, on which the numerals themselves are about 4 cm high; at the top left is the name of the tribute payer.

In these tribute lists the numbers always refer to payments of money; in this case the numbers after the semicolons are obols and the others drachmas. The monetary scale was 1 talent (T) = 60 minae, 1 mina = 100 drachmas (⊦), 1 drachma = 60 obols (|); in addition, the coin called the *stater* (Σ) was worth 4 drachmas.

Fig. 98 Athenian tribute list with early row numerals 4 cm high. Fragment of a square stele, almost 4 m high; 5th century B.C.
The inscription reads:
(AP)O THRAIKES PH(OROS) — "Payments from Thrace":
skiathioi 66 drachmai 4 oboloi; then
500 50 8;2 16;4 600 25.

[268]

The method of indicating the coin denomination was quite interesting (see Fig. 99). Either the coin was not specified, in which case it was understood to be drachmas, or it was placed on the left side of the number (top row in figure), or it took the place of the units in the numeral (middle row), or else it was incorporated into the number symbol itself (as in the bottom row).

In this method of specifying the coin in which an amount of money was stated, we have a beautiful documentation of the early stage of development in which numbers were still tied to the things they counted: "2 talents," "2 staters," or "2 obols" all have different symbols for the number 2.

ͰΔΙΙ
ΔͰͰ Δ𐅚𐅚 ΔͰͰΙΙΙ
ΤΤ ͰΓ

Fig. 99. Greek statements of amounts of money: From top to bottom: 12 drachmas — 12 staters — 12 drachmas 3 obols — 2 talents 105 talents

Fig. 100. Report of the treasurer of Athens giving the city-state's income and expenditures in the year 415 B.C. At the end is the total (*kephálaion*) of 327 talents. This inscription measures 50 × 40 cm.
British Museum, London.

We can read the total amount on an inscription in which the Athenian treasurer in 415 B.C. made a public accounting of his four-year term of office (Fig. 100):

KEPHALAION AN(alómatos toû epì tès) ARCHES
"head-number" (= total, Greek *kephalé*, "head"), which in this treasurer's term of office came to 327 talents.

Such were the numerals used by the city of Athens. The Boeotians, on the other hand, for 100 would write the first two letters *HE(katón)*, and for 500 would place a *Π(énte)* in front (see Fig. 101). For 1000 they used the arrow-form of the letter *chi* (instead of X — see Fig. 95, column 7, row 25, and Fig. 133 on p. 304).

100	500	1000	5623

Fig. 101 Boeotian numerals.

Although we see very clearly that the Greeks used these signs to write numbers, it is very hard to understand how they made computations with them. And in fact, as we shall see presently, they did not; they used counting boards instead. Using this system of written numerals, Archimedes would have been unable to compute the value of π or bracket it, by means of a 96-sided polygon, between 3-10/70 and 3-10/71. For such purposes numbers would have to be understood and visualized much more precisely and organized on some basis other than just the awkward grouping. These needs were met by

The Greek Alphabetical Numerals
As the table of Fig. 95 shows, the twenty-seven letters of the Greek alphabet (including the special forms) were arranged for use as numerals as follows:

	1	2	3	4	5	6	7	8	9
Units	A	B	Γ	Δ	E	F	Z	H	θ
	α	β	γ	δ	ϵ	ς	ζ	η	ϑ
Tens	I	K	Λ	M	N	Ξ	O	Π	Q
	ι	κ	λ	μ	ν	ξ	o	π	ς
Hundreds	P	Σ	T	Y	Φ	X	Ψ	Ω	λ
	ρ	σ	τ	υ	φ	χ	ψ	ω	λ
Thousands	$,\alpha$	$,\beta$	$,\gamma$	$,\delta$	$,\epsilon$	$,\varsigma$	$,\zeta$	$,\eta$	$,\vartheta$

For this purpose the Greeks added to their normal alphabet three Semitic letters for which they had no sounds (or no longer had them by the 5th century B.C.). These (see column 8 of Fig. 95) were: for 6 the *digámma* F with its later form ς, which was also called *st-igma* because it served as the abbreviation for -*st*-; for 90 the *koppa* ς; and for 900 the *sampi* λ, probably so named for the Semitic *sade* that comes after *pi*, through which the Greek alphabetical numerals from 90 upward displaced the Semitic numerals (compare column 5 with column 8; *hōs án* π, "like π"). Bishop Ulfilas adopted the arrow form of the *sampi*, 900, for use in his Gothic Bible (see Fig. 91).

Examples: PIA is 111, written in descending order of magnitude, but sometimes also in ascending order as AIP and more rarely without any order as IAP. In manuscripts the alphabetical numerals were topped with a short horizontal stroke, like $\bar{\epsilon}$, 5, and $\overline{\sigma\lambda\delta}$, 234, or placed between periods or lines of dots $\cdot E\cdot$ or $\vdots E\vdots$ or even sometimes turned on end like $\sqcap$. A prime to the right of the numeral meant an ordinal number, such as ε', "the fifth," or a unit fraction, $\frac{1}{5}$. The thousands were written just like the units, but with a short vertical stroke at the bottom left: $_{,}\varepsilon$, 5000 or $_{,}\alpha\omega\nu\varsigma$, 1856.

The next numerical rank, the *myrioi* = 10,000, could be written in three different ways:

The first used the symbol M from the early row numerals, but with multiples of *myrioi* no longer shown by repeating the numeral; thus 30,000 was not MMM but was written as in a place-value notation (see Fig. 102):

(1)   (2)   (3)

Fig. 102 The myriads were not repeated in the Greek alphabetical numerals, but their number indicated by placing the symbol for the appropriate unit in front as in 1*a* or on top (1b above). Here $\Gamma = 3$:

(1) 30,000   (2) 32,000   (3) 30,002

The second method used a dot after the myriads instead of the letter M, as did Diophantus of Alexandria (3rd century B.C.):

$1507284 = 150'7284$ $\rho\nu.,_{,}\zeta\sigma\pi\delta$ and $_{,}\alpha\tau\lambda\alpha.,_{,}\varepsilon\sigma\iota\delta$ 1331 5214;

The third method used two dots placed over the myriads, as in later Greek manuscripts:

$50,000 = 50000$ $\ddot{\varepsilon}$ or $50,000,000 = 5000$ myriads $_{,}\ddot{\varepsilon}$.

Now for the advantages and disadvantages of these Greek alphabetical numerals. Their advantage over the old row numerals can hardly be missed; they represent an enormous simplification, since they use only one sign for each unit (rank). Thus in inscriptions we generally find alphabetical numerals used wherever there was little room, for example to indicate numbers on sacred offerings, like the

Fig. 103 Greek coins with alphabetical numerals. The first two are from Alexandria, with the numerals A = 1 and IB = 12, the years of the respective rulers' reigns (about 250 B.C.); the third coin, at right, comes from Byzantium and shows the year (right to left) $ZI\Phi$ = 517 of the so-called Pontinian Era (which started in 297 B.C.); in other words, A.D. 220. The L on the coin at left indicates that the number following is a year.
State Numismatic Collection, Munich.

ΨΝΔ, 754, and the *ΣΟΓ*, 293, on two Greek vases from the 4th century B.C., which are among the oldest preserved documents with these numerals. In Alexandria they first appeared on coins about a century after Alexander the Great, and in this way came to be commonly used. Figure 103 shows two such Alexandrian coins and one from Byzantium.

On a sundial dating from the 9th century A.D. the hours are numbered in alphabetical numerals (see Fig. 104), and today in Athens there is still a *Γ*-September Street.

Fig. 104 Sundial with Greek alphabetical numerals, from a church in Boeotia, 9th century A.D.
Photograph by E. Delp.

And now another adantage of these alphabetical numerals, in fact their most important advantage: With these numerals it was possible at long last to make computations in writing, without having to use an abacus.

To us, however, who are thoroughly familiar with the Indian system of numerals, this very quality may seem to be questionable because of the lack of graphic identity of the units, tens, and hundreds. Thus $4 \times 10 = 40$ in the Indian system becomes $\delta \times \iota = \nu$ in the Greek alphabetical numerals, and the (to us) almost identical computation $4 \times 100 = 400$ is $\delta \times \rho = \mu$. The multiplication table and the addition table become obscured by the plethora of relationships.

Nevertheless Archimedes and Diphantus made their computations with Greek alphabetical numerals. But how?

At first glance this looks more difficult than it really is. To begin with: people read and memorized tables not as $\delta \times \iota = \mu$, but in terms of number words, "four times ten equals forty." Thus the ear could discern the similarity of $4 \times 100 = 400$ even if the eye could not. Moreover prepared multiplication tables were commonly used, as an arithmetician of the 14th century, Nicholas Rhabdas Artavazdos of Smyrna, has testified. The following multiplication table shows the beginning of this system, for $2 \times 1 = 2, 2 \times 2 = 4$, $2 \times 3 = 6$, etc., and correspondingly for $20 \times 10 = 200$ and $200 \times 100 = 20,000$:

β	α	β
	β	δ
	γ	ς

κ	ι	σ
	κ	υ
	λ	χ

σ	ρ	β
	σ	δ
	τ	ς̈

. . . .

$2 \times 1 = 2$ $20 \times 10 = 200$ $200 \times 100 = 20{,}000$
$\quad\;\; 2 \quad 4$ $\qquad 20 \quad 400$ $\qquad\; 200 \quad 40{,}000$

. . . .

For example, let us calculate 25×43, which we have already done according to the Egyptian method (see p. 219)

κ	ε	25
μ	γ	43
ω	ξ	800 60
σ	ιε	200 15
,α	οε	1000 75 = 1075

The Greeks began with the highest rank: $20 \times 40 = 800$; then they went on to $20 \times 3 = 60$; then $5 \times 40 = 200$; and finally $5 \times 3 = 15$; final answer: 1075. Each intermediate multiplication, such as 20×40, was broken down into a computational problem, $2 \times 4 = 8$, (the so-called *pythménes*, "root problem," from *pythmén* = "root, stem, base"), and the gradational or rank problem of placing the resulting 8 in the correct order of magnitude. This computation in two steps corresponds to the procedure on a modern slide rule, except of course that finding the right order of magnitude was harder for the Greeks than for us, because the numerals κ and μ have no zeros, as do 20 and 40, from which the numerical rank of the final answer 800 can be easily read. For the old rule for finding the numerical rank, which was derived from the counting board, see p. 316.

The study of practical computations, which the Greeks termed "logistics" in contrast to "scientific arithmetic," developed and flourished in Alexandria, the capital city of Hellenistic learning and culture, and were used from then on throughout the Hellenistic world and the Byzantine Empire. As the Indian place-value notation penetrated deeper and deeper in the West (from the 15th century on), it gradually displaced computations with alphabetical numerals. But not completely. The two systems of numerals finally reached an "equilibrium": the new numerals provided a place-value system in which *the Greek alphabetical numerals for the units were used as "Indian numerals."* The zero, which had the form of a dot according to the Eastern Arabic fashion, was added to the units α to ϑ:

αθες 1956   β.   20   β. .   200   β.β   202.

A Greek manuscript of a 15th-century textbook on arithmetic gives the following two-stage problem in computation involving fractions (as an easier method of using alphabetical numerals):

$12\frac{2}{7} \times 15$; rearranged as $86/7 \times 15$; without denominators 86×15 $= 430 + 860 = 1290$.

Transcription:		Translation:	
$\alpha\varepsilon$	$\alpha\beta\,\frac{\beta}{\zeta}$	15	$12\frac{2}{7}$
	$\eta\,\varsigma$		86
	$\alpha\;\varepsilon$		15
	$\delta\,\gamma\,\cdot$		430
	$\eta\,\varsigma$		86
	$\alpha\,\beta\,\vartheta\,\cdot$		1290

Fig. 105 A Greek computation carried out after the "Indian" manner. From a 15th-century manuscript. The digits are the Greek letters *alpha* through *theta* and the zeros are represented by dots. At right, next to the problem in fractions, is the division $1290 \div 56 = 23$–$1/28$.
Courtesy of Prof. K. Vogel, Munich.

Some Other Connections Between Letters and Numbers

The *Katapaya system* of Southern India arranged the 34 consonants of the Sanskrit alphabet to form the ten numerals 1 through 9 plus 0. Some digits thereby acquired three, and others four, different phonetic values; 1, for example, could be *k-t-p-y*, whence the name "Katapaya." Zero has only the sounds *nj* and *n*. In general, each consonant was sounded together with *a*, although it could also be combined with the other vowels *e i o u* without changing its numerical value. Thus the number 111 could be expressed, to use examples from German, not only by the word *Kette* but also by the word *Paket*. The Indian astronomers used to call the lunar cycle *anantapura* = n-n-t-p-r = 00612; this number, reading from right to left, gave the number of minutes of the half-month (15 × 24 × 60 = 21,600). This simple example shows the method and the advantage of designating a number by a word which itself has some quite different meaning, to be used as a secret means of communication, as a mnemonic device, or perhaps merely to clothe computations in poetic dress, as we saw earlier with the Indian verbal number symbols (see p. 120). The impression one gets from these Indian numbers, especially if we also recall the Indian number towers, is of the great joy that the Indians found in playing with numbers and their easy familiarity with place-value notation. Unlike the Greek alphabetical numerals, this Katapaya system of course had no practical significance.

The ancient *Irish Ogham characters*, which are found on inscriptions dating from the 4th century A.D. but are certainly older, are just the reverse of the Katapaya: here numerals are used as letters! The vowels are represented by up to five dots and the consonants by up to five perpendicular or slanting strokes on one side or the other of a long vertical line (see Figs. 106 and 107). The corner of a grave stele might serve as the vertical line on which the Ogham characters were aligned. These curious number symbols made "on the line" are strongly reminiscent of tally sticks. Northerners, it seems, generally loved secret writing.

The *old Germanic runes*, originally 24 in number, were probably derived from the old Northern Italic alphabets some time during the first three centuries A.D. The once variable forms became fixed by being carved into wood; all the horizontal strokes were drawn at an angle, as in the runic A and F, so that they would not run parallel to the grain of the wood, and the curved lines were either made straight or broken, as in O (see Fig. 95, column 12, rows 1, 6, and 16). There were two Germanic peculiarities: the old "sacred" order of the letters was changed (from this alphabet were formed the positional numbers shown in column 13), and the old names of the letters were changed to symbolic names like *f-e*, "possession, property"; *u-ruz*, "aurochs"; *þ-urs*, "giant"; *a-nsuz*, "god"; *r-aido*, "ride"; *k-enaz*, "torch." This leads us to the original point of the runes: these letters were (initially) not intended or used for writing, but for casting lots and telling the future, and were thus "written" (Old Saxon *writan*) or "inscribed" (Latin *scribere*) on small wooden

Fig. 106 Irish Ogham characters, expressing letters in the form of primitive numerals.

Fig. 107 Irish gravestone from Aglish, with Ogham characters, around A.D. 300. These are read from bottom to top beginning on the right and then continuing at left from top to bottom (beginning and end are lost; X is sounded as *k*):
"... *godika maqimaq*..."
"(grave of Lu)dudex, son of *maq*..."
Size: 88 × 25 × 5 cm.
Irish National Museum, Dublin.

[275]

sticks, as Tacitus describes in the passage from his *Germania*, already quoted (see p. 224; Old Norse *run* = "secret" or "counsel").

The important thing for us here, however, is that the runes had no numerical values. On the other hand, they were often replaced by numerals as a form of secret code. The 24 = 3 × 8 runes were divided into three groups of eight each (or "genders"), quite possibly because the old Germanic tribes likewise divided the rim of the heavens into eight parts. The number 8 was sacred. Thus the rune A was the fourth rune in the first gender. The two ordinal numbers (4 and 1) were written on a stick, as the memorial tablet from Rök, Sweden (Fig. 108), shows. This tablet bears the longest known inscription in runic characters: a passage in which a father curses the enemies of his slain son.

The secret runes are to be read clockwise beginning at the top left: 3 − 2, 1 − 5, then 2 − 2, 2 − 3, and 3 − 5, 3 − 2 — that is, *ol ni-röþr,* "as an old man of 90 years." Thus the first and third genders have been interchanged. After about A.D. 800 the runic characters, which were originally used by all the Germanic tribes, appear only in Scandinavia. Runic inscriptions, which are quite rare, continued to be carved until some time in the 14th century.

Finally, let us deal with the *Persian siyaq* script, which we have already encountered earlier (see p. 74; from Arabic *siyaq* = "order"). Persian is an Indo-European language, a descendant of Avestan. It was once written in cuneiform characters borrowed from the Babylonians. Since the Arab conquest of Persia in the 7th century, the Arabic alphabet has reigned there along with Islam, and with it many Arabic words have entered into the old Persian language.

As for the Persian numerals, the Arabs wrote out the number words for a long time even in mathematical writings. The *siyaq* script which made numerals out of abbreviated or eroded number words, probably goes back to this time (Fig. 109). Because they employ some

Fig. 108 Runic stone from Rök, Sweden, around A.D. 850 At the top are secret runes replaced by number symbols. More than 2 m high; now in the National Park, Stockholm.

| 5 | 50 | 5000 | 500 | 5005 |

Fig. 109 Persian *siyaq* numerals. The two parallel strokes in the number at the right stand for missing digits (zeros). This script is read from right to left; 5, represented by a loop, appears at the beginning of each of these numerals.

features of a place-value notation together with special signs for higher numbers, the *siyaq* numerals represent a unique intermediate stage between numerals and fully written-out number words.

To mention only one feature: the units (5), the tens (50), and the thousands (5000) all have the same beginning (5) but are differentiated by their terminal symbols; 500, however, is written differently, although the sign for 5 can be discerned in it as well. Another very interesting trait is the presence of a sign denoting the absence of a digit, but its use was not absolutely mandatory; its form however resembles that of the ancient Babylonian symbol for a missing digit (see Fig. 228). In writing a number, the units and the tens were transposed, as we have already seen (see p. 74).

The *siyaq* numerals were used for many centuries in Persian and Turkish financial and official documents. They had the advantage of not being familiar to everyone, and also of being hard to alter by anyone who might wish to falsify the accounts. In the bazaars of Iran they are still used by the old merchants for keeping their own records, but always only in terms of amounts of money and always expressing

Fig. 110 Computation written out in *siyaq* characters.

Fig. 111 The same computation as Fig. 110, written in Eastern Arabic numerals. Figures 110 and 111 were written in a "bilingual" manner, by a Persian. The numerical ranks are in the top row, and the numbers beneath.

Siyaq:	1250	550	2400	900;	(below) 50 (?)	(1)5750
East Arabic:	1.25	0.55	2.4	0.9	15	15.75

Kindly provided by H. Horst, Teheran.

the new *rial* currency in terms of the old *dinar* currency (1 *rial* = 1000 old *dinars*, whence the difference between the two sets of numbers in Figs. 110 and 111). The *siyaq* characters are used only for writing down the numbers; computations are made on the Russian abacus, the *ščët*, which in Persian has been and is still today called *tshoke* (see Figs. 148 and 149).

In the last century the *siyaq* characters were officially replaced by the Eastern Arabic style Indian numerals, i.e., the oriental forms of the numerals we ourselves use. Today the Iranian merchant must make out his tax declaration in the modern Indian-Arabic numerals (see Fig. 234).

In addition to the *siyaq* and the Indian numerals, however, Iranians still make use of the Greek alphabetical numerals, which the Arabs first learned in Syria. They arranged the numerals in Hebrew fashion according to their *abujdad* (so named after the first four letters, *a–b–j–d*), and for centuries used them for writing the figures in their astronomical tables. Today in Iran they are used only to number the pages of a book's preface, just as we commonly use the Roman numerals for the same purpose.

Fig. 112 The Arabic alphabetical numerals. These are arranged in the same order as in Hebrew (see Fig. 95, column 3). The numbers were written right to left, in the opposite direction from the digits in the Indian-Arabic numerals (Fig. 234). In the number 1955 we see the four numerals 5'50'900'1000 combined.

Now let us look back over our material: we have seen the co-existence of three systems of numerals that not only represent three different intellectual stages of development, but also indicate the three different ways in which numbers can be associated with letters. We have penetrated deeply into the fabric out of which time has woven the tapestry of cultural and ethnic history. We have — to mention only the main strands — seen the alphabet of the Phoenicians and the alphabetical numerals which the Greeks derived from them, and we have recognized the Arabs' custom of writing out number words and the Indians' method of writing numerals. Finally, one more important point: The *siyaq* characters as a peculiar intermediate link between number words and numerals suggest that the Indian numerals (from 4 on) may have started out as abbreviations of the corresponding number words.

With this insight we leave the alphabetical numerals behind. Gothic, Greek, Indian, and Persian numerals — truly a rich harvest has been reaped from the 22 seeds of the first Phoenician alphabet. Now we must return from the Orient to our western world.

The alphabetical numerals did not spread from the Goths to the other Germanic tribes in the West. This was probably because they were not a popular, but a "learned" set of numerals that might have been suited for use in the old and highly developed culture of Byzantium, with its army of scribes, but not by young and primitive peoples who were in many cases settling down for the very first time .after centuries of nomadic wandering.

But these Teutonic tribes were able to comprehend the primitive system of the Roman numerals which they adopted without hesitation and treated as their own.

The "German" Roman Numerals

Roman Numerals in Cursive Form

*Ich habe diss rechenbüchlein, dem
Leyen zu gutt vnnd nutz, dem die
Ziffernzale am ersten zu lernen schwer,
durch die gewöhnlich teutsch Zal
geordnet.*

"To make this little book of
computations agreeable and useful
to its reader, who will find numerals
hard to learn at first, I have used
the common German numbers
throughout."

A 16th-century arithmetician

Over the course of time many different variants of the old Roman numerals came into use in the West. The Romans themselves, for instance, never wrote 1000 as M or 2000 as MM, but as (I) and (I)(I), or at the most occasionally as IIM, using the M as an abbreviation for the number word *mille* (see Figs. 10 and 135). In the Middle Ages, however, M became quite common as a symbol for 1000: MMCXII = 2112.

Just as often, however, the M was replaced by the 1000-stroke above the number, which was known as the *titulus* in the Middle Ages and as *vinculum* in classical Latin, and which the later Romans also used occasionally; one inscription has $\overline{VI}XXXIX$ (= 6039) and likewise Iɔɔ∞XXXVIIII; see also Fig. 44. This 1000-stroke above the numeral is to be sharply differentiated from the stroke which merely distinguishes numerals from letters and likewise appears in classical Roman times: $\overline{V}$ = 5; $\overline{III}VIR$ = *triumvir*. This gave rise to our habit of enclosing Roman numerals between horizontal strokes: $\underline{\overline{III}}, \underline{\overline{V}}, \underline{\overline{X}}$.

The Roman numeral for 1 million was an X within a frame: $\boxed{X}$; the reason for the X has already been shown (p. 244).

As the art of writing spread, and came into increasing use by merchants and managers of various kinds of enterprises, the abstract but letter-like Roman numerals were written more and more as letters. Thus X became an *x* and V a *v* or, as often happened in the Middle Ages, a *u*; thus whoever "made an X for a U" tallied up double the amount, 10 instead of 5, and thus by extension figuratively cheated someone. After the so-called "uncial" script using only capital letters developed into the medieval form of writing with lower-case letters, a number such as 763 would then also be written as *dcclxiij*. And in the "Gothic" script which became common in Germany as well as the "German" script that developed from it, the numbers 19 and 7 came to look like ɪɪʀ and ʋɪɉ (see Fig. 114). To prevent falsification of a number, the last digit *i* was always lengthened and written as *j* (see Fig. 122).

The page from the Rüsselsheim book of accounts reproduced in Fig. 114, which dates from the middle of the 16th century, is an excellent example of the medieval "German" Roman numerals in their fully cursive form. Observe the fluency of the whole writing, especially in the number 19 in the fourth line from the bottom, which was required of bookkeepers in writing numbers (see p. 427). Compare also the beautiful examples illustrated in Figs. 121–123.

The transformation of the Roman numerals into letters of the Latin alphabet then also made possible the fascinating riddles in which a Latin verse would be concealed within the numerals of the date of an occurrence or vice versa (the so-called "chronograms" or "date riddles"):

LVtetIa Mater natos sVos DeVoraVIt,
"Mother Lutetia has devoured her own children."

This cryptic saying refers to the massacre of St. Bartholomew's

[281]

Fig. 113 The beginning of two multiplication tables from 13th-century monastic manuscripts: *semel* once, *bis* twice, *ter* 3 times, etc. Both are still fully "Roman." Above the second is written: *et ad scientiam utilissima*, "... and especially useful for learning ...." The resemblance to multiplication tables of other peoples and times is interesting (see Index).
State Library, Munich.

Fig. 114 "German" Roman numerals, as used in a book of receipts by an official of Rüsselsheim, 1554.
Hessian State Archive, Darmstadt.
The page reads:

> *Gelltt unstending.*
> *Vor verkaufft Strohe*————————————————37 *Gulden (f)* 11 *Albus*
> *Von wegen Johan Schellerts alte Regesschulden zum andern*
> *Ziel* ————————————————————————50 *Gulden*
> *Inn gemein von Juden, Wasenmeistern und vor Spraw*
> ————————————————————————32 *Gulden* 12 *Albus* 3 *Heller*
>   *Latus* 1ᶜ 19 *f* 23 *alb* 3 *hlr*
> *Summarum aller unstendigen Einnahm geldt*
>   6ᴹ 7ᶜ 13 *Gulden* 15 *Heller.*

> "Moneys received and paid.
> "From sale of straw————————————————36 Gulden (f) 11 Albus
> "Transfer of Johan Schellert's old debt to another purpose————50 Gulden
> "Total received from Jews, harnessmakers and for hay
> ————————————————————————32 Gulden 12 Albus 3 Heller
>   "In all, 1ᶜ 19 f 23 alb 3 hlr
> "Totals of all variable moneys received
>   6ᴹ 7ᶜ 13 (= 6713) Gulden 15 Heller."

Night — Lutetia Parisiorum was the old name for the capital of the Celtic tribe known as the Parisi — by incorporating the Roman numerals for the year 1572 in which the Huguenots were slaughtered in Paris.

MEDIEVAL NUMBERS

Sebastian Franck (1499–1542) wrote in his *Spiegel und bildtnis des gantzen erdbodens* ("Mirror and Image of the Whole World"):

Zimmet kumpt von Zailon . CC. vn LX teutscher meil von Calicut weyter gelegen. Die Nägelin kummen von Meluza/für Calicut hinausgelegen vij. c. vnd XL deutscher meyl.

Cinnamon comes from Ceylon, which lies .CC. and LX (= 260) German miles beyond Calicut. Cloves are brought from the Moluccas, which are vij. c. and XL (= 740) German miles from Calicut.

What strikes us here? The first time Franck writes 200 as CC in the old manner; but a few lines further down he writes 700 as vij. c. and not as DCC. Thus the unit numeral VII appears as the number in front of the rank C, to which the seven C-groups have been condensed. We have seen exactly the same thing happen with spoken numbers (see p. 45). With this change the Roman numerals have embarked on the path to a place-value notation. Yet no positional principle was ever fully formulated in terms of Roman numerals, because medieval usage placed numerals in front of the rank symbols only for the numbers 100 and up but clung to the serial principle for the tens below (see Fig. 114). No one took the mental step of realizing what had actually been done with what appeared as mere abbreviations.

A further obstacle to the development of the primitive Roman numerals into a "mature" form of numerals is revealed in the way the number 123456789 was written in the year 1792:

C	*MM*	*C*	*M*	*C*	
i	*xxiij*	*iiij*	*lvj*	*vij*	*lxxxix*
1	23	4	56	7	89

This is the same difficulty inherent in pronouncing or spelling out large numbers: the problem of breaking down the numerical ranks that come after the thousands. Scribes were led into the habit of again always using C and M instead of the higher ranks and thus destroying the original clarity of the system. MM, for example, stood for million, 1000 × 1000 (see p. 143).

Fig. 115 A document written on vellum in the year 1229, with seals. In the last line the year appears in ordinal form in Latin: *Anno d(o)m(ini) millesimo ducentesimo. .* abbreviated to the then common form *m°cc°*, etc.
Hessian State Archive, Darmstadt.

Now for some more examples, whose instructive features the reader will not fail to observe:

$\bar{c} \cdot \overline{lxiiij} \cdot ccc \cdot i$	164301	time: 1120
II · DCCC · XIIII	2814	1220
cIɔ · Iɔ · Ic	1599	1600
① D C X L	1640	1640
IIII milia · ccc · L · VI	4356	13th century
CCCM	300 000	1550

In the Rüsselsheim book of accounts the sum of 1859 Gulden is written sometimes as 1^M viij^c lix Gulden and sometimes as xviii^c lix Gulden.

In Papal Bulls 20 was written as $\overset{x}{\mathsf{X}}$ and 8 might appear as $\overset{=}{\mathsf{V}}$. Furthermore,

XV . Cet : II   1502
XIIII^c ain jar   1401

and Köbel (Fig. 117) wrote

MVI^{c}XII   1612.

One arithmetician distinguished CM ($= 900$) from C^M ($= 100,000$). There are vigesimal groupings:

IIII$_{xx}$ et huit	88
VII . XX . VII	$7 \times 20 + 7 = 147$
XIIII . XX . XVI	296
mil . IIII^c IIII^{xx} et V	1485.

In one church there is inscribed the date: I · V^c · V 1505; whoever inscribed this number was evidently already familiar with the new place-value notation; it is interesting that he gets around the zero by using the lower-case c.

The fraction $\frac{1}{2}$, which appears almost exclusively in money calculations, was thought of as half of unity and was thus written and printed as a 1 with a stroke through the middle (see Fig. 116), as in the

       $12^1/_2$ $76^1/_2$ $4^1/_2$ $9^1/_2$

Fig. 116 Medieval handwriting, showing various forms of the fraction $\frac{1}{2}$.

table of silver weights in the Bamberg book of computations (see Fig. 189). The letter *S* was used in medieval manuscripts as an abbreviation for the Latin word *semis*, "half," superimposed on V and X and then read as "five less half" ($4\frac{1}{2}$) or "ten less half" ($9\frac{1}{2}$). (We have already discussed these peculiar word-forms, p. 77.)

Truly a wonderful variety of modes of organization — sometimes just sequence and grouping, then the use of digits as coefficients crops up, and at still other times we see the old vigesimal grouping and the physically truncated whole used to represent the "half" — but always, despite all the many differences in form, easily understood.

The very multiplicity of such number-letter connections suggests the indigenous, popular nature of these numerals. Anyone could work with them. This also applies to the calendars, genuinely popular books, which up to 1500 were for the most part written with the "German numbers" (unless they used the old peasant numerals; see Fig. 85). And in the year 1514 the arithmetician Köbel put out a very popular and often consulted short book of arithmetic computations, which was printed entirely, even including the fractions, in Roman numerals: *Ich habe disz rechenbüchlein, dem Leyen zu gutt vnnd nutz, (dem die Ziffernzale am ersten zu lernen schwer) durch die gewönlich teutsch Zal geordnet* ("To make this little book of computations agreeable and useful to its reader, who will find numerals hard to learn at first, I have used the common German numbers throughout").

Even as late as the middle of the 18th century the French spoke of Roman numerals as *chiffres des Finances, dont les Financiers se servent ordinairement*, the "numerals of the (state) accounts, which financial officials ordinarily use."

Besides their clear and easily understood organization, the visual quality of the individual Roman numerals also contributed to their ready adoption by the common people. Just as the Germans now say of a man who is forty years old, that he has *genullt* (zeroed) four times, in the Middle Franconian dialect the expression was that he has *gekreuzelt* (crossed) four times.

Es war das concilium gehalt
zu Costnitz, das geschah als man zahlt
ein Ringgen, 4 Rosseisen dabei,
ein Schlaifen, Haggen und ein Ay,

The council was held
in Costnitz; that happened when we counted
a ring, 4 horseshoes too,
a ploughshare, a hook, and an egg,

reads a rhymed monastic chronicle of about 1500. The *Ringge* (= *Rinke*) is a buckle, the ploughshare means actually the iron tip of the plough, the horseshoes and the hook are self-explanatory, and the egg clearly means "one": the result is ⅮCCCCXVI, which is thus 1416, the date of the Council of Constance.

Date riddles of this kind first appeared around the end of the 14th century and continued until well into the 16th. The destruction of Basel in the year 1356 by plague and fire is recorded in the jingle:

Ein ringge und sin dorn
Trü rossyssin verkorn
Ein zimmeraxst und der gelten Zahl
Da (ver)fyel Basel überal.

A ring and its thorn,	Ⅾ	
Three horseshoes chosen,	CCC	
A carpenter's axe and the gelten number,	L	VI
That's when Basel came to an end.		

Fig. 117 Fractions in "German" Roman numerals, from Köbel's *Rechenbüchlein* (Short Book of Arithmetic), 1514.

Gelten here refers to jugs or pitchers, of which there were six at the Wedding of Cana.

THE GRADUAL PENETRATION OF THE NEW INDIAN NUMERALS

The fact that the Roman numerals were so deeply rooted in the customs and affections of the people at first made it exceedingly difficult for the new Indian numerals, the "figure numbers," to replace the old familiar Roman numerals. We shall learn later that in northern Europe the Indian numerals first began to be used by ordinary people about 1500. This date, the change from the fifteenth to the sixteenth century, is the great intellectual watershed of modern history, the time when all the new movements generally came to the fore. A variety of developments were also preparing the ground for the common adoption of the new system of numerals. Trade and commerce were expanding rapidly, and the merchant's control over his transactions had to grow accordingly. Money was replacing barter; the merchant princes in Lübeck or Nuremberg could no longer simply carve tally sticks: now they increasingly had to keep proper accounts, write numbers and make computations. The need for knowledge was universal. Now it was the towns and cities with their recently acquired prosperity which supported and required education, not just the cloistered monasteries. Schooling became an important issue, which was also aggravated by the problems of the Reformation and was thenceforward an urgent question. To be sure, the mental framework of the Middle Ages still strongly asserted itself and continued to live on for a long time to come; Humanism respected the fresh, new way of life of the cities and, instead of stepping out boldly as a pacemaker, tended to follow in the footsteps of foreign fashion. But above all printing extended and spread all these new trends: now almost anyone could learn and master things that had formerly been accessible to but a few.

At the turn of the sixteenth century all these various intellectual ingredients, the old and the new, the wished-for and the known, both properly comprehended and misunderstood, were boiling vigorously in the same pot. The following numerals may serve to symbolize this confusion and multiplicity:

M · CCCC · 8II	1482	With zero: I · 0 · VIII · IX	1089
I · 5 · IIII	1504	IVOII	1502
CC2	202	ICC00	1200
15×5	1515	I · II · τ · τ	1200
Cδ	104 (!)	15000 · 30	15030

On his Erasmus Altar at Louvain the painter Dirk Bouts placed the number MCCCC4XVII. In one book of that period, the pages were numbered as follows: 100, 100-1, 100-2, . . ., 200-4, 200-5, etc.

Let us see if we can characterize more precisely the period in which the new numerals were introduced. The stonecutter carving the year of Zeugwart Hallenburg's wife's death on his gravestone outside the Frauenkirche in Munich in the form M · DC · Z4 showed that he was about a hundred years behind the times as far as using the new numerals was concerned (see Fig. 118). On the other hand, his

Fig. 118 M·DCZ4 — that is, 1624 — carved on a tombstone (shown at top) outside the Frauenkirche in Munich.

compatriot and fellow guild member, who produced the beautiful knights' grave shown in Fig. 119 and carved the year of one knight's death in the form *m·cccc·lxxxij* at its head, put the date when his brother died on the left side of the tombstone in the new form, .94. — in the same breath, so to speak. The time when these two knights died and when knighthood died out generally was one of transition for it was also the epoch in which the merchant flourished.

And in the merchants' account books the new way of writing numbers took hold slowly but inexorably, at least where it was actually used. Unfortunately the books kept by the House of Fugger in Augsburg go back only as far as 1494; thus we must look instead at the account books of the city of Augsburg, which provide an almost unbroken record of the town's incomes and expenditures from the year 1320 on. These examples from the Augsburg municipal archives at the same time offer some valuable glimpses into the history of record-keeping, bookkeeping, and coinage (see Figs. 120–124).

THE ACCOUNT BOOKS OF THE IMPERIAL FREE CITY OF AUGSBURG

In the beginning both the transactions and the contracts were written in Latin (Fig. 120); later it was only the amount involved, invariably incorporated into the running text, until one day it was "displayed" separately (Fig. 121). The new numerals at first were confined to quoting the year (Fig. 123) until they finally dared tackle the running amounts. But even then the scribe protects himself by writing the money amounts in Roman numerals within the text before displaying them in Indian numerals (Fig. 124). This was in A.D. 1470. After this the ice was broken, although it still took more than half a century before the amounts were written entirely in the new Indian numerals. In the books kept by the progressive and sophisticated Fugger family, in contrast, as early as 1494 (the Fugger books are extant from then on) we find amounts recorded exclusively in the new numerals. In the year 1533 Anton Fugger had an inventory of all his property

Fig. 119 Tombstone of the two knights Ludwig and Hans von Paulsdorf. The first of these brothers died *Anno d(o)m(ini) mcccclxxxij* (1482) *am pfinztag vor vith,* "in the Year of Our Lord 1482 on the day of Pentecost" (date carved at the top); the second died *darnach im .94. am freitag vor liechtmessen,* "afterwards, in the year '94, on the Friday before Candlemas." Bavarian National Museum, Munich.

prepared, and in this we can already see the elements of "modern" bookkeeping (Fig. 125).

A remark or two is in order on the currency and the methods of indicating denominations of money:

The amounts were stated in pounds (*lib*, *lb*) of 240 pence each (*denarius*, abbreviated *d* with a flourish) as well as in shilling (*solidus*, *s* with a flourish) pennies, that is shillings of 12 pence each, and shilling heller, that is shillings of 12 heller (*h*) each. Besides these ordinary "short" shillings of 12 pence each, there was also the "long" Bavarian shilling of 30 pence, with eight "long" shillings to the pound. The shilling and the pound, which incidentally was originally a weight of around 367 grams according to Carolingian law, were not actual coins, but only accounting or computational pounds and shillings — larger abstract units into which the minted penny coins were grouped. Another coin was the Rhenish gulden, abbreviated *rf* (both letters with flourishes), which like the pound initially contained 240 pence but by about 1475, in Augsburg at least, was worth only 168 pence. (See pp. 343 ff. for a further discussion of medieval European money.) The joint use of the pound and the gulden side by side is also to be seen in the Dinkelsbühl counting tables and the counting cloths of Munich (see Figs. 178 and 180). Because of the varying standards of currency it is sometimes very difficult to convert the amounts of money recorded at various times into modern currency.

1. In the year 1320 the Augsburg city clerk put down everything in his elegant handwriting in Latin in the oldest municipal account books, called *Baumeisterbüchern* ("master builders' books") because they deal principally with payments to master builders (the Roman units are reproduced here as *i* and *j*):

Fig. 120
  Wertachprugg. Anno Domini M°CCC XX° recepti sunt pro dictis bumaisteren Heinricus Bache et Heinricus Bitschlin.
    Domine in tua (anima) de tribus septimanis recepimus de antiquo thelonio — iii lib et v.s.d.
    item de frumento — xxx s.d.
    factus est i lib minus ii d.

"Wertachbrücke (city tower). In the Year of Our Lord 1320 receipts for the said builders Heinrich Bache and Heinrich Bitschlin were:
God, in thy (soul). [This is the introductory formula which was repeated at the beginning of each new entry or section, and which is the ancestor of the phrase *Mit Gott* that appears in German account books of past generations.]

In three weeks we have received from the old customs collector —
3 pounds and 5 shilling pennies (= 3 × 240 + 5 × 12 = 780 pence).
In addition, for cereals — 30 shilling pennies.
This gives 1 pound less 2 pennies." [At that time 1 shilling was worth around
8 denarii, for the reckoning here is clearly based on the "long" shilling of
30 pence.]

NOTE: The word *minus*, abbreviated here, was in later times represented only
by a short horizontal stroke at the top. When this stroke was written alone, it
evolved into our present minus sign.

2. In the year 1400 the transactions were written in German, but
the sums in Roman numerals and clearly set off:

Fig. 121
Item an dem nächsten Samstag vor dem Sunnentag Oculi anno quo supra haben
wir den obengenannten Pumaister empfangen. Von den eingeschriebenen einemern
der Ungelt. Von dem Ungelt
ii^c lb lxxij lb d
vnd xxxiiij guld.

"Besides, on the next Saturday before the Sunday of Oculi in the afore-
mentioned year we received hired the aforenamed master builder. Of the
recorded moneys received (customs, duties, taxes, etc.). Moneys are
272 pound pennies
and 34 gulden."

The official of Rüsselsheim some hundred and fifty years later still
kept his accounts in much the same manner (see Fig. 114).

3. In the year 1410 we are given the sum of all amounts received,
which the city clerk wrote down majestically and with flourishes
(Fig. 122). Here we see the thousands and the hundreds marked off

Fig. 122
Suma Sumarum alles einemens in toto
x^M guld iij^c guld lxvij guld
iiij^M lb vij^c lb lxxxxviij lb x s ij d.

"The full sum of all receipts in toto is
10367 gulden and 4798 pounds 10 shillings 2 pence."

from the groups of tens and units, and written with the unit designations — in other words, with C used to indicate rank, as in 3C, and not just in sequence, as CCC. This separation of the hundreds and thousands was still taught by Adam Riese for reading numbers (see p. 142). Another surprising feature is the appearance side by side of gulden and pounds in the final sum, without any corresponding conversion. From this method of writing it becomes clear that such sums could have been arrived at only through computations on the counting board.

4. In the year 1430 the amounts of money were still given only in Roman numerals, but the date itself in Indian numerals stands above the entry (Fig. 123).

Fig. 123 *Anno* 1430. *Wertachbrücken-Zoll. Herr Geistpach git ain jar* 400 *und* 70 *Pfund tertius sabat bis Galli.*

"In the year 1430. Toll at the Wertachbrücken gate. For one year Herr Geistpach hands over 400 and 70 pounds on three Saturdays (that is, in three installments)."

5. In the year 1470 there is an entry that refers to the *kaiserlich Steur* ("Imperial taxes"). The transaction is recorded in Roman numerals and then the amount is repeated in the new Indian numerals (Fig. 124, top).

6. In the year 1500 the amounts hitherto also written in Roman numerals (for security) were now commonly recorded in Indian numerals at the side, each monetary denomination in its own column (Fig. 124, bottom).

7. In the year 1533 Anton Fugger, then one of the wealthiest merchants in all of Europe, made an inventory of all his possessions. It is worth taking a brief look at the way his accounts were drawn up and also reading the total of one section and the first few entries in his exchange book, in gulden, shillings, and heller (Fig. 125).

I shall leave the reader to search out other interesting points for himself, such as the manner in which the abbreviation of *libra,* "pound," developed from *lib* through ~~lib~~ and *lb* into the present symbol for pound, £ or ℔ or how *x* by infinite repetition as in Fig. 122 ultimately became ȝ:

Fig. 124 Entries of 1470 and 1500 in the Augsburg ledgers.

Top:

von desselben Herrn Rudolfs wegen an der kaiserlichen Stür, so auf Martini anno 70 verfallen wirdet, auf ain k. vnd auch sein selbs quitantzen geraicht und bezahlt iii^c und xviij lib Münnicher und iij s Münchener, tut in Gold

iii^c und lxiii r. guld, vnd iii lib —f 363 lb 3 s 0 h 0.

"For the same Herr Rudolf, an Imperial receipt was drawn for his payment of the Imperial tax which fell due on St. Martin's Day of the year '70, and a receipt was also given to Herr Rudolf, in the amount of iii^c and xviii lib Munich and iij shillings Munich, which is in gold —

iii^c and lxiii r. guld, and iii lib — f 363 lb 3 s 0 h 0."

Bottom:

item xxv lb vj s freitag vigilia uldalrici			f 0	lb	25	s	6 h	0
item liij lb sundag vor Margareten			f 0	lb	53	s	0 h	0
item xlviij lb xviiij s samstag vor Magdalenen			f 0	lb	48	s	19 h	0
		Summa	f 4	lb	1222	s	18 h	4

"then 25 pounds 6 shillings on the eve of St. Ulrich's Day
then 53 pounds on the Sunday before St. Margaret's
then 48 pounds 19 shillings on the Saturday before St. Magdalen's"

[292]

Fig. 125 An excerpt from Anton Fugger's inventory of his total worth, which he had made in 1533. In his *Wexelbuech* (exchange book) the name Welser appears in the first item; this was the name of another rich merchant of Augsburg. "The interest thereon" refers to amounts of interest due on loans made by Fugger.
Reproduced with kind permission of the Fürstlich Fuggersche Foundation, Augsburg.

Suma *Dukaten* 389 352 *die thuendt zu* 40 *Kreuzer*
rheinische Gulden   f 545092.–.–
Suma f 547019 —
Was für Haussrat vorhanden ist für nichts anzuschlagen
                                   Wexelbuech.

Bertholme Welser *cepa* (*compagnia*) ————————————— f 8000. –.–
Das ynteresse darauf ———————————————— f 316. 13. –4
Mergemelle ————————————————————— f 7000. –.–
Das ynteresse darauf ———————————————— f 262. 10.–
"Total ducats (*d* with a flourish) 389 352, each of 40 crowns
Rhenish gulden   f 545092.–.–
Total f 547019 —
Amounts on hand for household expenses are not to be included in total worth
                            Exchange book.

Bartholomew Welser Co. ————————————————— f 8000. –.–
The interest therefrom ————————————————— f 316. 13. –4
Items variously mentioned ———————————————— f 7000. –.–
The interest therefrom ————————————————— f 262. 10.–

WRITTEN NUMERALS AND COMPUTATIONS

By now the reader may be wondering why so much fuss is made over the mere fact that a new set of numerals replaced an old one. Whoever asks such a question shows that he does not understand the essence of the Indian system of numerals.

The primitive rules of ordering and grouping have now been superseded. At our present stage we can scarcely imagine the extreme difficulty of abandoning the highly satisfactory visual quality of the number groupings in order to adopt the more sophisticated principle of place-value. At first glance one is likely to notice that the new system is easier to write — but this was true only for those who were already familiar with it. Arithmetic texts of the 16th century go to unbelievable lengths, both in verbal exposition and in tables *der Verglychung tudtscher vnd ciferzal* ("for comparing the German numbers and the cipher numerals"), to make the new numerals and their use clear to people; we shall revert to this again later (see pp. 334 and 433).

"Numbering" — that is, the mere writing of the numbers — was in itself a leading subject in schools and books, and even in catechisms, which are among the oldest books of popular instruction. In a typical catechism printed in 1525 in Wittenberg, for example, we find, along with the *teyn Bade Gades* ("Ten Commandments of God"), the *vade unse* (the Lord's Prayer), the *Döpe* (Baptism) and the *Bycht* (Confession) also a section on *de düdesche tall mit den cifern* (the German numbers and the new ciphers).

But though at first glance one merely notices the greater brevity brought about by the new numerals, a second glance lets one see a little deeper: with the new digits we can now for the first time make computations!

With this remark we finally realize that writing numerals and making computations are two entirely different things; up to now we have generally had nothing to say about computations, although we quite thoroughly discussed spoken numbers and written numerals. But didn't people make calculations with Roman numerals? No, they did not! The fairly simple multiplication

325×47	in Roman numerals	CCCXXV · XLVII
2275		MMCCLXXV
13000		↻MMM
15275		↻ↁCCLXXV

looks clearly impossible to the uninitiated reader.

The how did people in the Middle Ages make the necessary computations — not to mention the Romans themselves, who developed the system of numerals used throughout the Middle Ages? This question about computations, and thus about the real purpose of written numbers, brings us to a fascinating subject of cultural history which has now largely been forgotten — the counting board or abacus.

Fig. 126 A table for learning the new system of numerals, from Köbel's book of arithmetic, 1524.

The Abacus

The Nature of the Counting Board

Die edle Rechenkunst Gott gab
In seiner Gnad von oben hrab.
Daraus wir dann erkennen gring,
Wie Gott allmechtig alle Ding
in zal vnd Mass vnd in Gewicht
durch seine Weisheit hat gericht.

"In His gracious providence from on high
God granted us the noble art of reckoning.
With this we then could recognize
How God almightily in His wisdom
Has established all things in terms of
Numbers and measures and weights."

From an old poem in praise of computation.

When an aborigine wants to add 6 coconuts to the 17 he already has and then give the total number, how does he arrive at the answer? Just as we do when we recite the alphabet: he counts 17 coconuts and then counts 6 more. He does not "compute" at all; he merely keeps on counting until he reaches 23. Yet in a sense he *has* made a calculation by combining two numbers into one. Then he can replace the coconuts by a supplementary quantity of pebbles or sticks, as the Wedda tribesman did (see p. 33). Thus he has physically "grasped" his answer, or even "written" it, if you will. Arab traders supposedly often carry important numbers about with them, in the form of such "pebble writing," in a small satchel. If we disregard their different face values, the coins we use are also supplementary quantities that give us the numbers we need, and hence are numerals of a sort: take 743, for instance. We do not carry this number about with us in the form of seven hundred forty three separate pennies — rather we group or combine it into a five-dollar bill, two one-dollar bills, four dimes and only three copper pennies. Why? Because, apart from other obvious advantages, by grouping it in this manner we can read the number more easily.

How are computations made in primitive system of numerals that does not use these "pebble numbers"? Let us begin by seeing how a peasant adds 60 and 457 in three separate steps (see Fig. 83). As a first step, he writes down the initial number 457 in his own symbols: he then puts down the number 60 quite independently of the first number; last of all he groups together the two numbers, first in V-groupings and ultimately in O-groupings, so that he can finally read off the answer quite easily. This example shows step by step what happened all at once in putting down the initial number: the peasant eliminated all the rows successively and replaced them with groupings, just as Gottfried Keller's old woman junk-dealer did (p. 249).

Peasant numbers are essentially no different from Roman numerals. But was the same procedure followed in the counting house of a great merchant: writing down and calculating the sums in chalk on a tablet or board in Roman numerals, and then erasing them and entering only the final results in the account book? Of course not. In making their computations, merchants, government officials, and counting house clerks abandoned the Roman numerals and once again calculated with pebbles — or counters — no longer just by lining them up in rows as the native did, however, but on a counting board.

We can understand why: For one thing, the counting board or abacus helps to break numbers down into groups so that they can be handled more easily; then too, it makes the numbers movable so that they can be combined in various ways without constant erasing and rewriting. How does it do that? A large number like 6238 could be broken down into groups represented by different kinds of pebbles, large and small, or of various colors. But this is still not the way it is done: the counters are given their grouping values by their positions. Thus a board or tablet is divided into strips or columns, one each for the units, the tens, the hundreds, and the

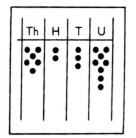

Fig. 127 Basic form of the counting board, showing the number 6238.

thousands, into which the number 6238 is readily broken down and "written" by means of the counters (Fig. 127). Thus the nine or less counters that appear in any one column can be visualized without any trouble at all.

Is this really how it was done in the Middle Ages? The question is quite pertinent, for this would mean that a mature form of complete place-value notation was used in computation, whereas in actual writing people clung blindly and stubbornly to the primitive rules of ordering and grouping numbers.

But it was indeed just so: in computation people used a mature place-value system with gradations based on 10; in writing the old numerals were retained. The two went along hand in hand for many centuries. The price paid for the "popular" simplicity of the Roman numerals was their uselessness in computation, but this disadvantage was again nicely circumvented by using the counting board with its more advanced gradational arrangement. The mutually complementary use of the numerals and the counting board thus created a fully adequate and convenient tool for simple computation, which people were therefore extremely reluctant to part with. For what is in fact the essence of the new Indian number system? That it combines the two aspects, writing numbers and making computations, into one single procedure, by extending the more advanced principle of the counting board — its place-value notation — to the numerals as well. The separation between computation and the writing of numbers was never realized during the Middle Ages; people wanted utility, not intellectual or spiritual perfection. It never occurred to anyone even to try to take the step which the Indians had taken. Cultural history shows over and over again that regardless of how close a subject matter may be to its ultimate and perfect expression, only the intellectual readiness of a culture to take this step will ensure its development to complete maturity. A striking example of this statement is the place-value notation we have been talking about: not only did Medieval Europe possess it for many centuries, but it was thoroughly familiar to people even in antiquity — on the counting board.

Now let us take up the history of the abacus.

The Counting Board in Ancient Civilizations

Little is known about the manner in which computations were made by ancient peoples — not only the Babylonians, the Egyptians, and the Indians, but also the Greeks and the Romans. The problems and questions in Egyptian papyri offer some intriguing glimpses into the mathematical thought of Pharaonic times, but the actual operations by which the Egyptians found or attempted to find the solutions must be laboriously deduced from the rules, and can often be only guessed at. We know the numerals used by all of these ancient cultures. These support the hypothesis that computations were performed on counting boards. But whenever the written, pictorial, or archaeological documents are lacking, we find ourselves groping in the dark.

And such documents are, unfortunately, missing throughout the ancient world except for Greece and Rome; only a few actual counting boards and representations or descriptions of counting boards used by the Greeks and Romans are still preserved. No wonder, then, that these are especially treasured as rare and precious documents of cultural history. The following pages will discuss and illustrate a few rare specimens, some perhaps for the first time.

THE SALAMIS TABLET

The only ancient Greek counting board that has been preserved is a tablet of white marble found about the middle of the last century on the island of Salamis (Fig. 128). Its exact date is not known.

Two groups of parallel lines have been chiseled into this tablet, one group of eleven lines crossed by a perpendicular line through their middle, and another group of five shorter ones at some distance from the first group. Along the two long sides of the tablet and across one of the shorter sides are letters, which can be identified as early Greek numerals and also as denominations of ancient coins (cf. Fig. 99).

T ℾX ℿH ℾᐃ ℾⱶ ICTX

Talent = 6000	1000	100 Drachmas	10	1	1	$\frac{1}{2}$	$\frac{1}{4}$	$\frac{1}{8}$
						Obol		

The three symbols at the right are those for the half, the quarter, and the eighth obol, which were respectively called the *hemiobólion* (represented by the half-symbol C because in Boeotia O signified a whole obol), the *tetartemórion* (T for "quarter-piece") and the *chalkós* (X, "copper" or "ore"): thus the last two are not to be confused with the symbols X for *chilioi* and T for *tálanton*, which stand at the left end of the row of numerals. When separated from their currency values, the fractions of the drachma become abstract fractions like the Roman:

1 drachma = 6 obols = 12 half = 24 quarter = 48 eighth obols.

These symbols for numerical values testify that the tablet was actually used for computations, very likely in government financial

[299]

Fig. 128 The Salamis Tablet, the only large ancient Greek counting board preserved today. Size: 149 × 75 × 4.5 cm (7.5 cm thick at the edges). National Museum, Athens.

offices, with whose public accounting we have already become familiar (see Fig. 100). Then if we read the following passage from the historian Polybius (2nd century B.C.), we shall see that the Salamic Tablet is essentially like the counting board already described in general terms.

The courtiers who surround kings are exactly like counters on the lines of a counting board, for, depending on the will of the reckoner, they may be valued either at no more than a mere *chalkós*, or else at a whole talent! (See also the quotation from Aristophanes on p. 211).

The *chalkós* and the *talent*, of course, are the lowest and the highest values on our tablet, at the extreme right and left ends of the scale. The counters took their values in computation from these: if they were placed under the symbol for the *chalkós* they had the value of an eighth of an obol, but if they were moved to the space beneath the talent sign, their value was increased 300,000 fold; a more graphic example of place-value notation could scarcely be found.

As early as the end of the seventh century B.C. Solon, the wise law-giver of Athens, compared a tyrant's favorite to a counter whose worth depended entirely on the whim of the person who pushed it from one place to another. This simile lived on through the centuries, as long as counting boards were in use: we shall meet it once again, this time in 18th-century disguise.

The Greeks called the counting board or table *abákion*, and the Romans *abacus*; the Greek word *ábax* means "round platter" or "stemless cup" and thus also a "table without legs." It is unlikely that this word was derived from the Semitic *abq* (dust), although the Semitic peoples also, especially later on, had similar sand-covered tablets which they used for drawing figures and making computations (see p. 331). *Abacus* was also the name for the ornate Roman display tables with costly inlays of marble or silver, for the smooth wall panels covered with plaster of marble-dust and for the top plate in the capital of Doric columns. In ancient times this word thus had the general meaning of a flat table or board. The Roman name *abacus* was used for the counting board throughout the Middle Ages. Greeks today still say of someone who is resourceful *ksérei tòn ávakon*, "he knows his abacus."

The counters were called *pséphoi* (pebbles). Since the word *pséphizein* (literally, "to pebble") is also the general word for "to compute" or "to calculate," the Greek language itself provides suggestive evidence as to where and how computations generally were made. As late as the 14th century the Byzantine scholar Maximus Planudes entitled his textbook on computations with the Indian numerals *Psēphophoría kat' Indôus*, "Pebble-placing in the Indian Fashion," showing how deeply rooted the ancient expression was in the abacus.

In addition to placing them in the columns of the counting board, the Greeks also arranged counters independently to form "geometrical numbers" — triangles, right angles, squares, etc. These "pebble-numbers" (counter-numbers), which turn up later in the history of mathematics as "figurative numbers," thus represent the oldest association of numbers with geometrical shapes. With the aid of these pebble-numbers the Greeks discovered quite a few important laws of number theory, for example the rule that the difference between two consecutive square numbers is always an odd number, the "corner" (*gnómon* in Greek) that is added to the previous square in a series of squares, as shown in the margin.

On the other hand, no information has been preserved concerning the actual operations performed on the abacus itself; nevertheless we can infer these with a high degree of probability from the general arrangement of the Salamis Tablet, from the remaining fragments of other Greek counting boards with columns drawn or incised on them, and above all from the medieval counting boards, about which we are plentifully and adequately informed.

To use the Salamis Tablet the reckoner would stand facing it on the right-hand long side. His *pséphoi* (counters) would be piled in the middle between the two groups of parallel lines. The group of lines at the left end of the board as he faced it were used for the integers,

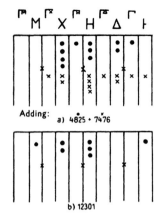

Adding: a) 4825 · 7476

b) 12301

Fig. 129 4825 + 7476 = 12301 on the Salamis Tablet.

Fig. 130 3509 − 1847 = 1662 on the Salamis Tablet. *Top:* putting down 3509 (●) and 1847 (.); *middle:* changing 1 thousand into two five-hundreds (○), one five-hundred into five hundreds (+) and one hundred into ten tens (×); *bottom:* answer 1662.

the group at the right end for fractions. The small crosses indicated the columns: the one at the right distinguished the units and the one at the left marked off the talents, or the myriads if the computation did not involve money.

Let us say, for example, that the quantity 7476 is to be added to 4825. First the reckoner forms the number 4825 (here indicated by the symbol ●) with pebbles in the appropriate columns, thereby observing the rule of quinary grouping which is always followed in each even column (the 2nd, 4th, etc.), as shown in Fig. 129. Then he simply "gives" the second number 7476 (indicated by ×) to the first, as expressed precisely by the Greek technical term *syntithénai* and the Latin *addere* (from *ad-dare*), which became our word *addition*. The first step of this procedure thus consists of "writing" or putting down the initial number, the second step of addition proper, and the third step of grouping, in which 5 units make up one 5-group and two of the latter a 10-group, or one unit in the next higher column. After the calculation has been "purified," as this grouping was called in the Middle Ages (the Romans called it *purgatio rationis*), the reckoner can then read off the final number, the answer 12301. The result may also have been placed over the intermediate numbers of the computation as a "head number" (perhaps above the middle line), as suggested by its Greek name *kephálaion* (from *kephalé*, "head"; see Fig. 100) and its Latin name *summa* (from *summus*, "the highest"). Thus the very same steps which we saw in peasant calculations are also essential to the reckoning table.

Subtraction, for example 3509 − 1847, is performed in the same graphic manner. The first number is put down (represened by ● in Fig. 130) and then the second number is "taken away" from the first: Greek *aphaireîn* and Latin *subtrahere*, "to draw away," whence comes our word *subtract*. In this process the second number must be either remembered or recorded elsewhere. On the Salamis Tablet this may have been done in one of the rows of numbers at the side, or arranged so that the first number appeared above and the second below the middle line.

It makes no difference with which column the subtraction begins. One can start with the highest rank just as well as with any of the others. The important thing is that one must be able to "take away"; in other words, the number of *pséphoi* of the first number in a column may not be smaller than that of the second number. If this happens, the ranks must be "changed" to lower ranks. Contrary to the procedure in addition, the numbers are "de-grouped," one ten being converted to two fives and one five into five units in the right-hand column. Then the "taking away" can be done quite easily, and the "rest" of the *pséphoi* that remain on the counting board are the answer, in this case 1662.

What was the purpose of the rows of number symbols carved along three sides of the Salamis Tablet? Evidently they served as a "supplementary" record of numbers to be placed on the abacus. In multiplying $874 + \frac{5}{6} + \frac{1}{12} + \frac{1}{24} + \frac{1}{48}$ by 93, the calculator had to keep these initial numbers constantly before his eyes (see Fig. 131). He now had to use the columns to build up the answer little by little (an example

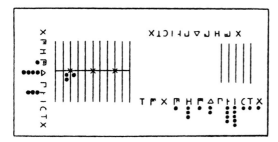

ig. 131 $(874 + \frac{5}{6} + \frac{1}{12} + \frac{1}{24} + \frac{1}{48}) \times 93$;— that is, 874 drachmas 5 obols
nd 1 half, 1 quarter and 1 eighth obol, times 93, multiplied on the Salamis
Tablet.

f the way multiplication was performed on the Roman abacus will be
xplained on p. 316). Perhaps in these operations a second reckoner
tood on the left side of the table and made calculations on his side
f the middle line. As counterparts of this in the Middle Ages, we
ave the Dinkelsbühl reckoning table and the counting board of the
three gentlemen of Basel" (see Figs. 177 and 178).

Besides the Salamis Tablet and the passage quoted from Polybius,
ve also have, by chance, a third witness, in the form of a picture of
uch a reckoner at work, on the famous

DARIUS VASE

his magnificent ceremonial vase in the red-figured Attic style, which
ommemorates the Persian Wars, was found in an ancient grave near
Canosa (the ancient Canusium) in Apulia and is now in the Museo
Nationale in Naples (Fig. 132). Southern Italy, including Apulia, was
art of the realm of Greek culture in ancient times.

In the top row of the three rings of figures, Zeus and the other gods
ake Hellas, menaced by the Persians, under their protection. In
ne middle row, the Persian King Darius is seated on his throne, and
Persian standing in front of him seems to be warning him against
ndertaking his proposed expedition against the Greeks. Arranged
bout Darius are figures symbolizing the members of his court and
fficials of his government, including, in the bottom row, the
evealing representation of Darius's royal treasurer (see Fig. 133).

A man dressed in Persian costume stands in front of the table with
sack of coins. He is bringing in the tribute paid by the various
onquered peoples, who are represented by the individual figures
neeling behind him. The Persian treasurer sits at a small table on
hich are the signs MY (Boeotian for X 1000), HΔΓ and 0 (Boeotian
r obol), ⟨ (or C, the sign for the half-obol) and T, which has the
ame meaning as on the Salamis Tablet (see p. 299). A striking
ature is the lack of the five-group, for the symbol Γ here clearly
tands for the units. Thus apart from the fractions, only the decimal
ystem is used. In contrast to the Salamis Tablet, the counte.s or
ebbles here lie, without inscribed columns, directly beneath the
ymbols for the numerical ranks. Possibly the pebble seen above

Fig. 132 The Darius Vase with its
depiction of a calculator at his
reckoning table (bottom row), one of
the only two known ancient
representations (see Fig. 134 for the
other). Height 1.30 m, greatest
circumference 2 m. Probably 4th
century B.C.
Museo Nationale, Naples.

Fig. 133 The treasurer on the Darius Vase, seated at his reckoning table, which has counters scattered upon it.

(below in this picture, which is upside down) the H stands for the five-group, after the Roman custom (see p. 317). Thus the number 1731–4/6 can be read on the Darius Vase; of course this whole representation is symbolic.

In his left hand the reckoner holds a diptych or double tablet, on which we read the word *talanta H*, "100 talents." On this tablet are posted the amounts which he successively finds on the abacus. Thus a living representation of computation in ancient times has been preserved through sheer good luck.

THE ETRUSCAN CAMEO

There is a surprising similarity between the Darius Vase and a tiny but exquisitely carved cameo of Etruscan origin (Fig. 134). On this gem, in almost exactly the same position and attitude as the treasurer on the Darius Vase, a man is seated at a small, three-legged table on which the counters are represented by three minute spheres. Like the reckoner on the vase, he holds a tablet in his left hand, this time with Etruscan numerals, which leave no doubt as to the man's activity. Perhaps the tablet alludes to the double rows of the signs for the numbers 1, 5, 10, 100, and certainly also 1000, which the reckoning table also has. The symbols ∧ and × are familiar to us from the Etruscan coins we have already seen (see Fig. 67).

The similarity between the representations on two works of art of such divergent origins, which were also certainly made at different times, bears witness to the universality of the reckoner, whether as merchant or as money-changer. Even his names in different languages, Greek *trapezitēs* and Latin *mensarius*, "tabl-er" (from the Latin *mensa*, "table") show unequivocally that computations were made on tables, just as the present "bank-er" ultimately derived his name from the medieval "counting bank" which was his tool (see Fig. 250).

With the Etruscan cameo we have entered the world of the Romans.

Fig. 134 Etruscan cameo, showing a reckoner at his table. 1.5 cm high. Cabinet des Médailles, Paris.

THE ROMAN HAND ABACUS

The only two specimens still preserved show that the Romans arranged the counting board very cleverly for the purpose of making smaller computations. One of the two is in the great collection known as the Cabinet des Médailles in Paris (see Fig. 135, which is a photograph of a plaster cast); the other, which once belonged to the famous

Fig. 135 Roman hand abacus. Plaster cast of the specimen in the Cabinet des Médailles, Paris. Almost full size. Between the two rows of grooves are the Roman number symbols for

1 million		10	1	1	½	¼	⅓
	Integers					Unciae	

Jesuit Athanasius Kirchner (died 1680), is now in the Museo delle Terme in Rome. To emphasize the "hand" size of this Roman abacus, the writer photographed this specimen held in his own hand (Fig. 136). The Augsburg scholar Marcus Welser also possessed a Roman hand abacus which is now lost, but which he fortunately illustrated and described precisely (in his *Opera historica et philologica*, 1682); Welser's abacus was the same as the other two, except that it had three small grooves at the right end (instead of only one). On the basis of these three, we can give a reliable account of the Roman hand abacus:

On a small bronze tablet are carved 8 long grooves (*alveoli*) with the same number of short grooves at the top; at the right end is another long groove (three more on Welser's abacus) without the short groove above. On the raised surface between the two rows of grooves are the Roman symbols for the whole numbers, for the *unciae* and along the extreme right groove for the half, quarter and third uncia called *semuncia*, *sicilius* and *duella* in Latin; see p. 161). In these grooves run little spheres (called *claviculi*, "little nails"), four in each long groove (5 in the case of the unciae) and one in each short groove above.

And how were the computations carried out? These dealt only with the simplest problems, for the most part probably the addition of two

Fig. 136 A real hand abacus, the second preserved specimen, now in the Museo delle Terme in Rome.

or more numbers. We shall run through a computation with integers; the reader will find the essential principles of operations with fractions on p. 158. To "insert" the first number 5328, the counters are pushed toward the raised bridge near the center. The rows above the bridge are the five-groups. Thus to form the 5, in the short groove at the top, the counter is pushed downward; to form the next digit three counters are pushed up in the next unit groove, as the diagram shows; the reader can follow the rest of the procedure himself.

Now, if a second number is to be added to the first, it cannot be placed separately on the hand abacus as it is with the free counters on the reckoning table, and the answer arrived at by combining the two. Here the combining must take place simultaneously — that is, the second number, which the computer keeps in his head or writes down elsewhere, must be immediately blended step by step with the first.

We now have only the two above-mentioned actual hand abaci, but we also have an invaluable Roman pictorial representation of a computer at work (see Fig. 138). In the left corner of the sculptured

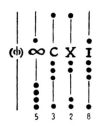

Fig. 137 5328 on the hand abacus.

Fig. 138 A Roman *calculator* (from *calculi*, "counters") with his hand abacus, calculating at the "dictation" of his master. From a 1st century A.D. Roman gravestone in the Museo Capitolino, Rome.

gravestone of a merchant stands the calculator of the household with his portable abacus, which is here portrayed symbolically. His stance and posture speak for themselves and are very indicative of the proceedings: the master "dictates" the amounts, the *calculator* adds them up. Every orderly head of a household kept a *codex accepti et dispensi*, a "book of receipts and disbursements," and in the *tablinum* of his Roman house he had a special counting room.

THE HAND ABACUS IN ASIA
The small Roman reckoning instrument did not take hold in the West. This is understandable, for it presupposed a skill in working with numbers which only the merchant would generally possess. On the other

hand, it penetrated to a surprising degree through all of the Far East, especially China and Japan, where it is now universally used.

Thus we shall interrupt our passage through antiquity at this point, to examine the Asiatic counterpart of the Roman abacus with its attached counters or buttons (*claviculi*), before we go on to discuss the operations procedures with loose counters (*calculi*).

The Japanese Soroban

This is an exact analogue of the Roman hand abacus. Within a rectangular frame, on each thin rod (Japanese *keta*, "row, wire") slide 6 small beveled beads, the sixth being separated from the other five at the bottom by a long strip perpendicular to the vertical wires or rods. The differences between the Japanese *soroban* and the Roman hand abacus are the following (see Fig. 139): instead of four counters at

Fig. 139 A Japanese *soroban* with 17 wires or rods (*keta*), typical of those which may be bought anywhere in Japan today. The number 231 is shown in the middle, and 1956 at the right. Size: 21 × 5.5 cm. This is a specimen in the present author's possession, through the kindness of Nobuko Yokota, Tokyo.

the bottom there are generally five; the vertical rods or wires have no numerical symbols, which are unspecified; the *soroban* has more vertical columns, as many as seventeen; and it has no columns especially set aside for fractions. They do have in common, however, the quinary and decimal grouping of the counters which run along fixed rods (or grooves). The Japanese might well wish, "If I only had one more *keta* salary!" when he dreams of multiples of ten decimals. On the *soroban*, as on the Roman hand abacus, only the counters pushed against the bridge or separating bar have value; in Fig. 139 the numbers 231 and 1956 are formed on the soroban (see also Fig. 141).

The *soroban* was introduced into Japan from China, probably in the 16th century. The origin of its name is unexplained. In the 1870's it was almost replaced by written computations with the "western" (that is, Indian) numerals, which had by then come into universal use throughout the world, but since about 1930 it has again been in great demand as a result of the growth of Japanese industry and trade. The Japanese chambers of commerce and industry offer yearly examinations and competitions in computations on the *soroban* in which, to name only one year, in 1942 about 40,000 participants competed for the prize.

As he writes, the author has before him a Japanese textbook, in its 4th edition of 1954, which provides instruction "using the latest methods in the mastery of the *soroban*, from the basic elements to the highest levels of proficiency." This book has been "registered and

Fig. 140 A Japanese calculating on the *soroban*, adding up amounts taken from a ledger.
Photo by Wolfram Müller.

accepted by the Soroban League of the Chamber of Industry and Commerce." It clearly and thoroughly teaches the four fundamental operations of arithmetic, using many examples and practice exercises, and every aspect is thought out and presented fully, down to the positions of the fingers in moving the beads (see Fig. 141).

Fig. 141 Excerpts from a Japanese *soroban* textbook published in 1954. *Top:* showing the proper technique for moving the counters along a *keta. Bottom left:* placing the number 5218 on the *soroban;* observe that numbers today are written in Indian numerals, but that computations are still (generally) made on the *soroban. Bottom right:* how the merchant adds the posted amounts together. The left hand turns the pages of the account book, while the right hand calculates. Compare the similar representations of reckoners at work in other cultures and at other times (Figs. 133, 134, 151, and 195).

There is an art even to pushing the counters. On the author's *soroban*, which is the type generally used in commerce, one *keta* with its beads is 5 cm long. This, however, includes the cross-bar, so that there is only a bare 4 mm of room left to move a counter in! And two adjacent *ketas* are only about 12 mm apart, while the sharp edges of the counters on them are separated by only 1 mm. Moreover the counters slide along the *ketas* almost completely without friction — all these circumstances call for a degree of dexterity and precision of movement which our European fingers, at least at the beginning, simply do not possess. The first few attempts to use the *soroban* strongly resemble the proverbial bull in the china shop, which breaks more dishes with its rear end than with its head: the beginner always finds himself moving more counters than he intends to.

All four operations of arithmetic are taught, especially multiplication and division. The many *ketas* or wires permit all the numbers in a